Statistical Mechanics
of
The Liquid Surface

Statistical Mechanics
of
The Liquid Surface

Clive A. Croxton

Department of Mathematics,
University of Newcastle,
New South Wales,
Australia.

A Wiley–Interscience Publication

JOHN WILEY & SONS
Chichester · New York · Brisbane · Toronto

British Library Cataloguing in Publication Data:

Croxton, Clive Anthony
  Statistical mechanics of the liquid surface.
  1. Liquids, Kinetic theory of
  2. Surfaces (Physics) 3. Statistical
  mechanics
  I. Title
  532'.6   QC175.3   79-40819
  ISBN 0 471 27663 4

Filmset in Northern Ireland at The Universities Press (Belfast)
Limited and printed at Pitman Press, Bath.

To
Karl Walter

Preface

If, as I once suggested, the bulk liquid could be regarded as a kind of purgatory between solid and gas, the nomination may be yet more appropriately applied to the interfacial region between liquid and vapour, having been the scene of expiation of many a theoretician over the last decade.

Whilst the equilibrium description of simple, homogeneous fluid systems is now relatively well understood, any physical excursion into more complex molecular or anisotropic systems has to follow largely uncharted paths in a rapidly expanding jungle of mathematical intractability. Progress in these more difficult areas has and is being made; reviews are beginning to appear in the literature and investigators are increasingly turning their attention to these more challenging areas of research.

The treatment of inhomogeneous fluids, in particular the free liquid–vapour interface, has followed just such a route, although the journey is far from complete. Nevertheless, it was felt that a detailed appraisal of the progress made so far might serve a useful purpose, both for those who have just arrived, and for those who have been there for some time already.

The result, this book, is substantially longer than both I and the publishers expected; my determination to present a comprehensive account became a point of obsession, and the manuscript grew uncontrollably, being the first monograph devoted exclusively to the subject. However, I felt I had strayed too far and decided to restrict myself to systems with which I had personal experience. Accordingly, the book presents a personal standpoint, and contains a good deal of original work which has not been published elsewhere. Undoubtedly my selection of chapters will exclude certain readers' favourite interfacial systems—electrolytes and critical fluids are cases in point—I can only encourage those working in the relevant fields to publish reviews which will complement the material presented here. In some cases they already exist—for example, The Critical Surface by B. Widom in C. Domb and M. S. Green (Eds), *Phase transitions and Critical Phenomena*— Ch. 3. Vol. 2. Academic Press London & New York (1972), and simply to repeat their content here would be pointless.

Of course, production of a book such as this depends heavily upon the forebearance, generosity, and support of many others. I wish to record again my gratitude to Professor Chandrasekhar and the Liquid Crystals Division at the Raman Research Institute, Bangalore, India for their stimulating and gracious hospitality; to Dr Paul Barnes and his wife Desirée for their generosity and understanding whilst working at the Department of Crystallography, Birkbeck College, University of London; to my wife, Sarah, for her forebearance and participation in countless appraisals of the manuscript on a subject in which she has no interest whatsoever, to Michael Claridge

viii

for his assistance in checking the manuscript, and finally to Karl Walter to whom I extend my sincere personal appreciation, and to whom the book is dedicated.

University of Newcastle,
New South Wales
June 1979

Prologue

Thermodynamically, the interfacial density transition between liquid and vapour may be conveniently regarded as if it were an infinitely sharp, structureless, density discontinuity separating the homogeneous coexisting liquid and vapour phases: from this essentially macroscopic standpoint, thermodynamic consistency completes the description. Of course, such an interface can only be an idealized abstraction—a caricature of the true interfacial structure. At the triple point the liquid density typically changes by a factor of perhaps $\sim 10^3$ over no more than a few atomic diameters, for simple fluids at least, whilst as the critical point is approached the transition zone assumes macroscopic proportions. Our task is to relate the interfacial structure and its associated principal thermodynamic functions to the nature of the interparticle potential. And, as for bulk isotropic fluids, the statistical mechanical connection will be effected primarily in terms of the anisotropic one- and two-particle density distributions, $\rho_{(1)}(\mathbf{r}_1)$ and $\rho_{(2)}(\mathbf{r}_1, \mathbf{r}_2)$. For equilibrium homogeneous bulk fluids the single-particle distribution is, of course, largely redundant, whilst the two-particle distribution is essentially isotropic. Our interest will therefore necessarily centre on the specification of both the single-particle density transition profile $\rho_{(1)}(\mathbf{r}_1)$ and the hierarchically-related anisotropic two-particle function $\rho_{(2)}(\mathbf{r}_1, \mathbf{r}_2)$.

The mathematical expedient of reducing the density transition profile to a surface of density discontinuity enabled Fowler[1] and Kirkwood and Buff[2] to present the first statistical thermodynamic description of the interface in terms of the *bulk* liquid radial distribution function. Whilst the essential inhomogeneity of the interfacial zone was recognized, the associated anisotropy of the pair distribution proved difficult to specify, and indeed remains a central issue as we shall see: nevertheless, their model embodied many of the qualitative features of the interfacial region.

With increasing temperature and interfacial delocalization, the step model becomes increasingly inappropriate, as might be expected. However, at the triple point where, presumably, the step model is at its most representative, reasonable numerical agreement with the principal experimental quantitities such as surface tension and surface energy may be achieved. Numerical coincidence of experimental and theoretical estimates of interfacial quantities provides an insensitive criterion, however, for the acceptability or otherwise of any given interfacial structure, the majority of the principal thermodynamic quantities of interest arising as *integrals* over the transition zone. Only the most general *a priori* features may be anticipated on the basis of the thermodynamic data. It is precisely for this reason that interest has centred primarily on the statistical specification of the single-particle or density profile $\rho_{(1)}(\mathbf{r}_1)$. Consequently, in each of the following chapters we shall regard the determination of the density profile as our principal task.

From our experience of isotropic bulk fluids we know that adjacent orders of particle distribution are hierarchically related, and, correspondingly, the specification of $\rho_{(1)}(\mathbf{r}_1)$ demands a knowledge of the anisotropic function $\rho_{(2)}(\mathbf{r}_1, \mathbf{r}_2)$. Quite how to describe this complex interfacial function has exercised the imagination of numerous investigators since Kirkwood, Buff, and Fowler were forced to adopt the isotropic bulk radial distribution. Presumably this anisotropic function must be a functional of the local densities at each of the particles constituting the pair. However, this raises a further question, for surely these local densities intermediate between the coexistence values of liquid and vapour represent essentially *unstable* regions located on the loop of the isotherm. In fact, the existence of density gradients may be shown to stabilize these transitional states. Nevertheless, debate regarding the specification of the anisotropic function continues, recently heightened by the introduction of one further seemingly innocuous feature: surface capillary waves.

That capillary waves represent a long wavelength undulation of the surface is not in question. However, whether they represent an *intrinsic* feature which our theories of surface structure should incorporate or whether they represent a subsequent undulation of a precalculated profile is currently a centre of considerable controversy. The specification of the pair function $\rho_{(2)}(\mathbf{r}_1, \mathbf{r}_2)$ in the two cases will differ fundamentally. If the horizontal, long wavelength capillary fluctuations have to be incorporated, then our present approaches will have to be substantially revised. The anisotropic interfacial pair distributions are generally compounded from short range isotropic bulk liquid and vapour distributions: of course, neither has any significant long wavelength component. Inclusion of capillary fluctuations would demand a long wavelength horizontal pair function having a correlation range of *thousands* of atomic diameters! Such long wavelength behaviour is generally associated with near-infinite compressibility, and suggests that the intermediate states mentioned above relate to the *tie-line* on the phase diagram, rather than stabilized states on the sinuous branch of the isotherm. If this is the case, then the intermediate densities between the two coexistence values can only be realized as a weighted combination of pure liquid and vapour states: in other words, as a spatial fluctuation in the location of a relatively sharp dividing surface, rather than a continuous distribution of intermediate densities. Whether these two standpoints may be reconciled is not yet clear; however, we shall discuss them at some length in the course of the book.

For a variety of reasons direct experimental investigation of the transition zone is not easy, and computer simulation has assumed an important role. It has become clear that the relatively weak coupling between liquid and vapour and their slow equilibration enforces runs considerably extended in comparison with their bulk fluid counterparts. Again, the application of horizontal boundary conditions, essential if wall effects are not to become significant, appears to suppress capillary fluctuations at the surface. Thus, whilst machine simulation has much to offer, the data may be deceptive and its interpretation has to be circumspect. Indeed, many of the results are capable of more than one interpretation, and careful consideration must be given to the length of the run, the equilibration period, wall effects, the area of the simulated interface, and suppression of capillary fluctuations, to mention but a few of the sources of confusion.

In the case of complex liquid systems—those having nonclassical, noncentral, anisotropic, or saturating interactions—the difficulties compound rapidly. To consider these various systems in any detail here would be to anticipate the forthcoming discussion. Suffice it to say that the nature of the molecular interaction, combined with the structural inhomogeneity, produces a complexity of interfacial behaviour which exceeds their homogeneous counterparts by a considerable margin. Even when the bulk fluid is reasonably well understood, for example the liquid alkali metals, the interfacial region may still present substantial theoretical difficulties: in the present case, selfconsistent ionic and electronic density profiles have to be separately determined, whilst the pseudopotential itself develops an anisotropy quite unfamiliar in the homogeneous liquid metal, not to mention specifically interfacial phenomena such as the development of surface plasmon contributions.

Water and polymeric systems present particularly formidable problems. Indeed, it could be argued that a discussion of their interfacial properties is premature, given that the description of their isotropic liquid phases is far from complete. Nevertheless, we include chapters on these as we do on a number of other complex liquid systems which, whilst at varying levels of theoretical development, represent an overall assessment of the current level of statistical mechanical description of liquid surfaces in general.

References

1. R. H. Fowler, *Proc. Roy. Soc.*, **A159,** 229 (1937).
2. J. G. Kirkwood and F. P. Buff, *J. Chem. Phys.*, **17,** 338 (1949).

Principal Symbols

Symbols which occur in a limited context and which are clearly defined where they are used are not included in this index. The figures in brackets indicate the pages of the text where the symbols are further defined or discussed. Except where otherwise shown, * denotes a reduced quantity.

Roman

Chemical elements are denoted throughout by the conventional symbols in roman type.

Å	Ångström unit
BGY	Born–Green–Yvon 42
KBF	Kirkwood–Buff–Fowler 27
HNC	Hypernetted chain 72
PY	Percus–Yevick 63

Italic and bold

a_0	Bohr radius 142
$a^{\dagger}$	Helmholtz free energy density 40
$a(z)$	Local Helmholtz free energy per particle 16
$a^{(0)}$	Reference Helmholtz free energy per particle 37
a_L, a_V	Helmholtz free energy per particle in homogeneous bulk liquid and vapour phase 16
a_1, a_2	Delocalization parameters in tanh density profile 37
A	Helmholtz free energy 6, 11
A_S	Surface excess Helmholtz free energy 7
$\hat{A}_S$	Surface orientational excess free energy 253
A_0	Intrinsic mean field Helmholtz free energy 77
A_{lm}	Spherical harmonic coefficients 54
$A^{(0)}, A^{(1)}, \ldots$, etc.	Reference and higher order contributions to Helmholtz free energy 93
c_N	Number of configurations of N-mer lost per unit area on account of boundary 222
$c_{\alpha\beta}$	Direct correlation between particles of species α, β 64
$c(r), c_{HNC}(r), c_{PY}(r)$	Direct correlation; in HNC and PY approximations 67
$\mathbf{e}_i$	Internal set of Euler angles specifying orientation of molecule i 88
$E_1, E_2, \ldots$, etc.	Energy eigenvalues 179
E_F	Fermi energy 150
$\mathbf{E}(z)$	Electric field 211
$f_{(N)}$	N-body phase distribution function 4
$f(r)$	Mayer f-function 41
$f(z)$	Spatial distribution about intrinsic Gibbs surface 80
$F(r)$	Indirect correlation 41
$\mathbf{F}_i$	Force on particle i 296
g	Acceleration due to gravity 78, 176

g_K	Kirkwood screening parameter 215
$g_N(\nu)$	Number of configurations allowed to N-mer with ν occupying surface sites 222
$g_{(2)}^{(1)}(r)$	Attractive component of radial distribution function 34
$g_{(2)}^{(0)}(r)$	Reference or repulsive component of radial distribution function 34
$g_{(2)}^L(r_{12}), g_{(2)}^V(r_{12})$	Bulk liquid, vapour radial distribution function
$g_{(2)}^{\dagger}(\mathbf{r}_\lambda, \mathbf{r}_\mu)$	Radial distribution between particles of species λ, μ 128
$g_n(\mathbf{r}^n)$	n-body configurational probability distribution 6
$g_{(1)}(\mathbf{r}_i), g_{(1)}(z_i)$	Single particle configurational distribution function for particle i
$g_{(2)}(\mathbf{r}_i, \mathbf{r}_j), g_{(2)}(z_i, \mathbf{r}_{ij})$	Two particle configurational distribution function for particles i, j
G	Gibbs free energy 7, 11
h	Subset of h particles 6
$h_{\alpha\beta}$	Total correlation between particles of species α, β 64
$h(r)$	Total correlation $= g_{(2)}(r) - 1$
$h, \hbar$	Planck and Dirac $(h/2\pi)$ constants
H	Heaviside step function 80
$^\kappa\mathbf{I}$	Internal state of κth chain
$\mathbf{I}_i$	Internal state of molecule i
I_P	Principal moment of inertia 288
$\hat{\mathbf{k}}$	Surface normal vector 240
k_F	Fermi wave number 142
$\mathbf{k}, k$	Wave number
k, k_B	Boltzmann constant 3
l_N	Number of configurations of N-mer lost per unit area on account of boundary 222
l_0	Interfacial thickness 76
l_x, l_y, l_z	Dimensions of rectangular parallelopiped 14
L	Interfacial thickness 18
L_+, L_-	Width of ionic, electronic density transition profile 159
$m; m_i$	Mass; mass of particle i
$\mathbf{M}; M_x, M_y, M_z$	Dipole moment of molecule; components of dipole vector 200
$\hat{\mathbf{n}}$	Director field vector 240
$n(\lambda)$	Refractive index 168
n_{λ_s}	Surface excess molar composition of species λ 7
$n_\alpha, n_\beta, \ldots,$ etc.	Molar composition of system containing species $\alpha, \beta, \ldots,$ etc 6
N	Number of particles in system
$\mathbf{p}_i$	Momentum vector of particle i 4
$\mathbf{p}^N$	Set of momentum vectors $(\mathbf{p}_1, \ldots, \mathbf{p}_N)$ 4
p_λ	Partial pressure of species λ 118
$p, p(z)$	Fugacity 34
P_l^m	Associated Legendre functions 54
$P; P_\perp, P_\parallel$	Pressure; normal, tangential components at surface 7, 11, 45
P_{HS}	Pressure of hard sphere fluid 32
$\mathbf{P}(z)$	Polarization density 211
q_i	Electrostatic charge of particle i 200
$Q(\mathbf{r}), Q_\mathbf{k}$	Amplitude of capillary wave 176
$\mathbf{Q}; Q_{xx}, Q_{xy}, \ldots,$ etc.	Quadrupole moment of molecule; components of quadrupole tensor 200

$\tilde{Q}(k), \tilde{Q}(-k)$	Fourier transform of uniform phase direct correlation	68
Q	Reduced quadrupole moment	101
r_c	Cutoff radius, ionic radius	144
r_s	Mean electronic spacing	142
$\mathbf{r}^N$	Set of position vectors $(\mathbf{r}_1, \dots, \mathbf{r}_N)$	
$\mathbf{r}_{ij}$	Separation vector between particles i, j	
$\mathbf{r}_i, \mathbf{r}_j$	Position vectors defining location of particles i, j	
R	Gas constant	
R	Truncation range of pair potential	266
R	Interfacial radius of curvature	72
R, R^*	Amplitude of reflectivity and its complex conjugate	168
$s(z)$	Local entropy per particle	16
S_s	Surface excess entropy	7
$\hat{S}, \hat{S}_s$	Orientational excess entropy, surface excess	251
S	Entropy	7, 11
S_β, S_σ	Bulk, surface entropy	163
$\bar{S}^\parallel(z, \mathbf{k})$	Mean transverse structure factor	292
t	Interfacial thickness	77
T	Absolute temperature	
T_i	Inversion temperature	163
T_c	Absolute critical temperature	2, 17
T_{NI}	Nematic–isotropic transition temperature	239
T_{SN}	Smectic–nematic transition temperature	255
u_{es}	Electrostatic energy	143
u_c	Correlation energy	142
u_{ke}	Kinetic energy	142
u_{ex}	Exchange energy	142
u_b	Single particle bulk energy	146
U_s	Surface excess energy	9, 11
$u(z)$	Local single particle energy	16
$U(\mathbf{r}_{ij})$	Total interaction energy between particles i, j	
U_{DD}	Dipole–dipole interaction energy	209
U_{DQ}	Dipole–quadrupole interaction energy	210
U_{QQ}	Quadrupole–quadrupole interaction energy	210
$\hat{U}, \hat{U}_S$	Orientational excess energy, surface excess	251
v_0	Volume of penetrable sphere	75
v_B, v_S	Bulk, surface transverse wave velocities	166
V	Volume of system	7
V_S	Surface excess volume	7
w	Interchange energy	120
W	Work associated with surface undulations	77
$x; x_{\lambda_s}$	Mole fraction; surface excess of species λ	123
z_G	Location of Gibbs surface	58
z_D	Difference between successive iteratively determined density profiles	60
Z	Location of intrinsic Gibbs surface	80
Z	Valency, ionic charge	144
z_S	Surface of tension	305
Z_Q	Configuration projection of N-body partition function	6
Z_N	N-body partition function	4
Z_P	Momentum projection of N-body partition function	6

$Z(r_{ik} \mid N)$ Spatial probability distribution of segments i, k in N-mer   231

Script

$\mathscr{A}$ Area of interface   6
$\mathscr{E}$ Ellipticity of scattered light   76
$\mathscr{H}$ System Hamiltonian   4
$\mathscr{H}^{I}$ Internal Hamiltonian   88, 235
$\mathscr{N}$ Avogadro's number
$\mathscr{S}$ Slater determinant   190

Sans serif

T   Tensor representation of dipole–dipole interaction   201
U   Tensor representation of dipole–duadrupole interaction 201
W   Tensor representation of quadrupole–quadrupole interaction   201

Greek

α Relaxation parameter   48
$\alpha, \beta, \gamma, \ldots$, etc.   Constituent species of system   115
β $(kT)^{-1}$
$\beta(\lambda)$ Imaginary part of refractive index   168
γ Surface tension   7, 11
γ Activity   16, 34
γ_0 Surface tension associated with planar interface or at zero point   78, 175
$\gamma^{(0)}, \gamma^{(1)}, \ldots$ etc.   Reference and higher order contributions to surface tension   94
γ_λ^0 Surface tension of pure component λ 123
γ_P, γ_S Surface tension of pure polymer, solvent   226
$\gamma_\parallel, \gamma_\perp$ Surface tension for parallel, perpendicular molecular orientations   247
$\hat{\gamma}$ Orientational contribution to surface tension   251
Γ_λ Surface excess concentration of species λ 117
$\Gamma_\lambda^{(\mu)}$ Relative adsorption of species μ with respect to species λ 117
Γ Gamma function   176
Γ_{i_p} Principal torque on particle i 295
Γ_N N-particle density matrix   180
δ Overlap parameter   110
δ^2 Mean square capillary displacement   176
$\delta(\lambda)$ Real part of refractive index   168
$\delta(\mathbf{r}_1 - \mathbf{r}_2)$ Kronecker delta function   73
ΔV Surface electrostatic potential   216
Δz_G Small displacement at Gibbs surface   58
ΔA_m Free energy of mixing   121
∇_i Gradient operator with respect to particle i
∇_{Ω_i} Angular gradient or torque operator with respect to particle i 107
ε Well depth of interaction potential   31
ε Dielectric constant   205
ε_0 Static dielectric constant   215
ζ Zeta function   176
$\zeta(x, y)$ Capillary wave displacement of surface about equilibrium plane   77

$\Phi^{(0)}(r_{12})$	Repulsive or reference component of pair potential 34
$\Phi^{(1)}(r_{12})$	Attractive component of pair potential 34
Φ_{AB}	Interaction potential between particles A, B 120
$\Phi^{(1)}(\mathbf{r}_{12}; \mathbf{e}_1, \mathbf{e}_2)$	First order anisotropic perturbation about reference potential 93
$\Phi^*(\mathbf{r}_i, \mathbf{r}_j)$	Screened pseudopotential 135
Φ_{PS}	Ionic pseudopotential 144
$\Phi_{MC}(r), \Phi_{MD}(r)$	Effective Monte Carlo, molecular dynamic pair potential 266
χ	Angle
χ_T	Bulk isothermal compressibility 18
ψ	Electrostatic potential 206
Ψ	Potential of mean force 23
Ψ, ψ	Electronic eigenfunction 142
Ψ_B	Boson eigenfunction 185
Ψ_F	Fermion eigenfunction 190
$\bar{\psi}_L, \bar{\psi}_V$	Mean bulk liquid, vapour potential 206
ω	Oscillation frequency 11
$\omega(k)$	Dispersion relation
Ω	Solid angle 92
$\dot{\Omega}$	Angular velocity 297
$\ddot{\Omega}$	Angular acceleration 297

Contents

xxii

CHAPTER 1

Statistical Thermodynamics of the Transition Region

1.1 Introduction

The ultimate objective of a statistical theory of matter is, of course, to relate the microscopic observables to the details of the intermolecular potential operating between the particles constituting the assembly. This objective has been largely achieved for dense, isotropic simple fluids; subsequently, extensions have been made to include more complex systems—liquid mixtures, molecular liquids, liquid metals, quantum fluids, and so on. Generally the connection between the principal macroscopic thermodynamic functions of state and the atomic microstructure, expressed in terms of a hierarchy of particle–particle correlation functions, is made through the formal link of the canonical N-body partition function, Z_N. Of the hierarchy, the lower order distributions appear to provide a virtually complete description of the configurational and transport properties of simple fluids. The single-particle distribution $g_{(1)}(\mathbf{r}_1)$ is naturally of little interest in homogeneous fluids, and the contribution of three-particle and higher order distributions is, for most purposes, negligible: it is the two-body distribution $g_{(2)}(\mathbf{r}_{12})$ which proves to be of central importance and is, moreover, an experimentally accessible function. Whilst we shall assume and utilize various results of single-component, homogeneous, classical monatomic liquid theory, their detailed discussion is not within the scope of this book, and the reader is referred to the various texts which will provide an adequate background theory of homogeneous fluids.[1]

There nevertheless remains a number of major outstanding problems in the theory of liquids—transport phenomena, irreversibility, and phase transitions, for example. The latter problem may be further resolved into various aspects of coexistence: the solid–fluid phase transition, critical phenomena, and the problem which primarily concerns us here, the inhomogeneous interphasal region at the surface between coexisting liquid and vapour. We enquire to what extent the familiar thermodynamic functions of state, pressure, density, chemical potential, and so on, may be extended into inhomogeneous regions, and in particular to what extent these functions may be expressed in terms of the lower order correlation functions as they are in isotropic systems. One feels intuitively that it should be possible to define local or point thermodynamic functions for each component in the interfacial region, even though such point functions lie outside the scope of macroscopic thermodynamics. Nonetheless, the specification of such functions remains a useful concept, and proves relatively easy to do in terms of the anisotropic surface correlations.[2] It should perhaps be emphasized that by 'point function' we do not mean a function whose value is determined by the properties at that point alone (such as temperature or density), but a function which is *well-defined* at that point and whose value is determined by the state of the local environment. The extent to which the point function is coupled to its environment will, of course, depend upon the range of the pair potential and local correlation involved in its specification. Any attempt to evaluate the point function in terms of the properties *at that point* necessarily reduces the analysis to a mean field approximation. As we might anticipate, the point functions, whilst conceptually useful, arise, as we shall see, as complicated integrals over the surface correlation functions.

Attempts to establish the principal thermodynamic functions of the free liquid surface (such as the surface tension and surface energy) in terms of the surface correlations by means of a direct statistical mechanical attack have been as spirited and as controversial as they have been unconvincing. The description of the structurally inhomogeneous region separating the two bulk phases is indeed a formidable problem, and initial euphoria based on numerical coincidence of calculated and experimental values of the surface thermodynamic quantities has now given way to a shift of emphasis more toward the determination of the structure of the transition zone itself, to which the thermodynamic functions are somewhat insensitive.

It is evident that if the pressure and temperature are chosen such that two coexistent phases are present, the existence of a weak gravitational field will spatially resolve the phases with a plane transition region, across which the local number density $\rho_{(1)}(z) = \rho_L g_{(1)}(z)$ varies from the bulk liquid value ρ_L along a z axis directed normally into the vapour of number density ρ_V. (Gravitational pressure gradients can be neglected both theoretically and experimentally for all temperatures such that $|T_c - T|/T_c \geqslant 10^{-5}$: see Reference 72 of Chapter 2.) However, the role of gravitational effects is coming under greater scrutiny (Section 2.11). Inhomogeneity in the local density (which, incidentally, is our first example of a point function) implies anisotropy of the two-particle distribution. Indeed, as the interfacial region is approached from either bulk phase we must expect the isotropic distribution to modify $g_{(2)}(r_{12}) \rightarrow g_{(2)}(z_1, \mathbf{r}_{12})$. The corresponding two-particle density distribution is related to the single-particle function as follows:

$$\rho_{(2)}(z_1, \mathbf{r}_{12}) = \rho_{(1)}(z_1)\rho_{(1)}(z_2)g_{(2)}(z_1, \mathbf{r}_{12}) \qquad (1.1.1)$$

by direct anology to the uniform fluid, $\rho_{(2)}(r_{12}) = \rho_L^2 g_{(2)}(r_{12})$.

The specification of the one-and two-particle distributions for the various liquid systems will occupy us throughout the rest of this book.

1.2   The Gibbs dividing surface

Many of the quantities of interest at the liquid surface are *excess* functions, representing the modification of a bulk quantity by transposing the same quantity of matter to the surface at constant V, T. The specification of an excess quantity implicitly assumes a knowledge of bulk values, both liquid and vapour, against which the superficial quantity may be compared: we shall presuppose these on the basis of our understanding of homogeneous liquid systems.[1]

In the calculation of these excess quantities it is convenient to introduce the artifice of a hypothetical dividing surface, first proposed by Gibbs.[3] This particular specification is not unique and has been criticized by a number of workers, notably Guggenheim[4] who has proposed a valid alternative exposition involving two mathematical planes isolating the surface transition zone. However, Gibbs' location of the dividing surface does have a number of advantages which will become evident as we develop the statistical thermodynamics of the liquid surface, and we shall adopt it here. (Indeed, the use of a dividing surface is indispensable for the treatment of curved surfaces in specifying the area and curvature of the interface.)

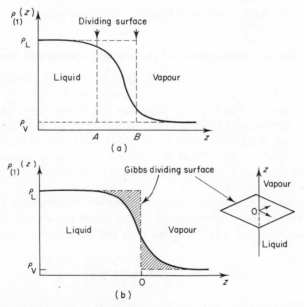

Figure 1.2.1 (a) The calculation of surface excess functions depends upon the location of the hypothetical dividing surface between homogeneous bulk liquid, density ρ_L, and homogeneous bulk vapour, density ρ_V. Clearly the two locations A and B will yield different excess values: the former being essentially the excess with respect to the vapour, whilst the latter yields the excess with respect to the liquid. (b) The Gibbs location of the dividing surface is chosen such that there is zero surface excess density of matter, the shaded areas being equal (equation 1.2.1)

The *surface excess* of a quantity is that excess of the real two-phase system over a hypothetical reference system in which the bulk properties remain constant up to a dividing surface separating the bulk liquid from the bulk vapour. The calculation of many, but by no means all, of the surface functions depends sensitively upon the location of the dividing surface. For example, if the hypothetical boundary of the two-phase system were located at A (Figure 1.2.1a) then clearly we would be calculating the excess of the quantity over that of an essentially isotropic vapour phase. This will be quite different to the location B in which the excess of the real function over the isotropic liquid phase is determined.

Gibbs' location of the dividing plane is chosen such that

$$\int_{-\infty}^{0} [\rho_{(1)}(z) - \rho_L]\,dz = \int_{0}^{\infty} [\rho_{(1)}(z) - \rho_V]\,dz \tag{1.2.1}$$

where the origin of coordinates has been located on the plane. This particular choice ensures that the superficial excess density of matter vanishes, that is, the shaded areas in Figure 1.2.1b are equal.

1.3 A formal development of the surface correlation functions

At sufficiently high temperatures such that the energy between adjacent quantum states $\Delta E \ll kT$, the laws of classical mechanics may be used in place of quantum mechanics, and although the latter provides the correct description of actual particle behaviour, there are many instances where it is simpler to use classical laws, and where their use leads to negligible error. This standpoint has to be revised for quantal systems of course, or for the distribution of molecules amongst vibrational or electronic states at normal temperatures for which $\Delta E \sim kT$.

For such quasi-classical systems the N-body distribution function $f_{(N)}$ $(\mathbf{p}^N, \mathbf{r}^N)$ describing the occupation of phase space is canonically related to the system Hamiltonian $\mathcal{H}(\mathbf{p}^N, \boldsymbol{\rho}^N)$ through the Boltzmann expression

$$f_{(N)}(\mathbf{p}^N, \mathbf{r}^N) = \frac{1}{Z_N} \exp\left\{ -\frac{\mathcal{H}(\mathbf{p}^N, \mathbf{r}^N)}{kT} \right\} \tag{1.3.1}$$

where we have abbreviated the momentum and configurational coordinates $(\mathbf{p}_1, \ldots, \mathbf{p}_N; \mathbf{r}_1, \ldots, \mathbf{r}_N)$ by $(\mathbf{p}^N, \mathbf{r}^N)$, in an obvious notation. The phase partition function Z_N enters as a normalizing constant and ensures that the probability of locating the N particles somewhere in the system is unity. We shall assume separable Hamiltonians of the form

$$\mathcal{H}(\mathbf{p}^N, \mathbf{r}^N) = \sum_{i=1}^{N} \frac{\mathbf{p}_i^2}{2m_i} + \Phi_N(\mathbf{r}^N) \tag{1.3.2}$$

where m_i is the mass of the ith particle. The total configurational potential Φ_N may be written as a sum of contributions

$$\Phi_N(\mathbf{r}^N) = \sum_{i=1}^{N} \Phi^{(1)}(\mathbf{r}_i) + \sum_{i>j}^{N} \Phi^{(2)}(r_{ij}) + \sum_{i<j<k}^{N} \Phi^{(3)}(r_{ij}, r_{jk}, r_{ik}) + \cdots \quad (1.3.3)$$

where $\Phi^{(1)}(\mathbf{r}_i)$ is an external potential producing an inhomogeneous spatial distribution of the particles, $\Phi^{(2)}(r_{ij})$ is the pair potential developed between two isolated particles at separation r_{ij} and $\Phi^{(3)}(r_{ij}, r_{jk}, r_{ik})$ represents the non-addititive three-particle potential developed between the isolated particles i, j, k, and so on. This latter and higher order terms will be neglected and we shall work in the so-called pair approximation which, incidentally, ensures that the thermodynamic and structural properties of the surface may be expressed simply in terms of the one- and two-particle distribution functions without reference to distributions of higher order. We point out that the pair potential $\Phi^{(2)}(r_{ij})$ (which we shall from now on designate $\Phi(r_{ij})$ without ambiguity) is simply a scalar function of the separation r_{ij}, and does not modify with density or under conditions of structural inhomogeneity. Such is not always the case, however; the electronic component of the interaction in liquid metal systems, for example, is a sensitive function of ionic density and hence structural anisotropy. We shall postpone discussion of these systems until Chapter 5. We shall assume the existence of a weak gravitational field in order to resolve spatially the liquid and vapour phases, but for our purposes the component $\sum_i^N \Phi^{(1)}(\mathbf{r}_i)$ will be negligible in comparison with the pair term, and in consequence will be neglected.

The configurational projection of (1.3.1) may be obtained simply by integrating over the momentum coordinates

$$\rho_{(N)}(\mathbf{r}^N) = \int f_{(N)}(\mathbf{p}^N, \mathbf{r}^N) \, \mathrm{d}\mathbf{p}^N \quad (1.3.4)$$

where for an inhomogeneous system $\rho_{(N)}(\mathbf{r}^N) = \rho_{(1)}(\mathbf{r}_1) \cdots \rho_{(1)}(\mathbf{r}_N) g_{(N)}(\mathbf{r}^N)$. Generally, we shall be interested in the low-order correlation functions $g_{(h)}(\mathbf{r}^h)$, $h \ll N$, where the h-body correlation function $g_{(h)}(\mathbf{r}_1, \ldots, \mathbf{r}_h)$ may be projected from the N-body correlation $g_{(N)}(\mathbf{r}_1, \ldots, \mathbf{r}_N)$ by performing a spatial integral average over the remaining coordinate variables:

$$\rho_{(1)}(\mathbf{r}_1) \rho_{(1)}(\mathbf{r}_2) \cdots \rho_{(1)}(\mathbf{r}_h) g_{(h)}(\mathbf{r}^h) = \frac{1}{(N-h)!} \frac{1}{Z_N} \int \cdots \int \exp\left\{ -\frac{\mathscr{H}}{kT}(\mathbf{p}^N, \mathbf{r}^N) \right\}$$

$$\mathrm{d}\mathbf{p}_1 \cdots \mathrm{d}\mathbf{p}_N \, \mathrm{d}\mathbf{r}_{h+1} \cdots \mathrm{d}\mathbf{r}_N$$

$$= \frac{1}{(N-h)!} \frac{1}{Z_Q} \int \cdots \int \exp\left(-\frac{\Phi_N(\mathbf{r}^N)}{kT} \right) \mathrm{d}\mathbf{r}_{h+1} \cdots \mathrm{d}\mathbf{r}_N. \quad (1.3.5)$$

where $Z_N = Z_P Z_Q$, and

$$Z_P = \frac{1}{h^{3N}} \int \cdots \int \exp\left\{-\sum_{i=1}^{N} \frac{\mathbf{p}_i^2}{2mkT}\right\} d\mathbf{p}_1 \cdots d\mathbf{p}_N$$

$$Z_Q = \frac{1}{N!} \int \cdots \int \exp\left\{-\frac{\Phi_N(\mathbf{r}^N)}{kT}\right\} d\mathbf{r}_1 \cdots d\mathbf{r}_N$$

We point out that (1.3.5) relates to a conservative, canonical, though not necessarily homogeneous, ensemble of N simple classical spherical particles. $g_{(h)}(\mathbf{r}^h)$ is dimensionless, and represents the relative probability of finding the subset of h particles in the configuration $\{\mathbf{r}_1, \ldots, \mathbf{r}_h\}$, relative that is to the probability of the uniform, uncorrelated (random) distribution. If we write down the $(h+1)$th distribution analogous to (1.3.5) a recurrence relation is obtained between adjacent orders of correlation:

$$g_{(h)}(\mathbf{r}_1, \ldots, \mathbf{r}_h) = \frac{1}{(N-h)} \int \rho_{(1)}(\mathbf{r}_{h+1}) g_{(h+1)}(\mathbf{r}_1, \ldots, \mathbf{r}_{h+1}) \, d\mathbf{r}_{h+1}. \qquad (1.3.6)$$

This hierarchical relation may be applied to the one- and two-particle distributions at the liquid surface yielding the selfconsistency condition

$$1 = \frac{1}{(N-1)} \int \rho_{(1)}(z_2) g_{(2)}(z_1, \mathbf{r}_{12}) \, d\mathbf{r}_2. \qquad (1.3.7)$$

Attention is drawn to the fact that whilst equations (1.3.5)–(1.3.7) apply specifically to spatially inhomogeneous regions, they nevertheless reduce to their homogeneous bulk forms, i.e. when the local point density function $\rho_{(1)}(\mathbf{i})$ adopts its uniform value ρ.

1.4 Thermodynamic functions of the liquid surface

Before attempting to develop a detailed statistical mechanical analysis of the principal thermodynamic functions of the liquid surface, it will perhaps clarify matters if we first of all consider the classical macroscopic thermodynamics of a planar transition region which the statistical theory must, of course, ultimately describe. There is no particular difficulty in generalizing the discussion to a curved interface, in which case the planar results are recovered in the limit of the radii of curvature becoming infinite; we direct the reader to Reference 9 for details. The interrelations between work and energy in its various forms at the liquid surface may be accounted for largely in terms of familiar bulk thermodynamic expressions generalized to include the interfacial region.

We consider the Helmholtz free energy of a classical multicomponent system, $A = f(T, V, \mathscr{A}, n_\alpha, n_\beta, \ldots, n_\nu)$ having a plane interface of area $\mathscr{A}$ and of the molar composition $n_\alpha, \ldots, n_\nu$ where $\alpha, \ldots, \nu$ represent the constituent species. A distinction should be made between the Helmholtz free energy A, which represents the total amount of reversible work which may be obtained from a given thermodynamic state *including* work of volumetric expansion and areal extension, and the Gibbs free energy G, which excludes these latter contributions.

The infinitesimal change in Helmholtz free energy dA arising from small changes in temperature, volume, area, and composition is given by the total differential

$$dA = \left(\frac{\partial A}{\partial T}\right)_{V,\mathscr{A},n_\alpha,\ldots,n_\nu} dT + \left(\frac{\partial A}{\partial V}\right)_{T,\mathscr{A},n_\alpha,\ldots,n_\nu} dV + \left(\frac{\partial A}{\partial \mathscr{A}}\right)_{T,V,n_\alpha,\ldots,n_\nu} d\mathscr{A}$$

$$+ \sum_\alpha^\nu \left(\frac{\partial A}{\partial n_\lambda}\right)_{T,V,\mathscr{A},n_\alpha,\ldots,n_\nu \,|n_\lambda} dn_\lambda \quad (1.4.1)$$

where by $|n_\lambda$ we mean this component is excluded. Since

$$\left(\frac{\partial A}{\partial T}\right)_{V,\mathscr{A},n_\alpha,\ldots,n_\nu} = -S \qquad \left(\frac{\partial A}{\partial V}\right)_{T,\mathscr{A},n_\alpha,\ldots,n_\nu} = -P \qquad \left(\frac{\partial A}{\partial \mathscr{A}}\right)_{T,V,n_\alpha\ldots n_\nu} = \gamma$$

and

$$\left(\frac{\partial A}{\partial n_\lambda}\right)_{T,V,\mathscr{A},n_\alpha,\ldots,n_\nu \,|n_\lambda} = \mu_\lambda \quad (1.4.1a)$$

we have immediately

$$dA = -S\,dT - P\,dV + \gamma\,d\mathscr{A} + \sum_\lambda \mu_\lambda\,dn_\lambda. \quad (1.4.2)$$

More particularly, for constant composition, temperature, and volume

$$\frac{dA}{d\mathscr{A}} = \gamma. \quad (1.4.3)$$

Since we are concerned here primarily with the thermodynamics of the interfacial region of the system we may assume that the bulk properties remain essentially unmodified by extensions of the surface area. If this is so, then the differential form of the Helmholtz equation (1.4.2) may be expressed in terms of surface *excess* quantities:

$$dA_s = -S_s\,dT + \gamma\,d\mathscr{A} + \sum_\lambda \mu_\lambda\,dn_{\lambda_s} \quad (1.4.4)$$

which is the fundamental Gibbs equation for the interfacial layer. (Guggenheim's objection[4] to the Gibbs treatment whereby the interfacial region is accorded zero volume arises from the possible inclusion of a term $-P\,dV_s$ in (1.4.4) corresponding to an extension of the thickness of the surface region, in which case there is ambiguity over the nature of the pressure within the zone. In retaining the Gibbs notation we tacitly assume the magnitude of the term is negligible and 'volumetric' terms in the general sense are restricted to extensions of area.)

Clearly, if we keep the temperature, composition, and height of the system constant and double the area, then the quantities A, V, $\mathcal{A}$ and the number of moles of each component will double: A is a homogeneous function of first degree in V, $\mathcal{A}$, and composition. It follows from application of Euler's theorem to (1.4.2) that

$$A = \sum_\lambda \mu_\lambda n_\lambda - PV + \gamma \mathcal{A}$$

whilst for the surface excess Helmholtz free energy we obtain from (1.4.4)

$$A_s = \sum_\lambda \mu_\lambda n_{\lambda_s} + \gamma \mathcal{A}.$$

We recall that A_s and n_{λ_s} depend upon the location of the dividing surface, although it may be chosen so that $\sum_\lambda \mu_\lambda n_{\lambda_s}$ is zero, in which case the above equation reduces to

$$A_s = \gamma \mathcal{A}. \tag{1.4.5}$$

For a single-component system this choice of location of the dividing surface coincides with the Gibbs equimolecular surface, and only for this choice is the surface tension equal to the superficial density of Helmholtz free energy.[9, 16]

Equation (1.4.5) has the total differential

$$dA_s = \gamma \, d\mathcal{A} + A \, d\gamma. \tag{1.4.6}$$

Combining (1.4.4) and (1.4.6) we obtain

$$\frac{d\gamma}{dT} = -S_s \tag{1.4.7}$$

for a pure liquid.

The surface excess entropy per unit area affords a certain amount of qualitative insight into the nature of the surface structure, and although we shall be considering possible interpretations of the temperature derivative $d\gamma/dT$ in more detail later, it is nevertheless appropriate to make some initial observations here. The orthodox form of the experimental $\gamma(T)$ characteristic and its derivative is shown in Figure 1.4.1 for liquid argon: a monotonic decreasing curve tending to zero as $T \to T_c$, its slope decreasing in a similar fashion. This appears fully consistent with our qualitative understanding of a progressively spatially delocalized liquid surface, a measure of whose disorder relative to the bulk liquid we may take as S_s, and which diminishes as the critical point is approached when the interfacial boundary disappears altogether. How, then, are we to account for $\gamma(T)$ characteristics which show *positive* slopes, which have been reported for a number of systems (Figure 1.4.2)? As we remarked earlier, it would appear that the temperature variation $d\gamma/dT$ is more sensitive to the structure of

9

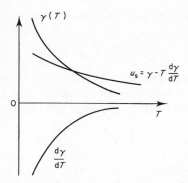

Figure 1.4.1 Schematic variation of the surface tension γ, the surface excess energy U_s and the surface excess entropy $-d\gamma/dT$ with temperature

the transition zone than is the surface tension itself. One possible conclusion which immediately comes to mind is that the molecular organization at the surface is somehow *greater* than it is in the bulk, at that temperature. Indeed, Faber[14] has calculated that on the basis of an entropy of fusion of $1.2k$ per atom, we should require the top three layers of the liquid to be crystalline to account for the positive slopes of Zn, Cd, and Cu (Figure 1.4.2). Of course, this degree of surface order is suggested only for the purposes of simple calculation; nevertheless it is one of a number of possible explanations which will be considered in the course of our discussion.

A straightforward extension of the Helmholtz equation for a pure liquid yields the following expression for the surface excess energy per unit area

$$U_s = A_s + TS_s \tag{1.4.8}$$

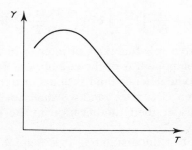

Figure 1.4.2 Schematic $\gamma(T)$ characteristic for a system exhibiting inversion of the surface tension with temperature

or alternatively, from equations (1.4.5) and (1.4.7),

$$= \gamma - T\frac{d\gamma}{dT}. \tag{1.4.9}$$

Rearrangement of (1.4.8) provides us with a considerable insight into the thermodynamic forces controlling the form of the liquid–vapour density transition profile for spherically symmetric interactions. Since

$$A_s = U_s - TS_s \tag{1.4.8a}$$

we see at once that if the bulk liquid properties extend unmodified up to a discontinuously sharp surface, then $S_s = 0$. Clearly isothermal minimization of the Helmholtz free energy favours a *broadening* of the transition zone as far as the excess entropy term in concerned, whilst the surface excess free energy contribution U_s favours a *sharpening* of the density transition, enabling the surface atoms to reside in the energetically most favourable region of the attractive surface field of the bulk fluid. It is the competition between these two agents which determines the equilibrium form of the profile, and it will be instructive to consider the results of the various calculations in these terms as the opportunity arises. From an experimental measurement of the $\gamma(T)$ characteristic U_s may be readily determined, and is generally a sensibly constant quantity (Figure 1.4.1). This has led a number of workers, particularly those working in the field of organic liquids, to consider U_s rather than γ to be the more fundamental parameter of the liquid surface.

The higher derivative, $dU_s/dT = -T\,d^2\gamma/dT^2$, may be identified as the excess specific heat per unit area, but its significance is obscure, and although an interpretation has been placed upon it by Einstein,[15] we shall consider it no further here.

We recall that the thermodynamic functions of classical homogeneous systems may be related to the structural properties of the fluid in terms of $-kT \ln Z_Q$, where Z_Q is the configurational partition function[1]

$$Z_Q = \frac{1}{N!} \int \cdots \int \exp\left(-\frac{\Phi_N(\mathbf{r}^N)}{kT}\right) d\mathbf{r}^N \tag{1.4.10}$$

The principal thermodynamic functions of interest in terms of Z_Q are listed in Table 1.4.1 for a component α of the equilibrium fluid. The expression of these quantities for homogeneous fluid systems in terms of particle correlation functions are given elsewhere[1] and we shall not reproduce them here. Instead, we shall consider their modification in the vicinity of the liquid–vapour transition zone.

At our present level of understanding it is helpful to regard the surface entropy as comprising two components: the thermal and the configurational. Disordering contributions from the former vanish at absolute zero, at least for classical systems, whilst the configurational term remains and may be related to the variety of geometric arrangements which may be adopted by

Table 1.4.1. Principal thermodynamic functions in terms of the configurational partition function Z_Q.

Helmholtz free energy	$A = -kT \ln Z_Q$
Pressure	$P = -kT \left(\dfrac{\partial \ln Z_Q}{\partial V} \right)_{T, n_\alpha}$
Chemical potential of species α	$\mu_\alpha = -kT \left(\dfrac{\partial \ln Z_Q}{\partial n_\alpha} \right)_{T, V, n_\beta, \ldots, n_\nu}$
Internal energy	$U = kT^2 \left(\dfrac{\partial \ln Z_Q}{\partial T} \right)_{V, n_\alpha}$
Entropy	$S = k \ln Z_Q + \dfrac{U}{T}$
Gibbs free energy	$G = -kT \ln Z_Q + kTV \left(\dfrac{\partial \ln Z_Q}{\partial V} \right)_{T, n_\alpha}$
Surface tension	$\gamma = -kT \left(\dfrac{\partial \ln Z_Q}{\partial \mathscr{A}} \right)_{T, V, n_\alpha, \ldots, n_\nu}$

the particles. Although we cannot in general totally separate the two regimes, it is nevertheless helpful to do so, and we show below that for an imperfect lattice, at least, the total entropy separates naturally into thermal and configurational parts.

The potential energy $\Phi_n(\mathbf{r}^n)$ of a lattice of n oscillators is given in the harmonic approximation as

$$\Phi_n(\mathbf{r}^n) = \sum_{i=1}^{3n} \tfrac{1}{2} m \omega^2 r_i^2 + \Phi_0 \qquad (\Phi_0 = \text{constant})$$

where m is the mass of the particle and ω is the Einstein frequency of the oscillator about the lattice site. Substitution in (1.4.10) yields a sum of terms each of which integrates to give $(2\pi kT/m\omega^2)$. If the lattice has $N(>n)$ sites, the integral may be taken over all N sites and will be approximately N times the integral over one site. Similarly, the second atom can occupy any of $(N-1)$ sites, whilst the nth can occupy any of $(N-n+1)$ sites. Indistinguishability of the particles requires us to divide by $n!$, whereupon we have

$$Z_Q = \exp \left(-\frac{\Phi_0}{kT} \right) \frac{N!}{(N-n)!\, n!} \left(\frac{2\pi kT}{m\omega^2} \right)^{3n}.$$

From Table 1.4.1 we see that the free energy $(-kT \ln Z_Q)$ will comprise three parts

$$A = \Phi_0 - kT \ln \{N!/(N-n)!\, n!\} - 3nkT \ln (2\pi kT/m\omega^2)$$

from which we may extract the entropy

$$S = k \ln \{N!/(N-n)!\, n!\} + 3nk \ln (2\pi kT/m\omega^2).$$

The entropy is clearly seen to be resolved into its configurational and thermal components. In principle, knowledge of the internal energy at constant volume as a function of temperature would allow the Helmholtz free energy, and hence the entropy, of a liquid to be determined by integration of the Gibbs–Helmholtz equation. At this stage such a calculation is prohibitively difficult, requiring a knowledge of the two-particle distribution as a function of P and T in the presence of the surface.

The form of the density transition profile $\rho_{(1)}(z)$ is subject to the usual constraints governing a stable two-phase coexistence: constancy of the chemical potential and normal pressure across the transition zone:

$$\mu_{\text{liquid}} = \mu(z) = \mu_{\text{vapour}}$$
$$P_{\text{liquid}} = P_{\perp}(z) = P_{\text{vapour}}. \tag{1.4.11}$$

The quasi-thermodynamic point functions $P_{\perp}(z)$ represents the normal component of the pressure tensor, and its continuity across the interface ensures mechanical stability at the free surface, whilst constancy of the chemical potential $\mu(z)$, or local Helmholtz free energy per particle, ensures the thermodynamic stability of the surface. Partial expressions apply in the case of multicomponent systems.

What we require is a rigorous definition and justification for the use of these conceptually attractive point functions in the inhomogeneous region between the bulk liquid and vapour phases, and we now consider whether the thermodynamic functions appropriate to homogeneous assemblies can be extended to the specification of local point functions. Hill[2] has adapted Kirkwood's equation[5] for the bulk chemical potential which applies for either homogeneous phase: here we have taken it for the liquid[7]

$$\mu = \mu_{\text{L}} = kT \ln \rho_{\text{L}} \Lambda^3 + \rho_{\text{L}} \int_0^1 \int_{\text{V}} \Phi(r_{12}) g_{(2)}^{\text{L}}(r_{12}, \xi) \, d\mathbf{r}_{12} \, d\xi \tag{1.4.12}$$

where $\Lambda = h/(2\pi mkT)^{1/2}$.

The integral term in (1.4.12) represents the work necessary to 'charge up' a new molecule at $\mathbf{r}_1$ by means of the coupling parameter $(\xi = 0 \rightarrow 1)$ within the homogeneous fluid: this work may be equated to $kT \ln \gamma$, where γ is the activity coefficient. Hill now eliminates the second term in the r.h.s. of (1.4.12) by taking Kirkwood's integral equation for the single-particle distribution function,[6] which is

$$kT \ln \rho_{(1)}(z_1) = kT \ln \rho_{\text{L}} + \rho_{\text{L}} \int_0^1 \int_{\text{V}} \Phi(r_{12}) g_{(2)}^{\text{L}}(r_{12}, \xi) \, d\mathbf{r}_{12} \, d\xi$$
$$- \int_0^1 \int_{\text{V}} \Phi(r_{12}) \rho_{(1)}(z_2) g_{(2)}(z_1, \mathbf{r}_{12}, \xi) \, d\mathbf{r}_{12} \, d\xi \tag{1.4.13}$$

where particle 1 is located at z_1. The result is, for any point in the system

$$\mu = \text{constant} = kT \ln \rho_{(1)}(z_1) \Lambda^3 + \int_0^1 \int_{\text{V}} \Phi(r_{12}) \rho_{(1)}(z_2) g_{(2)}(z_1, \mathbf{r}_{12}, \xi) \, d\mathbf{r}_{12} \, d\xi$$
$$= \mu(z_1) \tag{1.4.14}$$

which is seen, by comparison with (1.4.12), to reduce correctly to μ_L or μ_V in either bulk phase. The r.h.s. of (1.4.14) Hill *defines* as the local or point chemical potential $\mu(z_1)$. The integral represents, as in the homogeneous case, the work necessary to 'charge up' a new particle at z_1 and Hill relates this to a local activity coefficient $\gamma(z_1)$, where $kT \ln \gamma(z_1)$ is set equal to this work. The activity is then also constant across the transition zone.

Equation (1.4.14), which is exact to within the assumption of pairwise additivity of the potential, may be solved for the transition profile, its solution ensuring constancy of the chemical potential across the interfacial zone. As we shall see in the next chapter, the Born–Green–Yvon integro-differential equation for the single-particle distribution $\rho_{(1)}(z_1)$, developed on the basis of mechanical stability of the surface, may be shown to be equivalent to the requirement that the local chemical potential $\mu(z_1)$ remains constant in the transition zone.

The question arises as to whether these point functions can associate thermodynamically as their homogeneous counterparts do. It would appear, *a priori*, that we cannot expect the usual relations between partial derivatives of uniform point functions to hold: these latter functions depend upon only *one* intensive variable, say pressure or temperature, whereas in the present situation the point functions are coupled through the long-range parts of $\Phi(r)$ to the non-uniform regions of the density profile.

The local potential energy per particle $u(z_1)$ may be readily expressed in terms of the surface distributions as follows

$$u(z_1) = \tfrac{3}{2}kT + \frac{1}{2} \int_V \Phi(r_{12}) \rho_{(1)}(z_2) g_{(2)}(z_1, \mathbf{r}_{12}) \, d\mathbf{r}_{12} \qquad (1.4.15)$$

and is seen to be a straightforward generalization of the homogeneous result to the anisotropic surface region. Adopting the Gibbs coordinate frame (1.2.1), the surface excess energy U_s follows directly as

$$U_s(T) = \frac{1}{2} \int_{-\infty}^{0} \int_V [\rho_{(2)}(z_1, \mathbf{r}_{12}) - \rho_{(2)}^{L}(r_{12})] \Phi(r_{12}) \, d\mathbf{r}_{12} \, dz_1$$

$$+ \frac{1}{2} \int_{0}^{\infty} \int_V [\rho_{(2)}(z_1, \mathbf{r}_{12}) - \rho_{(2)}^{V}(r_{12})] \Phi(r_{12}) \, d\mathbf{r}_{12} \, dz_1 \qquad (1.4.16)$$

where $\rho_{(2)}^{L}(r_{12})$ and $\rho_{(2)}^{V}(r_{12})$ represent the homogeneous liquid and vapour pair density distributions, respectively. It is apparent from (1.4.16) that the surface excess energy is highly sensitive to the location of the origin of coordinates.

Kirkwood and Buff[8] adopt a mechanical definition of surface tension in terms of the stress transmitted across a strip of unit width, normal to a Gibbs dividing surface, and find the surface tension to be the integral of a surface excess tangential pressure

$$\gamma = \int_{-\infty}^{\infty} (P - P(z_1)) \, dz_1 \qquad (1.4.17)$$

14

in terms of the local tangential pressure point function, $P_{\parallel}(z_1)$. Instead, we shall determine the surface tension as the isothermal work of formation of unit area of interface (Table 1.4.1), in accordance with the thermodynamic definition, incidentally obtaining the result (1.4.17).

Consider a column of pure fluid of volume $l_x l_y l_z$ extending from the bulk isotropic phase, across the transition zone, into the bulk isotropic vapour (Figure 1.4.3). Now, suppose we perform the two following isothermal reversible processes:

(i) extend the volume a distance dl_x
(ii) compress the volume in the liquid and vapour phases by a total distance dl_z.

These two isothermal processes are performed in such a way that the final volume and pressure are identical to their initial values, the only difference being an isothermal increase in cross-sectional area, $d\mathscr{A}$. We point out that with the equimolecular choice of the Gibbs dividing surface (1.2.1) its location remains invariant to processes (i) and (ii).

Now, $\gamma = -kT(\partial \ln Z_Q/\partial \mathscr{A})_{V,T,N}$ and in terms of the two processes above, the first process

$$kT \left(\frac{\partial \ln Z_Q}{\partial l_x} \right)_{l_y, l_z, T} dl_x \tag{1.4.18}$$

represents the incremental change in the Helmholtz free energy by an isothermal *volumetric* expansion with the sides of the fluid column l_y, l_z held fixed. Similarly, the second process

$$-kT \left(\frac{\partial \ln Z_Q}{\partial l_z} \right)_{l_x, l_y, T} dl_z \tag{1.4.19}$$

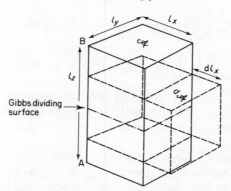

Figure 1.4.3 Symmetrical, reversible, isothermal distortion of a fluid column about the Gibbs surface yielding an increase in interfacial area $d\mathscr{A}$

represents the effect of an isothermal volumetric compression but with the sides l_x, l_y held fixed. From Table 1.4.1 it is apparent that (1.4.18), (1.4.19) represent nothing other than the x and z components of the pressure tensor. Whilst these components are unequal in regions of molecular anisotropy, mechanical stability requires that the z component $P_\perp$ be constant and equal to either of the bulk phase pressures, $P_L = P_V = P_\perp$. No such constraint applies to the tangential or x-component of the pressure, and so we anticipate that this term will vary along the z axis.

We may now write, for the entire column,

$$-kT\left(\frac{\partial \ln Z_Q}{\partial \mathscr{A}}\right)_{V,T} d\mathscr{A} = kT \int_A^B \left(\frac{\partial \ln Z_Q}{\partial l_x}\right)_{l_y,l_z,T} l_y\, dl_x\, dl_z - kT\left(\frac{\partial \ln Z_Q}{\partial l_z}\right)_{l_x,l_y,T} l_y\, dl_z.$$
(1.4.20)

The isochoric increase in surface area of the volume requires $dl_z = l_z\, dl_x$, whereupon (1.4.20) may be written

$$-kT\left(\frac{\partial \ln Z_Q}{\partial \mathscr{A}}\right)_{V,T} d\mathscr{A} = kT \int_A^B \left(\frac{\partial \ln Z_Q}{\partial l_x}\right)_{l_y,l_z,T} l_y\, dl_x\, dl_z$$

$$-kT \int_A^B \left(\frac{\partial \ln Z_Q}{\partial l_z}\right)_{l_y,l_x,T} l_y\, dl_x\, dl_z \quad (1.4.21)$$

since $(\partial \ln Z_Q/\partial l_z)_{l_y,l_x,T}$ is necessarily independent of z and $l_z = \int_A^B dl_z$. Moreover, since $d\mathscr{A} = l_y\, dl_x$, we have immediately

$$\gamma(T) = kT \int_A^B \left\{\left(\frac{\partial \ln Z_Q}{\partial l_x}\right)_{l_y,l_z} - \left(\frac{\partial \ln Z_Q}{\partial l_z}\right)_{l_x,l_y}\right\} dl_z \qquad (1.4.22)$$

and we regain the Kirkwood–Buff pressure tensor result (1.4.17)

$$\gamma(T) = \int_{-\infty}^{\infty} (P - P_\parallel(z))\, dz.$$

The range of integration has been extended to $\pm\infty$ since in either isotropic bulk phase the integrand rapidly becomes zero: the surface tension evidently may be represented as an excess pressure integral over the transition zone.

From a knowledge of the homogeneous expression for the pressure in a classical, dense isotropic system of structureless particles[1]

$$P = \rho kT - \frac{2}{3} \int_V \nabla_1 \Phi(r_{12})\rho_{(2)}(r_{12})r_{12}\, d\mathbf{r}_{12} \qquad (1.4.23)$$

we may construct the x and z components of the local pressure point functions, and obtain

$$\gamma = \frac{1}{2} \int_{-\infty}^{\infty} \nabla_1 \Phi(r_{12})\rho_{(2)}(z_1, \mathbf{r}_{12}) \frac{x_{12}^2 - z_{12}^2}{r_{12}}\, d\mathbf{r}_{12}\, dz_1. \qquad (1.4.24)$$

(Other expressions for $P_\parallel(z)$ do exist, for example that of Harasima,[10] modifying the functional form of the above expression for γ, for the local Helmholtz free energy and the local entropy.) Since the surface tension must obviously be a positive quantity, it follows from the pressure–tensor representation (1.4.17) that $P_\parallel(z)$ has the overall character of a *tension* across the transition zone. The point should also be made that, unlike certain of the other excess quantities, the various expressions do not depend upon the location of the dividing surface since the normal component of the pressure $P_\perp = P$, a constant (against which the excess pressure is determined). However, only for the equimolecular Gibbs choice of dividing surface do we have from (1.4.21) the simple relation (c.f. 1.4.14)

$$A_s = \gamma \mathscr{A} \tag{1.4.25}$$

where A_s is the surface excess Helmholtz free energy.[9] (A unique location of the equimolecular dividing surface can only be specified for a single-component system, of course. The more general relation (1.4.3) cannot be simplified in this way.) Indeed, in terms of the point function $a(z)$—the local Helmholtz free energy per particle—we are able to express A_s in a precisely analogous way to the specification of the surface excess energy (1.4.16):

$$\gamma = \int_{-\infty}^{0} [\rho_{(1)}(z_1)a(z_1) - \rho_L a_L]\, dz_1 + \int_{0}^{\infty} [\rho_{(1)}(z_1)a(z_1) - \rho_V a_V]\, dz_1$$
$$\tag{1.4.26}$$

where the single-particle distribution $\rho_{(1)}(z_1)$ has the characteristic that it minimizes the free energy of the system subject to the constraint that the number of particles remains constant, and is equivalent, of course, to the demand that γ is also minimized. Equation (1.4.26) provides a variational approach to the determination of the density distribution function, and examples of its use will be discussed in Chapter 2.

The local or point Helmholtz free energy per particle $a(z)$ in the interfacial region may be defined by the equation[2]

$$a(z_1) = \mu(z_1) - [P_\parallel(z_1)/\rho_{(1)}(z_1)] = \mu - [P_\parallel(z_1)/\rho_{(1)}(z_1)] \tag{1.4.27}$$

which, from (1.4.5),

$$= kT \ln \rho_{(1)}(z_1)\Lambda^3 + \int_{0}^{1} \int_{V} \Phi(r_{12})\rho(z_2)g_{(2)}(z_1, \mathbf{r}_{12}, \xi)\, d\mathbf{r}_{12}\, d\xi - [P_\parallel(z_1)/\rho_{(1)}(z_1)]$$
$$\tag{1.4.28}$$

or, in terms of the local activity,

$$= kT \ln \rho_{(1)}(z_1)\Lambda^3 + kT \ln \gamma(z_1) - [P_\parallel(z_1)/\rho_{(1)}(z_1)] \tag{1.4.29}$$

where, as before, $\Lambda = h/(2\pi mkT)^{1/2}$.

We go on to define a local entropy per particle in the transition zone, since $A = U - TS$

$$Ts(z_1) = u(z_1) - a(z_1) \tag{1.4.30}$$

the detailed expression following from equations (1.4.6) and (1.4.27)–(1.4.29).

It is quite apparent from the foregoing expressions that the selfconsistent specification of the density transition profile $\rho_{(1)}(z_1)$ and the pair correlation function $\rho_{(2)}(z_1, \mathbf{r}_{12})$ is central to the statistical thermodynamic description of the liquid surface. In the absence of an accurate and explicit theory of these functions various expedients have been adopted in an effort to circumvent the quite considerable difficulties involved in their specification. Fowler,[11] for example, as an initial assumption and Kirkwood and Buff,[8] for the purposes of numerical evaluation, resort to shrinking the transition zone to a mathematical surface of density discontinuity coincident with the Gibbs dividing surface, whilst the only concession to anisotropy of the pair distribution is the truncation of the isotropic function $\rho_{(2)}^L(r_{12})$ by the interfacial dividing plane. It is found, as we might expect, that in the Kirkwood–Buff formulation agreement between theoretical and experimental estimates of surface tension and surface energy deteriorates with increasing temperature. This is undoubtedly due to inadequate account being taken of the delocalization of the liquid–vapour interface as $T \rightarrow T_c$ in conjunction with the wholly inadequate representation of the pair distribution. Such an approach might be justified in the vicinity of the triple point, but becomes clearly untenable as the temperature rises. Linear and exponential profiles have been tried with moderate success, in some cases parametrically adjusted to bring the calculated surface tension and surface energy into agreement with experiment, but these cannot afford much physical insight into the basic problem which is, of course, the structural description of the interfacial transition zone.

More recently attempts have been made[12] to estimate the surface thickness or spatial delocalization of the transition zone by fitting model density profiles—linear, hyperbolic tangent, etc.—to the experimentally determined surface tension, but, again, this approach sheds virtually no light at all on the central issue. Again, simple approximate expressions for the excess free energy developed at the liquid surface may be readily obtained in calculating the entropy difference involved in replacing bulk vibrational modes with their surface counterparts. This, of course, neither requires nor yields any but the most general information as to the nature of the atomic structure of the liquid surface; indeed, it is a reflection of the relative insensitivity of the thermodynamic functions to surface structure. On the other hand, having agreed upon a dispersion relationship for the surface vibrational modes, the analysis may proceed by direct analogy with the familiar Debye theory. At low temperatures only long wavelength hydrodynamic capillary waves will be excited: the density of states function $\sigma(\omega)$ effectively suppresses the very long wavelength gravity waves, and these are neglected in such an approximation. The application of this analysis to obtain a limiting law for the temperature dependence of the surface tension of ^{4}He will be given in Chapter 6. At higher temperatures, however, the thermal excitation of nondispersive Rayleigh waves will gradually replace the capillary wave representation—this standpoint supported by suppression of the lower-frequency capillary waves by the density of states function. Of course, in such an analysis as this we need a 'characteristic surface temperature' Θ_s, corresponding to the highest vibrational frequency $\omega_{max} = k\Theta_s/\hbar$ which the surface can sustain, and Faber[13] suggests for an area of surface $\mathscr{A}$

$$\int_0^{\omega_{max}} \sigma(\omega)\, d\omega = \frac{\mathscr{A}}{a} \qquad (1.4.31)$$

where a is the area associated with a single atom, so that the number of normal modes is equal to the number of particles per unit area.

Instructive as these approaches may be, they lack much of the *a priori* objectivity of the truly molecular theory; we must eventually embark upon a full statistical mechanical analysis of the surface correlations before acceptable estimates of the thermodynamic functions are to be obtained, and this we shall proceed to develop in subsequent chapters.

Of more lasting appeal is the original phenomenological description of density fluctuations by van der Waals, more usually associated with the names of Cahn and Hilliard.[17] We know from fluctuation theory that the increase in free energy ΔA_1, associated with a number density fluctuation $\Delta\rho$ occurring in a subvolume V is

$$\Delta A_1 = cV(\Delta\rho)^2/\rho^2\chi_T \qquad (1.4.32)$$

where c is a constant, ρ is the bulk number density, and χ_T the bulk isothermal compressibility. If now we regard the superficial liquid–vapour density transition as the density fluctuation, then the excess free energy per unit area is

$$\gamma_1 = cL(\Delta\rho)^2/\rho^2\chi_T \qquad (1.4.33)$$

where L is simply a measure of the interfacial thickness. Clearly, γ_1 would be minimized for $L = 0$, a discontinuously sharp density transition, which is evidently incorrect. Cahn–Hilliard therefore further assume that there is a second contribution, γ_2, to the excess free energy associated with the inhomogeneity proportional to the square of the density gradient, yielding

$$\gamma = \gamma_1 + \gamma_2 = \frac{cL(\Delta\rho)^2}{\rho^2\chi_T} + bL\left(\frac{\Delta\rho}{L}\right)^2 \qquad b = \text{constant} > 0 \qquad (1.4.34)$$

clearly showing entropy and energy contributions. This may now be minimized with respect to the surface thickness L:

$$\frac{d\gamma}{dL} = \frac{c(\Delta\rho)^2}{\rho^2\chi_T} - b\frac{(\Delta\rho)^2}{L^2} = 0. \qquad (1.4.35)$$

The corresponding value for the surface tension then becomes

$$\gamma = \frac{2cL(\Delta\rho)^2}{\rho^2\chi_T}. \qquad (1.4.36)$$

Now, since $\Delta\rho = \rho_L - \rho_V \sim \rho_L$ we have

$$\gamma \sim 2cL/\chi_T. \qquad (1.4.37)$$

A statistical mechanical development of equation (1.4.34) is given in Section 2.4 (2.4.20 *et seq.*).

A number of authors[18, 19] have observed that the product of the surface tension and the bulk isothermal compressibility $\gamma\chi_T$ is virtually independent of the nature of the liquid at the triple point, and from (1.4.37) is seen to represent a characteristic length proportional to the surface thickness (Table 1.4.1). This empirical fact appears to have been known to Frenkel, and has since been rediscovered by Egelstaff and Widom.[19] It does illustrate that bulk and surface properties must, in many ways, be closely related. A comparison of liquid metals with nonmetallic systems is significant because inter-ionic forces in metals are density-dependent. Thus, had the density gradient at the interface not been sharp, it might have been expected to show up as a difference in the value of $\gamma\chi_T$ for metals and insulators, and no such difference is found. We remark here in passing that the density gradient approach has its counterpart in terms of electron densities as we shall see when we come to a specific discussion of liquid metal systems (Chapter 5).

The relation appears to hold for systems of quite dissimilar binding,[18] including molten salts, organic liquids, and aqueous solutions in addition to liquid metals and the liquid inert gases—indeed, $\gamma\chi_T$ for ^{4}He and ^{3}He extrapolated to 0 °K agrees with the classical liquids. For water, however, the product $\gamma\chi_T$ appears to pass through a shallow minimum over the range 20–40 °C before showing a conventional rise of interfacial broadening with increasing temperature.

References

1. C. A. Croxton, *Liquid State Physics—A Statistical Mechanical Introduction*, Cambridge University Press, London and New York (1974).
   G. H. A. Cole, *An Introduction to the Statistical Theory of Classical Simple Dense Fluids*, Pergamon, Oxford (1967).
   P. A. Egelstaff, *An Introduction to the Liquid State*, Academic Press, London (1967).
   J. S. Rowlinson, *Liquids and Liquid Mixtures*, 2nd Ed., Butterworth, London (1969).
   S. A. Rice and P. Gray, *The Statistical Mechanics of Simple Liquids*, Interscience, New York (1965).
   I. Z. Fisher, *Statistical Theory of Liquids*, University of Chicago Press, Chicago (1961).
   T. M. Reed and K. E. Gubbins, *Applied Statistical Mechanics*, McGraw-Hill Kogakusha, Tokyo (1973).
   C. A. Croxton, *Introduction to Liquid State Physics*, John Wiley & Sons Ltd, London (1975).
2. T. L. Hill, *J. Chem. Phys.*, **30,** 1521 (1959).
3. J. W. Gibbs, *Collected Works*, Vol. 1, Yale University Press, New Haven (1928).
4. E. A. Guggenheim, *Trans. Faraday Soc.*, **36,** 397 (1940); *Thermodynamics*,

20

North Holland, Amsterdam (1959). (The criticisms have been omitted from the 5th edition, 1967.)

G. Scatchard, *J. Phys. Chem.*, **66,** 618 (1962).

E. A. Guggenheim and N. K. Adam, *Proc. Roy. Soc.*, **A139,** 218 (1933).

G. Bakker, Handbook of Experimental Physics, Vol. 6. *Capillarity and Surface Tension*, Akadverlagsgesellschaft, Leipzig (1928).

J. E. Verschaffelt, *Bull. Sci. Acad. Roy. Belge.*, **22,** 373 (1936).

E. A. Flood, *Canad. J. Chem.*, **33,** 979 (1955); *The Solid/Gas Interface*, Vol. 1, Dekker, New York (1966).

D. W. G. White, *Metals, Materials and Metallurgical Reviews* (July 1968).

5. J. G. Kirkwood, *J. Chem. Phys.*, **3,** 300 (1935).

6. J. G. Kirkwood and E. M. Boggs, *J. Chem. Phys.*, **10,** 394 (1942).

7. In this, and the successive equations relating to the point chemical potential and point Helmholtz free energy we have, with Hill,[2] adopted the Kirkwood coupling parameter representation. An alternative method which avoids the use of the coupling parameter ξ, but to which it is nevertheless equivalent, is discussed by T. M. Reed and K. E. Gubbins, *Applied Statistical Mechanics*, McGraw-Hill Kogakusha, Tokyo (1973), p. 183.

8. J. G. Kirkwood and F. P. Buff, *J. Chem. Phys.*, **17,** 338 (1949).

9. S. Ono and S. Kondo, *Hand. Phys.* **10,** 134 (1960).

10. A. Harasima, *Adv. Chem. Phys.*, **1,** 203 (1958).

11. R. H. Fowler, *Proc. Roy. Soc.*, **A159,** 229 (1937).

12. D. D. Fitts and W. J. Welsh, *Chem. Phys.*, **26,** 379 (1977).

13. T. E. Faber, *An Introduction to the Theory of Liquid Metals*, Cambridge University Press, London and New York (1972), p. 86.

14. T. E. Faber, *An Introduction to the Theory of Liquid Metals*, Cambridge University Press, London and New York (1972), p. 90.

15. A. Einstein, *Ann. Physik*, **4,** 513 (1901).

16. R. H. Fowler, *Proc. Roy. Soc.* **A159,** 229 (1937).

17. J. W. Cahn and J. E. Hilliard, *J. Chem. Phys.*, **28,** 258 (1958).

18. R. D. Present, *J. Chem. Phys.*, **61,** 4267 (1974).

19. P. A. Egelstaff and B. Widom, *J. Chem. Phys.*, **53,** 2667 (1970).

CHAPTER 2

The Surface Correlations $\rho_{(1)}(z_1)$ and $\rho_{(2)}(z_1, \mathbf{r}_{12})$ in Simple Classical Fluids

2.1 Introduction

Although formal statistical mechanical expressions are available for the principal thermodynamic and quasi-thermodynamic point functions at the liquid surface (Chapter 1), *a priori* calculations require a knowledge of the one- and two-particle density distributions $\rho_{(1)}(z_1)$ and $\rho_{(2)}(z_1, \mathbf{r}_{12})$ throughout the transition zone. These surface modifications of the bulk distributions remain theoretically and experimentally elusive, and such information as we do have is indirect, circumstantial, and subject to considerable experimental uncertainty. The density transition curves reported so far are highly controversial in as far as both monotonic and oscillatory profiles have been suggested. Computer simulations have not helped since both monotonic and oscillatory transitions have been obtained (Chapter 10)! A consensus of opinion is developing, however.

The form of the transition profile, or single-particle density distribution $\rho_{(1)}(z)$ along a z-axis directed normally across a planar interface from liquid to vapour is characterized by constancy of the normal pressure $P_\perp(z)$ and chemical potential $\mu(z)$ across the interfacial zone, and together ensure the mechanical and thermodynamic stability of the interface. The specification of these quasi-thermodynamic point functions within the inhomogeneous surface region necessarily involves the anisotropic one- and two-particle distributions (Chapter 1) which, moreover, are hierarchically related (1.3.6) and combine self-consistently as follows:

$$\rho_{(2)}(z_1, \mathbf{r}_{12}) = \rho_{(1)}(z_1)\rho_{(1)}(z_2)g_{(2)}(z_1, \mathbf{r}_{12}) \qquad (2.1.1)$$

where $g_{(2)}(z_1, \mathbf{r}_{12})$ is the anisotropic radial distribution function. By selfconsistently we mean that the one- and two-particle distributions cannot develop independently but are intimately related through an expression such as (1.3.7). Clearly, any attempt to determine $\rho_{(1)}(z)$ or any of the thermodynamic functions of the surface explicitly or implicitly involves the specification of $\rho_{(2)}(z_1, \mathbf{r}_{12})$. It therefore appears that, from the outset, we are confronted with the description of the inhomogeneous pair function, an exceedingly difficult problem for which no entirely satisfactory prescription has yet been proposed. However, numerous approximate 'closures', as we shall call them, since they enable an iterative determination of $\rho_{(1)}(z)$ to proceed, have been suggested and have yielded single-particle distributions which may be subsequently used in the calculation of surface thermodynamic quantities. These latter, as we have mentioned before, are insensitive to the precise form of the single-particle distribution, generally arising as they do as integrals over the transition zone. Numerical coincidence of the experimental and theoretical values for a thermodynamic parameter is a necessary but insufficient criterion of acceptability of the calculated density profile $\rho_{(1)}(z)$.

To help fix ideas we may consider a general algorithm for the determination of $\rho_{(1)}(z)$ and thence the associated thermodynamic quantities. Few, if any, analytical profiles have been obtained, certainly none for realistic systems, and so an iterative procedure is necessarily adopted. This schematically illustrated method of solution (Figure 2.1.1) underlies virtually all the determinations of $\rho_{(1)}(z)$ yet reported. First, we have to specify a 'closure', a convenient representation of $\rho_{(2)}(z_1, \mathbf{r}_{12})$. One of the earliest and simplest is that of Green:[1]

$$\rho_{(2)}(z_1, \mathbf{r}_{12}) \sim \rho_{(1)}(z_1)\rho_{(1)}(z_2)g_{(2)}^{L}(r_{12}). \qquad (2.1.2)$$

We see that a degree of anisotropy is expressed through the single-particle distributions, whilst for the anisotropic radial distribution function (c.f. 2.1.1) Green adopts the *isotropic* bulk liquid form, $g_{(2)}^{L}(r_{12})$. We should realize that at this stage we have forfeited simultaneous mechanical and thermodynamic stability, or both: only the exact pair distribution will ensure both constraints on the profile are simultaneously satisfied. Given an equation relating the one- and two-particle distributions, and assuming an initial guess for the single-particle distribution $\rho_{(1)}^{(0)}(z)$ (perhaps a step function), solution may proceed.

The iteration continues, subject to certain boundary conditions ($\rho_{(1)}(-\infty) = \rho_L$), until stable convergence of the profile appears to have developed: an example of the intermediate and final stages of an iterated profile is shown in Figure 2.1.2.[2] Of course, the solution rarely, if ever, evolves as smoothly as the above algorithm might suggest; numerous complications and technicalities conspire to yield physically unacceptable or divergent solutions. Nevertheless, Figure 2.1.2 gives a schematic indication of the general procedure. Quite clearly, it is primarily in the specification of the closure device that approximation is introduced, although further numerical error is

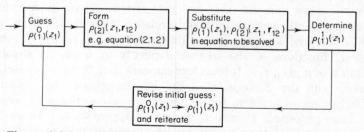

Figure 2.1.1 Algorithm for the iterative determination of the single-particle interfacial density profile, $\rho_{(1)}(z)$

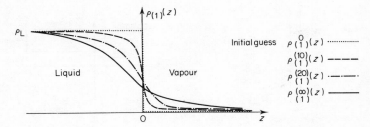

Figure 2.1.2 Schematic form of iterative solutions relaxing from an initial guess, a step function

incurred in the course of iteration, assuming the equation being solved is itself exact. Consequently, it will be necessary to examine the specification of $\rho_{(2)}(z_1, \mathbf{r}_{12})$ quite closely in assessing the reported single-particle distribution.

A useful general approach[3] is to express the density profile in Boltzmann form, relating it to a hypothetical single-particle potential of mean force, $\Psi(z_1)$:

$$\rho_{(1)}(z_1) = \rho_L \exp\left(-\Psi(z_1)/kT\right). \qquad (2.1.3)$$

The associated surface constraining field, $-\nabla\Psi(z_1)$, is another useful concept. This point function expresses the net force exerted on a particle in the vicinity of the surface and provides a means of qualitative assessment of the various closure devices, as we shall see in due course.

All closures for $\rho_{(2)}(z_1, \mathbf{r}_{12})$ so far proposed presuppose a knowledge of the *homogeneous* pair distribution $g_{(2)}(r_{12} \mid \rho, T)$ over a continuous range of densities from that of the bulk liquid to that of the bulk vapour—in some cases at densities for which a uniform fluid is unstable.† Indeed, the use of these isotropic functions underlies the simplification introduced in an attempt to represent the more complicated exact function $\rho_{(2)}(z_1, \mathbf{r}_{12})$. In the remainder of this chapter we shall therefore consider first of all the calculation of isotropic pair distribution functions, and go on to introduce the various integral equations for $\rho_{(1)}(z)$, illustrating their closure and solution for the simplest of systems, the classical single-component liquid inert gases.

2.2. The equation of state of a non-uniform fluid

The stable coexistence of two equilibrium bulk phases at the same temperature is, as we know, characterized by their having the same chemical potential and pressure: we require that each of these thermodynamic quantities be respectively identical at all points A and A′ throughout each interior phase in Figure 2.2.1a. The coexistence is achieved by virtue of the

† The surface field stabilizes inhomogeneous systems over regions of the (ρ, T)-plane inherently unstable for uniform fluids; however, this standpoint has recently been contested by Kalos *et al.*[93] amongst others. See Section 2.11.

24

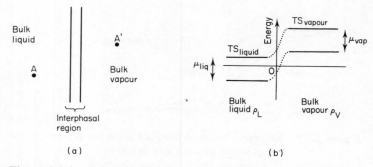

Figure 2.2.1 Coexisting liquid and vapour phases characterized by constancy of normal pressure and chemical potential μ across the interface

fact that at constant PVT the chemical potential, or free energy per particle, is simply the difference between the total U and the unavailable energy TS: in the bulk liquid phase the entropy term is relatively small, whilst in the vapour it is relatively large. Nevertheless, for two distinct values of density ρ_L, ρ_V identical chemical potentials are obtained, and they therefore form a stable coexistence. The two phases are, however, gravitationally resolved as distinct liquid and vapour regions. The locus of coexisting densities is readily determined from a consideration of the stability of the system with respect to spinodal decomposition, and analyses based on thermodynamic[83] and statistical mechanical[84] reasoning have been advanced. A similar line of reasoning enables us to understand the development of coexisting fluid densities in terms of the condition for mechanical stability—equality of the bulk pressures in either phase. In this case, the pressure represents the difference between a kinetic and a configurational term whose relative magnitude remains constant throughout the system but whose absolute magnitudes nevertheless differ in either phase.

For a single component fluid there can be two and only two subcritical coexisting values of the density ρ_L and ρ_V at any given temperature. How is it then that densities $\rho_L \geqslant \rho(z) \geqslant \rho_V$ develop in the interphasal region of Figure 2.1.1a, representing states which for a bulk isotropic fluid are inherently unstable (Figure 2.2.2)?

The answer, of course, is to be attributed to the essential *inhomogeneity* of the interphasal zone. In the immediately subcritical region $(T \leqslant T_c)$ the free energy per particle is of the form

$$a[\rho(\mathbf{r})] + C[\rho'(\mathbf{r})]^2 \qquad C = \text{constant} > 0 \qquad (2.2.1)$$

to first order in the density gradient, a result of Cahn and Hilliard[83] which derives from earlier work of van der Waals. More recently a statistical mechanical basis has been developed for this well-known 'squared-gradient' result, from which it appears that there are, not surprisingly, departures

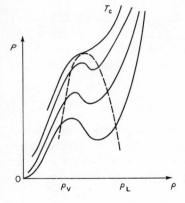

Figure 2.2.2 Phase diagram showing locus of stability of the coexisting bulk liquid and vapour phases as a function of temperature

from Cahn–Hilliard behaviour at lower temperatures.[84] The squared-gradient dependence nevertheless persists, at least to first order. A qualitative justification of the form of equation (2.2.1) for the per particle free energy in an inhomogeneous region follows from a consideration of the surface *excess* energy and entropy. The excess energy contribution is minimized for an infinitely sharp interphasal zone ($|\rho'| = \infty$) representing a clear resolution of the system into the bulk liquid and bulk vapour phases. The excess entropy, on the other hand, is maximized for a completely delocalized interface ($\rho' = 0$): the minimum interfacial free energy per particle is achieved for some intermediate density profile which, of course, is characterized by invariance of the chemical potential across the transition zone.

Now, in either isotropic bulk phase $\rho'(\mathbf{r}) = 0$, whereupon for stable coexistence $a[\rho_L] = a[\rho_V]$. In regions of structural anisotropy, however, $a[\rho(\mathbf{r})]$ or more particularly $\rho(\mathbf{r})$ is no longer restricted to one or other of the two coexistence values—the appropriate chemical potential may be achieved in conjunction with the product $C[\rho'(\mathbf{r})]^2$. The consequence is, of course, that the inhomogeneity or surface field effectively *stabilizes* the bulk isotherms within the region $[\rho_L, \rho_V]$, and for the purposes of the theoretical description of interfacial thermodynamics we shall necessarily require a knowledge of the equation of state and the associated molecular structure not only over the conventional liquid and vapour branches but also over the region $\rho_V \lesssim \rho \lesssim \rho_L$. Obviously experiment or machine simulation of bulk systems will be of no assistance here if interphasal description is to be made in these terms.[93]

Probably the most successful description so far of high density bulk Lennard–Jones fluids is afforded by the perturbation theories and their derivatives, details of which may be found elsewhere in the literature.[85] Suffice it to say that, in general, these treatments regard the attractive branch of the interaction as a perturbative departure from a hard sphere interaction, reflecting the role of the hard core as the primary agent responsible for the structural features of high density fluids. The agreement with experiment and machine simulation is striking, although the discrepancy at low densities is quite serious. For the purposes of interfacial description we shall, of course, require a knowledge of the equation of state and its associated structure over the *entire* density range. Osborn and Croxton[87] have proposed the linear interpolation

$$g_{(2)}(r) = \alpha \exp(-\Phi(r)/kT) + (1-\alpha)g_{(2)}(r)_{\text{WCA}} \qquad (2.2.2)$$

where $0 \leqslant \alpha \leqslant 1$ over the range $0 \leqslant \rho \leqslant \rho_0$, an arbitrary density above which the WCA representation is considered adequate. Such an interpolation between the Mayer and Weeks–Chandler–Anderson distribution over the density range $\rho_V \leqslant \rho \leqslant \rho_L$ would, on the basis of the virial equation

$$P = \rho kT - \frac{2\pi\rho^2}{3} \int_0^\infty \nabla_1 \Phi(r_{12}) g_{(2)}(r_{12}) r_{12}^3 \, dr_{12} \qquad (2.2.3)$$

yield the usual *sinuous* isotherm. The z component of the pressure tensor $P_\perp$ in a region of structural anisotropy is, however, given by (1.4.24)

$$P_\perp(z_1) = \rho_{(1)}(z_1)kT - \frac{1}{2} \int \nabla_1 \Phi(r_{12}) \rho_{(1)}(z_1) \rho_{(1)}(z_2) g_{(2)}(z_1, \mathbf{r}_{12}) \frac{z_{12}^2}{r_{12}} \, d\mathbf{r}_{12}$$
$$(2.2.4)$$

at a point z_1 in the transition zone. The $g_{(2)}(z_1, \mathbf{r}_{12})$ is appropriate to a density $\rho_V \leqslant \rho \leqslant \rho_L$, and for a correct specification of

$$\rho_{(1)}(z_1)\rho_{(1)}(z_2)g_{(2)}(z_1, \mathbf{r}_{12}),$$

$P_\perp$ *remains constant* across the density transition, as of course it must for mechanical stability.

Our concern regarding the specification of the structure and equation of state over the entire fluid range, in particular within the interval $\rho_L \geqslant \rho \geqslant \rho_V$, is not confined to the maintenance of thermodynamic and mechanical stability alone. Within the transition zone the anisotropic pair distribution $\rho_{(2)}(z_1, \mathbf{r}_{12})$ remains to be specified as we have seen from the various expressions developed throughout this book. Its expression, in anything other than purely formal terms, is generally based on isotropic distributions appropriate to the local density within the interphasal region: the various approximations or *closures* are discussed in detail in Section 2.8. Without pre-empting that discussion, we point out that an appreciation of the demands made upon the equation of state provides an important physical criterion for the acceptability or otherwise of the closure approximation. For example, one of the most widely used approximations is that of Green:[86]

$$\rho_{(2)}(z_1, \mathbf{r}_{12}) \sim \rho_{(1)}(z_1)\rho_{(1)}(z_2)g_{(2)}(r_{12} \mid \rho_L) \qquad (2.2.5)$$

where $g_{(2)}(r_{12} \mid \rho_L)$ is the isotropic bulk liquid distribution appropriate to the temperature and pressure in question. *A priori* considerations would suggest that (2.2.5) as it stands, whilst approximate, is nevertheless unobjectionable; it contains an angular dependence and is symmetrical with respect to exchange of particle coordinates. If, however, we determine the equation of state for such a fluid, that is evaluate (2.2.3) with fixed (density-independent) $g_{(2)}(r \mid \rho_L)$ and temperature, the virial equation of state simply becomes quadratic in density and consequently there can be no two coexisting fluid phases: obviously the density dependence of $g_{(2)}(r)$ is crucial in any adequate specification of the interphasal structure. An obvious refinement of (2.2.5) is to replace $g_{(2)}(r_{12} \mid \rho_L)$ by $g_{(2)}[r_{12} \mid \rho_{(1)}(z_1)]$. If this is substituted in (2.2.3) it does, of course, yield an equation of state which at least exhibits a liquid–vapour coexistence. Whether its insertion in (2.2.4) will yield a constant normal pressure $P_\perp$ is, however, another matter. Anyway, there are other grounds for rejecting this particular approximation, as we shall see.

Much more disturbing is the possibility that $g_{(2)}(z_1 \mid \mathbf{r}_{12})$ may differ *fundamentally* from either bulk phase distributions, whilst still permitting both mechanical and thermodynamic stability. Such a proposition has been made recently by Kalos et al.,[93] amongst others, who suggest that the correlation range of the pair distribution may extend over *macroscopic* dimensions parallel to the surface within the interphasal zone: clearly, there is no prospect whatsoever of constructing such distributions from either of the bulk functions (Section 2.11).

2.3. The Kirkwood–Buff–Fowler (KBF) step model of the liquid surface

In the absence of an accurate and explicit theory of the functions $\rho_{(1)}(z_1)$ and $\rho_{(2)}(z_1, \mathbf{r}_{12})$, both of which are required for the calculation of the surface tension and surface energy, Kirkwood and Buff,[4] and Fowler,[5] (KBF), by quite independent methods, considered a single-component fluid with the distribution functions

$$\rho_{(1)}(z_1) = \rho_L \qquad z_1 \leqslant 0$$
$$= 0 \qquad z_1 > 0 \qquad (2.3.1)$$

$$\rho_{(2)}(z_1, \mathbf{r}_{12}) = \rho_{(1)}(z_1)\rho_{(1)}(z_1 + z_{12})g_{(2)}^L(r_{12}).$$

Thus, the homogeneous fluid extends up to a surface of density discontinuity coincident with the Gibbs surface, beyond which the vapour is of zero density. $g_{(2)}^L(r_{12})$, the radial distribution appropriate to the homogeneous bulk phase, is retained right up to the surface boundary and for which Kirkwood and Buff fitted an analytic approximation to the scattering data of Eisenstein and Gingrich.[6]

For those regions shown in Figure 2.3.1a, where the region of integration is spherically symmetric, the contribution to the surface tension in equation (1.4.24)

$$\gamma = \frac{1}{2} \int_{-\infty}^{\infty} \nabla_1 \Phi(r_{12}) g_{(2)}(z_1, \mathbf{r}_{12}) \frac{x_{12}^2 - z_{12}^2}{r_{12}} \, d\mathbf{r}_{12} \, dz_1$$

is zero. However, when $r_{12} > z_1$, as in Figure 2.3.1b, transforming to

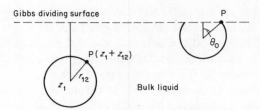

Figure 2.3.1 Geometry for calculation of the Kirkwood–Buff expression for the surface tension (equation 2.3.2)

spherical coordinates we have, for the angular integration in (1.4.24),

$$\int_0^{2\pi} \int_{\theta_0}^{\pi} (x_{12}^2 - z_{12}^2) \sin \theta \, d\theta \, d\varphi = \frac{\pi z_1 (r_{12}^2 - z_{12}^2)}{r_{12}}.$$

Equation (1.4.24) now becomes

$$\gamma = \frac{\pi \rho_L^2}{2} \int_{-\infty}^{\infty} z_1 \int_z^{\infty} g_{(2)}^L(r_{12}) \nabla_1 \Phi(r_{12}) (r_{12}^2 - z_{12}^2) r_{12} \, dr_{12} \, dz_1$$

which may be reduced to

$$\gamma = \frac{\pi \rho_L^2}{8} \int_0^{\infty} \nabla_1 \Phi(r_{12}) g_{(2)}^L(r_{12}) r_{12}^4 \, dr_{12} \qquad (2.3.2)$$

representing the surface tension at an interface between a liquid phase and a vapour phase of negligible density. For the surface excess energy they obtain

$$U_s = \frac{-\pi \rho_L^2}{2} \int_0^{\infty} g_{(2)}^L(r_{12}) \Phi(r_{12}) r_{12}^3 \, dr_{12}. \qquad (2.3.3)$$

With a knowledge of the effective pair potential $\Phi(r_{12})$ and an analytic fit to the experimental radial distribution $g_{(2)}^L(r_{12})$, surprisingly good estimates for the surface tension and surface energy of liquid argon at the triple point were obtained (Table 2.3.1). Buff[12] has extended the theory to the curved surface case.

Obviously the actual transition profile will have a more relaxed form than that assumed by Kirkwood and Buff, and presumably has a lower excess Helmholtz free energy per unit area whereupon we can understand the

Table 2.3.1. Surface tension and surface energy of liquid Argon at the triple point (90 °K)

	Kirkwood & Buff[4]†	Hill, modified by Hill[7] Plesner & Platz[8]	Plesner & Platz[8]	Shoemaker et al.[9]	Croxton & Ferrier[10]	Expt[108]
γ dyne/cm	16.84	6.91	21.6	16.55	15.6	13.48	13.45
U_s erg/cm^2	44.3	19.43	60.59	50.55	27.08	35.35	35.01

† Estimated from their 90 °K values by $\gamma = \gamma_0 (1 - T/T_c)^{1.28}$, $U_s = -T(d\gamma/dT)$.

inequalities $\gamma_{(step)} > \gamma_{(expt)}$ and $U_{s(step)} < U_{s(expt)}$ (See equation (1.4.8a) and subsequent discussion.) In the Kirkwood–Buff formulation, agreement between theory and experiment for the surface tension and surface energy naturally deteriorates with increasing temperature when the step model becomes progressively inappropriate with increasing surface delocalization.[99]

Harasima[13] has calculated the tangential and normal pressures $P_\parallel, P_\perp$ on the basis of the components of the pressure tensor

$$P_\parallel(z_1) = kT\rho_{(1)}(z_1) - \frac{1}{2}\int \nabla\Phi(r_{12})g_{(2)}(z_1, \mathbf{r}_{12})\frac{x_{12}^2}{r_{12}}\,d\mathbf{r}_{12}$$

$$P_\perp(z_1) = kT\rho_{(1)}(z_1) - \frac{1}{2}\int \nabla\Phi(r_{12})g_{(2)}(z_1, \mathbf{r}_{12})\frac{z_{12}^2}{r_{12}}\,d\mathbf{r}_{12} \qquad (2.3.4)$$

using the same approximations as Kirkwood and Buff (2.3.1). The results are shown in Figure 2.3.2, from which we can see that the system is far from hydrostatic equilibrium: the normal pressure $P_\perp$, instead of being constant as it should be, rises to a maximum value of 660 atm at $x = -2.4$ Å. $P_\parallel(z)$ shows a similar behaviour, although their difference yields a reasonable estimate when based upon the mechanical definition of surface tension in the form $\gamma = \int_{-\infty}^{\infty}[P_\perp(z) - P_\parallel(z)]\,dz$, but not, of course, on the basis of $\int_{-\infty}^{\infty}[P - P_\parallel(z)]\,dz$ in which the condition for hydrostatic equilibrium, $P_\perp(z) = P$, is assumed.

There have been attempts to compute the effect of an exponential type of transition zone using the Kirkwood–Buff analysis, but the discrepancy with experiment is increased rather than decreased. Exponential profiles with a single adjustable parameter, the 'relaxation length' of the transition zone, have been used by Berry et al.[14] in the Kirkwood–Buff approximation, and by fitting the experimental values for the surface tension conclude that for a number of systems (N_2, O_2, CH_4, Ne, Ar) at the triple point the transition is complete within two or three atomic diameters. Linear,[15,16] cubic,[16] and tanh[17] profiles have also been tried with moderate success, but such approaches cannot, of course afford much physical insight into the structural problem.

Shoemaker et al.[9] have utilized recent X-ray scattering determinations of the radial distribution function for the evaluation of the Kirkwood–Buff–Fowler relations for γ and U_s: these results, summarized in Table 2.3.1, may

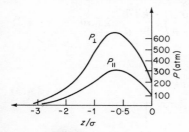

Figure 2.3.2 The tangential and normal components of the pressure tensor (equation 2.3.4) in the vicinity of the liquid–vapour interface

be taken as the best possible estimate of the thermodynamic parameters of the surface on the basis of the KBF model.

2.4 Quasi-thermodynamic approaches—the chemical potential and free energy.

The difficulties arising in the Kirkwood–Buff calculation of the surface properties (Section 2.3) arise from the over-restrictive assumptions (2.3.1) which take inadequate account of the delocalization of the liquid–vapour density transition $\rho_{(1)}(z_1)$ and the adoption of the isotropic bulk liquid radial distribution $g_{(2)}^{L}(r_{12})$ in place of the inhomogeneous function $g_{(2)}(z_1, \mathbf{r}_{12})$.

A statistical mechanical specification and justification for the use of quasi-thermodynamic point functions in the nonhomogeneous transition zone, such as the local chemical potential $\mu(z_1)$ and local Helmholtz free energy per particle $a(z_1)$ has been given by Hill.[18] And as we saw in Chapter 1, the constraints of chemical, mechanical, and thermodynamic equilibrium which characterize the stable transition zone yield integral equations (1.4.14), (1.4.26), (1.4.28) for the single-particle profile $\rho_{(1)}(z_1)$. These equations necessarily involve an explicit, or at least implicit, knowledge of the pair distribution $\rho_{(2)}(z_1, \mathbf{r}_{12})$ for their solution.

A purely quasi-thermodynamic attempt to determine the density profile $\rho_{(1)}(z_1)$ on the basis of the constancy of the chemical potential has been subsequently refined by several investigators and, despite a number of inconsistencies between assumptions concerning the contributions of energy and entropy to the chemical potential, remains one of the important *a priori* quasi-thermodynamic approaches to the surface structure.

The local density ρ is related to the local chemical potential μ in terms of the tangential component of the stress tensor $P_{\parallel}$ through the Gibbs–Duhem equation[11]

$$\left(\frac{\partial \mu}{\partial P_{\parallel}} \right) = \frac{1}{\rho}. \tag{2.4.1}$$

The simplest model of any value is that based on the van der Waals equation which allows $P_{\parallel}$ to be related to the density ρ. In such a 'smoothed potential' model both the potential energy per molecule and the free volume per molecule are functions of density only. It follows directly[19] by integration of the Gibbs–Duhem equation that the chemical potential μ is given as

$$\mu = \mu_0(T) + kT \ln \left(\frac{kT}{b} \right) + kT \ln \left(\frac{\theta}{1-\theta} \right) + \frac{kT\theta}{1-\theta} - kT\alpha\theta \tag{2.4.2}$$

where

$$\left.
\begin{array}{ll}
\theta = \dfrac{Nb}{V} = \rho b & \alpha = \dfrac{2a}{bkT} \\[3mm]
a = \varepsilon b & b = \dfrac{2\pi\sigma^3}{3}
\end{array}
\right\} \tag{2.4.3}$$

where the integration constant $\mu_0(T)$, a function of temperature only, is determined such that μ approaches the ideal gas value in the limit of infinite dilution. a and b are the usual van der Waals molecular constants, and $-kT\alpha\theta$ is the potential energy of interaction of a molecule with the rest of the fluid assuming the Sutherland interaction potential

$$\left.\begin{aligned}\Phi(r_{12}) &= -\varepsilon\left(\frac{\sigma}{r_{12}}\right)^6 & r_{12} \geqslant \sigma \\ &= +\infty & r_{12} < \sigma\end{aligned}\right\}$$
(2.4.4)

and the (inconsistent) radial distribution

$$\left.\begin{aligned}g_{(2)}(r_{12}) &= 1 & r_{12} \geqslant \sigma \\ &= 0 & r_{12} < \sigma\end{aligned}\right\}.$$
(2.4.5)

As it stands, the local point functions involved in (2.4.1) are determined solely in terms of the density *at that point*. To this extent the liquid transition zone may be imagined as divided into elemental strata each of which constitutes a thermodynamic entity, as shown in Figure 2.4.1.

Hill[20] generalizes the above treatment to take account of interactive coupling to the spatial variation of the density, rather than assuming, as we have in setting up (2.4.2), that it is the density at a point which entirely defines the potential energy and the entropy *at that point*. Clearly the specification of these local point functions involves its coupling to the neighbouring inhomogeneous regions of the fluid through the long range part of the pair potential (2.4.4). Hill is able to take partial account of this, and modifies the potential term in (2.4.2), $-kT\alpha\theta$, to read

$$kT\Psi(z_1) = \int \rho_{(1)}(z_1 + z_{12})g_{(2)}(r_{12})\Phi(r_{12})\,\mathbf{dr}_{12}$$
(2.4.6)

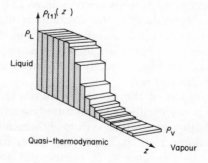

Figure 2.4.1 Subdivision of the transition zone into elemental strata for the quasi-thermodynamic description of the interface

whilst he assumes that it is the local density at each point which determines the entropy contribution to the chemical potential in (2.4.2).

On the basis of the interaction (2.4.4) and distribution (2.4.5) the bulk equation of state may be expressed as[21]

$$P = P_{HS}(\rho) - \frac{24\varepsilon}{kT}\,\eta^2 \qquad (2.4.7)$$

where $\eta = \frac{1}{6}\pi\sigma^3\rho$. The first term in (2.4.7) represents a purely density-dependent hard sphere equation of state whilst the second term represents the van der Waals correction arising from the attractive branch of the potential (2.4.4). For $P_{HS}(\rho)$ Hill utilizes the Tonks[22] equation of state, whilst a more recent determination by Plesner and Platz[8] uses the Reiss–Frisch–Lebowitz[23] equation for a hard sphere fluid. Both analyses, however, retain the distribution (2.4.5) for the specification of the particle potential $\Psi(z)$ (2.4.6) instead of the inhomogeneous function $g_{(2)}(z_1, \mathbf{r}_{12})$. Plesner and Platz obtain the following nonlinear equation for the density profile

$$\ln\frac{\eta(z)}{1-\eta(z)} + \frac{7\eta(z)}{1-\eta(z)} + \frac{15\eta^2(z)}{2(1-\eta(z))^2} + \frac{3\eta^3(z)}{(1-\eta(z))^3} + \Psi(z) = \text{constant}$$

$$(2.4.8)$$

in terms of the reduced density $\eta(z) = \frac{1}{6}\pi\sigma^3\rho_{(1)}(z)$, which may be solved numerically for the reduced profile, $\eta(z)$.

The results, as a function of reduced temperature T/T_c, are qualitatively similar to those of Hill and are shown in Figure 2.4.2. An obvious criticism is, of course the adoption of an inconsistent pair interaction and pair distribution, the latter of which is quite structureless and, moreover, isotropic, and could only be expected to yield monotonic profiles. One of the most controversial questions concerns the development or otherwise of *structured* transition profiles; clearly the present approach as it stands is incapable of resolving the dilemma. Again, the inconsistent methods of calculation of the energy and entropy contributions to the chemical potential are to be criticized. Nevertheless, the calculation of surface energy and surface tension may now proceed, and the results are shown in Table 2.3.1. Also shown in the table are Hill's original results, and a corrected version using parameters ε and σ which correctly reproduce the critical temperature and density, which Hill's did not.

But for the specification of the potential energy term $\Psi(z)$ (2.4.6), the preceding treatment is essentially quasi-thermodynamic, and more appropriate to low-density high-temperature systems in which the elemental strata, into which the transition zone is effectively subdivided, constitute thermodynamic entities. In this case it is meaningful to discuss them in a thermodynamic rather than a statistical mechanical sense. For low-temperature systems possessing a relatively sharp transition zone the elemental subdivision would have to be so fine ($\ll\sigma$) that it is no longer meaningful to regard them as thermodynamic systems, and the propriety of

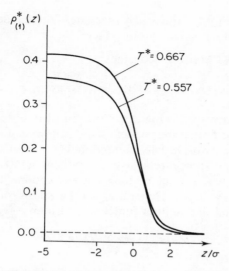

Figure 2.4.2 The reduced density transition profiles as a function of reduced temperature determined on the basis of constancy of the chemical potential across the transition zone[8] (equation 2.4.8)

the assumption that it is the properties at the point which determine the value of the point function (even with Hill's refinement for Ψ) becomes questionable. Plesner *et al.* have subsequently extended their treatment to include binary mixtures[81] (Section 4.4), and a similar analysis has been made by Buff and Stillinger[24] for the distribution of an electrolyte against a metallic surface.

The condition for mechanical stability of the free surface is, of course, *not* satisfied. The pressure within each elemental stratum is isotropic, and this has occurred simply because macroscopic thermodynamics were used to relate local pressure to local density. Either this, or there can be no surface tension. Such an approach may be justified in the vicinity of the critical point, however. Then the transition zone is sufficiently delocalized that the density is sensibly constant within each semi-macroscopic stratum, $P(z)$ is virtually isotropic, the density profile is almost certainly structureless, and the surface tension is virtually zero.

On the basis of Hill's[18] statistical mechanical specification of the quasi-thermodynamic point chemical potential in an inhomogeneous region (1.4.14)

$$\mu = \mu(z_1) = \text{constant} = kT \ln \rho_{(1)}(z_1)\Lambda^3$$

$$+ \int_0^1 \int_V \Phi(r_{12})\rho_{(1)}(z_2)g_{(2)}(z_1, \mathbf{r}_{12}, \xi)\, d\mathbf{r}_{12}\, d\xi$$

where $\Lambda = h/(2\pi mkT)^{1/2}$, Toxvaerd[25] rearranges the equation to express the constant fugacity $p(z_1)$ across the interface:

$$\mu(z_1) - kT \ln \Lambda^3 = kT \ln p(z_1) = \text{constant}$$

$$= kT \ln \rho_{(1)}(z_1) + \int_0^1 \int_V \Phi(r_{12}) \rho_{(1)}(z_2) g_{(2)}(z_1, \mathbf{r}_{12}, \xi) \, d\mathbf{r}_{12} \, d\xi. \quad (2.4.9)$$

A perturbation approach[26] is made for the anisotropic radial distribution in which it is assumed that the structure is determined essentially by the repulsive core of the two-particle interaction $\Phi(r_{12})$; realistic interactions are resolved into a short-range repulsive component $\Phi^{(0)}(r_{12})$ with associated structure $g_{(2)}^{(0)}(r_{12})$, and a weaker, long-range component $\Phi^{(1)}(r_{12})$ with associated structure $g_{(2)}^{(1)}(r_{12})$. Generally, on the basis of the perturbation technique, the radial distribution function is written

$$g_{(2)}(\mathbf{r}_{12}, \xi) = \sum_{i=0}^{\infty} g_{(2)}^{(i)}(\mathbf{r}_{12}, \xi)$$

where only the asymptotic, isotropic form of the first two terms ($i = 0, 1$) is known at present. However, the perturbation expansion is found to converge very rapidly since the hard-core term $g_{(2)}^{(0)}(r_{12})$ effectively establishes the principal structural features of the system. The perturbation expansion was originally derived by Zwanzig,[27] but has awaited approximations due to Barker and Henderson[26] before applications could be made to the theory of uniform fluids.

Ignoring angular dependence of the pair distribution in (2.4.9), Toxvaerd finally obtains

$$\text{constant} = kT \ln \gamma_0(z_1) + 2\pi \int_{-\infty}^{\infty} \rho_{(1)}(z_1 + z_{12}) \int_{|z_{12}|}^{\infty} \Phi^{(1)}(r_{12})$$

$$\times \left\{ g_{(2)}^0[r_{12}, \rho_{(1)}(z_1)] + \frac{\Phi^{(1)}}{2kT} g_{(2)}^{(1)}[r_{12}, \rho_{(1)}(z_1)] \right\} r_{12} \, dr_{12} \, dz_{12} + \cdots, \quad (2.4.10)$$

Since $kT \ln \gamma_0(z_1) = \mu_0(z_1) - kT \ln \Lambda^3$, this term may be determined from the uniform density coexistence conditions, $\mu(\rho_L) = \mu(\rho_V)$ and $P(\rho_L) = P(\rho_V)$.

We point out that the anisotropic pair distribution is of the approximate form of Green[1] (2.1.2), which is commonly adopted, and in which the angular dependence of the radial distribution is suppressed, although the radial distribution function is in this case taken to be density-dependent. More particularly, however, we note from (2.4.10) that the functions $g_{(2)}^{(0)}[r_{12}, \rho_{(1)}(z_1)]$, $g_{(2)}^{(1)}[r_{12}, \rho_{(1)}(z_1)]$ are *appropriate to the density at particle 1*, and independent of the location of particle 2. Clearly, configurations will occur in the course of integration, particularly in the vicinity of the transition zone, where particle 1 is, say, in the vapour region whilst particle 2 is in the liquid: in this case a vapour-like radial distribution will be adopted. If now the particles are interchanged, obviously a liquid-like distribution is assumed: the model is evidently not symmetrical with respect to particle exchange. (Such an approximation would be even more serious in the case of a two-component system.)

The consequences of this asymmetry are difficult to assess, although since the number density varies by a factor of $\sim 10^3$ across the transition zone in the vicinity of the triple point, it is clear that if particle 1 is located on the vapour side of the dividing surface with particle 2 on the liquid side then the two particles will correlate in a vapour-like manner, and vice versa. Of course the radial distribution has to be weighted by the product $\rho_{(1)}(z_1)\rho_{(1)}(z_2)$ in forming the approximation to $\rho_{(1)}(z_1, \mathbf{r}_{12})$. Perhaps a slightly less restrictive approximation

$$g_{(2)}(z_1, \mathbf{r}_{12}) \sim g_{(2)}(r_{12}, \bar{\rho}) \tag{2.4.11}$$

could be adopted where $\bar{\rho}$ is some average density, neither that at z_1 nor that at z_2, but some intermediate value. In the vicinity of the critical point an obvious effective or average density is $\frac{1}{2}(\rho_L + \rho_V)$, whilst near the triple point $\bar{\rho} \to \rho_{(1)}(z_1)$ since the overestimate of correlations in the vapour phase are relatively unimportant. Lekner[28] claims that some justification for this standpoint is afforded on the basis of functional differentiation techniques. However, it is worthwhile observing that the principal peak in the radial distribution, which largely determines the values of the integrals, is very often of greater amplitude in the *vapour* rather than the liquid phase! Difficulties nevertheless remain regarding asymmetry with respect to particle exchange and, unless $\bar{\rho}$ is chosen judiciously, a *smearing* of structural features, if any, will occur. For example, an effective density chosen on the simple basis (Figure 2.4.3)

$$\bar{\rho} = \frac{1}{2}[\rho_{(1)}(z_1) + \rho_{(1)}(z_2)]$$

would seriously *over*estimate the correlation in the fluid, whilst for a small displacement

$$\bar{\rho} = \frac{1}{2}[\rho_{(1)}(z_1') + \rho_{(1)}(z_2')]$$

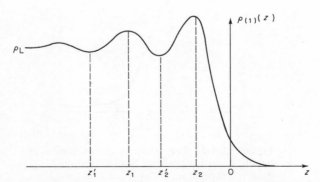

Figure 2.4.3 Calculation of effective density at the liquid surface in a structured region of the interface

would represent a considerable *under*estimate. An effective density of the form[3]

$$\bar{\rho} = \frac{\int_{z_1}^{z_2} \rho_{(1)}(z)\,dz}{|z_1 - z_2|}$$

might be a better choice which, moreover, is symmetrical with respect to particle exchange.

Nevertheless, Toxvaerd solves (2.4.10) for a system of square-well particles having the pair potential

$$\begin{aligned}
\Phi^{(0)}(r_{12}) &= +\infty & r_{12} &\leqslant \sigma \\
\Phi^{(1)}(r_{12}) &= -\varepsilon & \sigma &\leqslant r_{12} < 1.5\sigma.
\end{aligned} \tag{2.4.12}$$

For $g_{(2)}^{(0)}(r_{12}, \rho)$, Verlet and Weiss'[29] parametrized expression of hard sphere fluid computer results were used, whilst the tabulated results of Smith, Henderson, and Barker[30] also based on computer simulations, were used for $g_{(2)}^{(1)}(r_{12})$. The density profile at $T^* = \varepsilon/k = 1.00$ is shown in Figure 2.4.4, from which we see that the transition appears to be a monotonic, rapidly decreasing function at temperatures far from the critical point.

On the basis of a similar perturbation expansion of the pair distribution, and a closure subject to the same objections regarding structural asymmetry with respect to particle exchange and isotropy of the radial distribution, Toxvaerd[31] has, on the basis of (1.4.28), obtained a perturbation expansion of the point Helmholtz free energy in the vicinity of the liquid–vapour density transition. The free energy per particle may be expressed as a perturbation expansion

$$a(z_1) = \sum_{i=0}^{\infty} a^{(i)}(z_1) \tag{2.4.13}$$

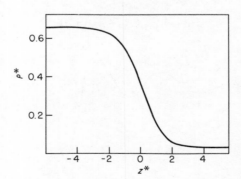

Figure 2.4.4 The density transition profile for a square-well fluid at the reduced temperature $T^* = \varepsilon/k = 1.00$ determined on the basis of constancy of the chemical potential[25] (equation 2.4.10)

where the first three terms of the series are written explicitly in (2.4.14):

$$a(z_1) = a^{(0)}(z_1) + \frac{1}{2kT} \int \Phi^{(1)}(r_{12})\rho_{(1)}(z_1 + z_{12})g_{(2)}^{(0)}[r_{12}, \rho_{(1)}(z_1)] \, d\mathbf{r}_{12}$$

$$- \frac{1}{4kT} \int [\Phi^{(1)}(r_{12})]^2 \rho_{(1)}(z_1 + z_{12})g_{(2)}^{(0)}[r_{12}, \rho_{(1)}(z_1)] \left(\frac{\partial \rho_{(1)}(z_1 + z_{12})}{\partial p}\right)^{(0)} d\mathbf{r}_{12}$$

$$+ \cdots \quad (2.4.14)$$

$a^{(0)}(z_1)$ is the free energy per particle in the hard sphere reference system of density $\rho_{(1)}(z_1)$ which is approximated by a Percus–Yevick hard sphere fluid with a temperature-dependent collision diameter.

On the basis of the approximation

$$g_{(2)}(z_1, \mathbf{r}_{12}) \sim g_{(2)}(r_{12}, \rho_{(1)}(z_1)) \quad (2.4.15)$$

Toxvaerd determines the density distribution $\rho_{(1)}(z_1)$ which minimizes the Helmholtz excess free energy (1.4.26) (and at the same time permits a calculation of the surface tension and surface excess free energy per unit area) subject to the constraint (1.2.1) for a Lennard–Jones[31] and a square well fluid.[25] Following Fisk and Widom,[32] Toxvaerd makes a variational determination for a trial function of tanh form:

$$\rho_{(1)}(z_1) = \{\rho_L \exp[a(b - z_1)] + \rho_V\}/\{\exp[a(b - z_1)] + 1\} \quad (2.4.16)$$

where

$$a = a_1 \quad (z_1 < b)$$
$$a = a_2 \quad (z_1 > b)$$

and

$$b = \left(\frac{1}{a_1} - \frac{1}{a_2}\right) \ln 2.$$

The two variable parameters of this profile, a_1 and a_2, are adjusted so as to minimize variationally the interfacial free energy, although as Toxvaerd points out,[31] the trial function (2.4.16) does not have the correct asymptotic form. There is also a discontinuity in the density gradient at $z = b$ if $a_1 \neq a_2$, which apart from Cahn–Hilliard considerations, is disturbing. An asymptotically correct trial function

$$\rho_{(1)}(z_1) = (\rho_L + \rho_V)/2 - (\rho_L - \rho_V)/\pi \tan^{-1} a(z_1 - b) \quad (2.4.17)$$

with

$$b = \frac{1}{\sqrt{3}} \left(\frac{1}{a_1} - \frac{1}{a_2}\right)$$

increases γ by 5–10%. Toxvaerd's results for the square-well and Lennard–Jones systems are shown in Figures 2.4.4 and 2.4.5a, respectively. The

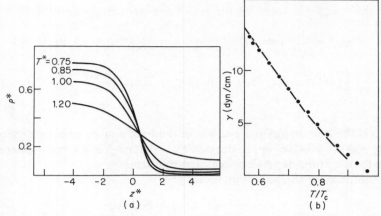

Figure 2.4.5 (a) Density transition profile for a Lennard–Jones fluid as a function of temperature determined as a variationally optimized tanh function,[31] yielding a minimum excess Helmholtz free energy. (b) Variation of surface tension of a Lennard–Jones fluid as a function of temperature on the basis of the tanh profiles shown in (a)

agreement between the profiles based on constancy of the chemical potential and minimization of the surface excess free energy is striking, although perhaps not altogether surprising since the qualitative features of the profile, in particular its range subject to the constraints $\rho_{(1)}(-\infty) = \rho_L$, $\rho_{(1)}(+\infty) = \rho_V$, are related more or less directly to the attractive branch of the pair interaction through the Boltzmann factor. And there remains the problem associated with the non-invariance of the two-body closure device with respect to renumbering of the particles.

In the case of the Lennard–Jones fluid, parametric adjustment of the tanh profile (2.4.16) to yield the minimum superficial Helmholtz free energy enables a determination of the surface tension to be made. Excellent agreement with experimental data for liquid argon is obtained (Fig. 2.4.5b), but unfortunately the insensitivity of γ to the structural details of the transition zone permits only the most general conclusions to be drawn regarding the nature of the density profile.

Based on the WCA perturbation analysis, but perhaps physically somewhat more transparent, is the recent expression of Upstill and Evans[80] who resolve the pair potential of a homogeneous reference system into a repulsive and an attractive component:

$$\Phi(r) = \Phi^{(0)}(r) + \Phi^{(1)}(r).$$

They go on to assume that the anisotropic pair distribution at the surface $\rho_{(2)}(z_1, \mathbf{r}_{12})$ may be represented by $\rho_{(2)}^{(0)}(z_1, \mathbf{r}_{12})$ the 'repulsive component', which is further approximated by the mean field expression

$$\rho_{(2)}^{(0)}(z_1, \mathbf{r}_{12}) \sim \rho_{(1)}(z_1)\rho_{(1)}(z_2)g_{(2)}^{(0)}\left[r_{12}; \rho_{(1)}\left(\frac{z_1 + z_2}{2}\right)\right]$$

which is likely to be accurate only for slowly varying profiles. For the purposes of calculation the 'repulsive' distribution $g_{(2)}^{(0)}(r)$ is further approximated by $g_{(2)}^{(0)}(r; \rho_L)$—that is by its bulk liquid form throughout, neglecting the z-dependence of the profile. This Upstill and Evans justify since most contributions to the surface free energy arise from the liquid side of the dividing surface. Finally, it is assumed that the free energy of the inhomogeneous reference system can be calculated in the usual local density approximation. Whilst more exact theories of the interfacial region are available, this analysis does provide some insight into the processes governing the overall form of the transition zone. For example, it is found in this simplification that the surface excess free energy may be resolved into two components:

(i) a hard sphere term which has the form of a surface entropy and favours a spatially delocalized profile, and
(ii) an attractive term having the form of a surface energy which favours a sharper transition zone, there being a shift in emphasis from component (ii) to component (i) with increasing temperature.

However, such a resolution is clearly apparent in the Cahn–Hilliard description of the critical interface. In more exact theories—that of Kirkwood–Buff, for example—the temperature dependence is concealed in the radial distribution. Upstill and Evans[80] adopt a trial transition profile of exponential form with a characteristic length M which is variationally adjusted to yield a minimum surface excess free energy. The numerical agreement with experiment and more sophisticated theories is quite good, although the model deteriorates badly with increasing temperature.

An interesting rearrangement of equation (2.4.14) has recently been given by Abraham[70] who notes that the total Helmholtz free energy of a macroscopic non-uniform system may be written in the form

$$A = \int \rho_{(1)}(z_1) a(z_1)\, dz_1$$

$$\int \rho_{(1)}(z_1) a(z_1)\, dz_1 = \int (a\dagger[\rho(z_1)] + \delta a_1(z_1) + \cdots)\rho_{(1)}(z_1)\, dz_1$$

(2.4.18)

where, on the basis of a Weeks–Chandler–Anderson perturbation approach

$$a\dagger[\rho(z_1)] = \rho_{(1)}(z_1) a^0[\rho_{(1)}(z_1)] + \frac{\rho_{(1)}(z_1)}{2kT} \int \Phi^{(1)}(r_{12}) g_{(2)}^{(0)}[r_{12}, \rho_{(1)}(z_1)]\, d\mathbf{r}_{12}$$

(2.4.19)

$$\delta a_1(z_1) = \frac{\rho_{(1)}(z_1)}{2kT} \int \Phi^{(1)}(r_{12})[\rho_{(1)}(z_1 + z_{12}) - \rho_{(1)}(z_1)] g_{(2)}^{(0)}[r_{12}, \rho_{(1)}(z_1)]\, d\mathbf{r}_{12}$$

where

$$\Phi^{(1)}(r_{12}) = -\varepsilon \qquad r_{12} < r_m$$
$$= \Phi(r_{12}) \qquad r_{12} > r_m$$

where r_m is the value for which $\Phi(r_{12})$ is a minimum and ε the depth at the minimum. $g_{(2)}^{(0)}$ is the radial distribution function of the reference fluid.

Similar, but somewhat more complicated, expressions follow on the basis of a Barker–Henderson expansion. Abraham notes that the resolution of the local Helmholtz free energy $a(z_1)$ into the terms $a\dagger[\rho_{(1)}(z_1)] + \delta a_1(z_1) + \delta a_2(z_2)$ is of the van der Waals/Cahn–Hilliard form

$$a(z_1) = a\dagger[\rho_{(1)}(z_1)] + A\left(\frac{\partial\rho_{(1)}(z_1)}{\partial z_1}\right)^2 \qquad A > 0 \qquad (2.4.20)$$

where $a\dagger[\rho_{(1)}(z_1)]$ is the analytic continuation of the function $a\dagger(\rho)$ in a uniform fluid into an inhomogeneous region, whilst the 'square-gradient' term, appropriate for slowly varying critical profiles, is replaced by term(s) requiring integration over the inhomogeneous density profile.

Abraham expresses the surface tension of the fluid as

$$\gamma = \lim_{h\to\infty}\left\{\int_{-h}^{h} a(z_1)\,dz_1 - h[a(-h) + a(h)]\right\} \qquad (2.4.21)$$

which he minimizes with respect to an assumed Fermi function profile.

This development is of interest in as far as it shows how the van der Waals/Cahn–Hilliard free energy function is to be modified from the square-gradient theory at temperatures well below the critical. However, the result is incorrect in that the radial distributions in (2.4.19), apart from not being invariant to particle exchange and angle independent, are *not* simply functions of the local density $\rho_{(1)}(z_1)$ but instead are complex functions of both $\rho_{(1)}(z_1)$ and $\rho_{(1)}(z_2)$, whereupon the terms $a\dagger$, δa_1, and δa_2 no longer bear comparison with the van der Waals/Cahn–Hilliard form (2.4.20) except, of course, near the critical point when $\rho_{(1)}(z_1) \sim \rho_{(1)}(z_2)$.

Again, the adoption of a class of Fermi functions for the transition profile affords little physical insight into the detailed nature of the transition zone, although, of course, the surface tension or surface excess Helmholtz free energy is relatively insensitive to the form of the profile.

A refinement of this free energy approach, based on the original van der Waals formulation, has been given by Buongiorno and Davis[71] who minimize their free energy expression with respect to the equilibrium profile, obtaining an integral equation for the single-particle density distribution and, of course, an expression for the surface tension. The anisotropic pair distributions arising in their expressions are approximated by an angle-independent local density function, $g_{(2)}(z_1, \mathbf{r}_{12}) \sim g_{(2)}(r_{12}, \rho_{(1)}(z_1))$, and they further assume that the density profile changes slowly over the range of the interparticle potential, enabling the integral equation to be reduced to a more amenable differential form. Transition profiles for a reduced Lennard–Jones system are then obtained over a variety of subcritical temperatures obtaining monotonic profiles. The assumption that the profile varies only slightly over the range of the molecular interaction is belied at low temperatures where the density is seen to vary by a factor of $\sim 10^2$–10^3 over two or three atomic diameters. It would appear that the model is more appropriate to critical interfaces[72] and does not bear legitimate extension to low temperatures, when it reduces virtually to a mean field approximation, at least in its present differential form. A similar analysis based on a Fermi profile function has been given by Singh and Abraham.[73] A more satisfactory approach[74] is to express (2.4.19) as a functional of *both* $\rho_{(1)}(z_1)$ and $\rho_{(1)}(z_2)$:

$$a^\dagger[\rho_{(1)}(z_1)] = \rho_{(1)}(z_1)a^{(0)}[\rho_{(1)}(z_1)] - \frac{1}{2}\rho_{(1)}(z_1)\int \Phi^{(1)}(r_{12})g^{(0)}_{(2)}(z_1,\mathbf{r}_{12})\,d\mathbf{r}_{12}$$

(2.4.22)

$$\delta a_1(z_1) = \frac{1}{2}\rho_{(1)}(z_1)\int [\rho_{(1)}(z_1) - \rho_{(1)}(z_1 + z_{12})]\Phi^{(1)}(r_{12})g^{(0)}_{(2)}(z_1,\mathbf{r}_{12})\,d\mathbf{r}_{12}.$$

For low-density, slowly varying interfaces we may write

$$[\rho_{(1)}(z_1) - \rho_{(1)}(z_1 + z_{12})] \sim -\left(\frac{d\rho_{(1)}(z_1)}{dz_1}\right)z_{12}$$

(2.4.23)

whilst, at critical densities, we may form a midpoint approximation to the radial distribution function

$$g^{(0)}_{(2)}(z_1,\mathbf{r}_{12}) = 0.5g^{(0)}_{(2)}(r_{12},\rho_{(1)}(z_1)) + 0.5g^{(0)}_{(2)}(r_{12},\rho_{(1)}(z_2))$$

(2.4.24)

which is, moreover, angularly dependent and invariant to particle exchange. We may expand (2.4.24) as a cluster expansion in the direct correlation function from the Ornstein–Zernike equation

$$g^{(0)}_{(2)}(r_{12},\rho_{(1)}(z_1)) = 1 + c(r_{12}) + \rho_{(1)}(z_1)\int c(r_{23})c(r_{13})\,d\mathbf{r}_3 + \cdots$$

(2.4.25)

with a similar expression for $g_{(2)}(r_{12},\rho_{(1)}(z_2))$. At high temperatures and low densities we may replace the direct correlation with the Mayer f-function and obtain from (2.4.24)

$$g^{(0)}_{(2)}(z_1,\mathbf{r}_{12}) \sim 1 + f(r_{12}) + \left[\rho_{(1)}(z_1) + \frac{1}{2}\frac{d\rho_{(1)}(z_1)}{dz_1}z_{12}\right]F(r_{12})$$

(2.4.26)

where

$$F(r_{12}) = \int f(r_{13})f(r_{23})\,d\mathbf{r}_3 \sim \int c(r_{13})c(r_{23})\,d\mathbf{r}_3.$$

Inserting (2.4.23) and (2.4.26) in (2.4.24) we obtain the only remaining terms which are

$$a^\dagger[\rho_{(1)}(z_1)] = \rho_{(1)}(z_1)a^{(0)}[\rho_{(1)}(z_1)]$$

$$da_1(z_1) \sim -\frac{\rho_{(1)}(z_1)}{4}\left(\frac{d\rho_{(1)}(z_1)}{dz_1}\right)^2\int \Phi^{(1)}(r_{12})F(r_{12})(z_{12})^2\,d\mathbf{r}_{12}.$$

(2.4.27)

Now the term $a^\dagger[\rho_{(1)}(z_1)]$ *is a function of* $\rho_{(1)}(z_1)$ only, and the square gradient term is multiplied by a weakly z-dependent positive constant

$$A = \frac{-\rho_{(1)}(z_1)}{4} \int \Phi^{(1)}(r_{12}) F(r_{12}) z_{12}^2 \, d\mathbf{r}_{12} > 0. \tag{2.4.28}$$

At lower temperatures A becomes very strongly z-dependent in the vicinity of the transition zone. We note that the above development does not yield the square-gradient term characteristic of the van der Waals/Cahn–Hilliard theory if we make the simplifying assumption $g_{(2)}(z_1, \mathbf{r}_{12}) \sim g_{(2)}[r_{12}, \rho_{(1)}(z_1)]$. The term $F(r_{12})$ could have been left in its original Ornstein–Zernike form, $\int c(r_{13}) h(r_{23}) \, d\mathbf{r}_3$ with the additional and reasonable assumption that this integral is only weakly dependent upon density.

2.5 The single-particle Born–Green–Yvon equation

Formal expressions for the one- and two-particle distributions in the vicinity of the liquid surface have already been developed:[33] it follows from (1.3.5) that, in the canonical ensemble, $\rho_{(1)}(z_1)$ and $\rho_{(2)}(z_1, \mathbf{r}_{12})$ are given by

$$\rho_{(1)}(z_1) = \frac{1}{(N-1)! Z_Q} \int \cdots \int \exp\left(\frac{-\Phi_N(\mathbf{r}^N)}{kT}\right) d\mathbf{r}_2 \cdots d\mathbf{r}_N \tag{2.5.1}$$

$$\rho_{(2)}(z_1, \mathbf{r}_{12}) = \frac{1}{(N-2)! Z_Q} \int \cdots \int \times \exp\left(\frac{-\Phi_N(\mathbf{r}^N)}{kT}\right) d\mathbf{r}_3 \cdots d\mathbf{r}_N \tag{2.5.2}$$

where the configurational partition function

$$Z_Q = \frac{1}{N!} \int \cdots \int \exp\left(\frac{-\Phi_N(\mathbf{r}^N)}{kT}\right) d\mathbf{r}_1 \cdots d\mathbf{r}_N.$$

We have assumed that for a plane, non-uniform zone $\rho_{(1)}$ depends only on z coordination and that the configurational potential may be expressed as a sum of scalar pair interactions. Differentiation of (1.3.5) with respect to z_1 yields

$$\nabla_1 \rho_{(1)}(z_1) = -\frac{1}{Z_Q kT} \int \cdots \int \sum_i \sum_j \nabla_i \Phi(r_{ij}) \exp\left(\frac{-\Phi_N(\mathbf{r}^N)}{kT}\right) d\mathbf{r}_2 \cdots d\mathbf{r}_N \tag{2.5.3}$$

or, using (2.5.2),

$$\nabla_1 \rho_{(1)}(z_1) = \frac{1}{kT} \int \nabla_1 \Phi(r_{12}) \rho_{(2)}(z_1, \mathbf{r}_{12}) \frac{z_{12}}{r_{12}} \, d\mathbf{r}_{12}. \tag{2.5.4}$$

This equation is a generalization of the Bogoliubov–Born–Green–Yvon[34] equation for the surface transition zone. Closely related and quite equivalent is Kirkwood's integral equation in terms of the coupling parameter[35] ξ (1.4.13):

$$kT \ln \rho_{(1)}(z_1) = kT \ln \rho_L + \rho_L \int_0^1 \int_V \Phi(r_{12}) g_{(2)}^L(r_{12}, \xi) \, d\mathbf{r}_{12} \, d\xi. \quad (2.5.5)$$

Both these equations relate the single-particle distribution $g_{(1)}(z_1)$ to the anisotropic pair function $\rho_{(2)}(z_1, \mathbf{r}_{12})$. Both equations are exact to within the assumption of pairwise additivity of the total potential, and in one or other of the two forms are basic equations in the statistical mechanics of the liquid surface.

Given the hierarchical nature of adjacent orders of distribution (1.3.6), it follows that a similar integro-differential equation may be set up for $\rho_{(2)}(z_1, \mathbf{r}_{12})$ although it, of course, would be a function of the triplet distribution $\rho_{(3)}(z_1, \mathbf{r}_{12}; z_2, \mathbf{r}_{23}; z_3, \mathbf{r}_{13})$. It is necessary to terminate the hierarchy of linked equations so as to reduce any given order of equation to a closed form for solution. For uniform fluids the two-particle distribution $\rho_{(2)}(r_{12})$ is of central interest, and the triplet distribution $\rho_{(3)}(r_{12}, r_{23}, r_{13})$ is often expressed in terms of pair distributions—the Kirkwood superposition approximation (KSA).[35] For non-uniform systems $\rho_{(1)}(z_1)$ and $\rho_{(2)}(z_1, \mathbf{r}_{12})$ are of primary interest, and although an inhomogeneous counterpart to the Kirkwood superposition approximation could in principle be used to generate $\rho_{(2)}(z_1, \mathbf{r}_{12})$, unfortunately its determination presupposes a knowledge of $\rho_{(1)}(z_1)$! Instead, various *ad hoc* representations of the anisotropic two-particle distribution have been proposed with varying degrees of physical justification and whose consequences are often difficult to assess: both oscillatory and monotonic BBGYK transition profiles have been obtained on the basis of different closures and different numerical procedures. Unfortunately, we have few experimental or theoretical guidelines regarding the specification of the anisotropic distribution, and although detailed examination of the two-particle distributions developed in the machine simulations would help immensely, no reports appear in the literature. (See, however, Section 10.10.)

Before considering the *ad hoc* closures in any detail, we first discuss the interrelation between the assumed closure prescription and the development of the single-particle distribution in terms of the BBGYK equation generalized to an inhomogeneous system. Integration of (2.5.4) subject to the boundary condition $\rho_{(1)}(-\infty) = \rho_L$ yields[3]

$$\rho_{(1)}(z_1) = \rho_L \exp \left\{ -\frac{1}{kT} \int_{-\infty}^{z_1} \int_V \nabla_1 \Phi(r_{12}) \rho_{(1)}(z_2) g_{(2)}(z_1', \mathbf{r}_{12}) \, d\mathbf{r}_{12} \, dz_1' \right\}$$

$$(2.5.6)$$

$$= \rho_L \exp \left(\frac{-\Psi(z_1)}{kT} \right) \quad (2.5.7)$$

where $\Psi(z_1)$ is an effective single-particle potential of mean force developed in bringing particle 1 up from $z = -\infty$ to its current location. Apart from the temperature-dependence of Ψ (through the two-particle distribution) it is clear that for classical systems the profile $\rho_{(1)}(z_1)$ is related to the constraining surface potential through a Boltzmann factor, and that it will exhibit spatial delocalization with increasing temperature.

The iterative solution of (2.5.6) will generally involve the adoption of an approximate closure for $\rho_{(2)}(z_1, \mathbf{r}_{12})$ or $g_{(2)}(z_1, \mathbf{r}_{12})$, and an initial guess at the transition profile—usually a step function—which are inserted in the algorithm shown in Figure 2.1.1. As the centre of integration z_1 moves towards the surface a particle located at this point experiences an anisotropic distribution of force

$$-\nabla_1 \Psi(z_1) = -\int \nabla_1 \Phi(r_{12}) \rho_{(1)}(z_2) g_{(2)}(z_1, \mathbf{r}_{12}) \, d\mathbf{r}_{12} \qquad (2.5.8)$$

which is directed into the bulk fluid. The specification of the closure clearly specifies the surface constraining field $-\nabla_1 \Psi(z_1)$, and we speculate that an 'over constrained' closure device induces oscillations in the transition profile, whilst 'under constraint' results in a monotonic density transition.[3] The exact constraint remains unknown, of course and will vary from system to system. Nevertheless, Osborn and Croxton[36] have recently shown that a given closure device is capable of yielding both oscillatory *and* monotonic transitions on the basis of the same input quantities. A more detailed discussion follows in Section 2.7: suffice it to say at the moment that in the course of solution of the BGY equation a shift in the location of the Gibbs surface is associated with the non-equilibrium, intermediate stages of iteration. The precise method of relocation of the dividing surface appears to underlie the different forms of solution.

Again, the kernel of the integral appearing in (2.5.6) is extremely sensitive to the details of the pair potential and two-particle distribution, and as Borstnik and Azman[37] have recently observed, the solution of the BGY equation can strongly depend upon the details of the input quantities in the case of realistic systems.

In view of the complexity of the mathematical formulation of the problems discussed above, it is interesting to consider the effect of an extreme constraint such as that imposed by an ideal infinitely high potential wall for which a full and rigorous solution is possible.[38,39] A detailed development of Fisher and Bokut's analysis[39] is not appropriate here, except to say that for a high-density system of hard spheres adjacent to an infinitely high potential wall the development of a long-range layered structure occurs, as one might intuitively expect. The correlation range extends over many hard sphere diameters and, for this athermal case, depends solely upon packing density. The analysis has recently been extended to include hard spheres against a soft wall by Singh and Abraham.[79] Such results have been obtained in simulation for Lennard–Jones systems by a number of investigators, and will be considered in some detail in Chapter 10. Bernal[40] has also simulated hard sphere systems against both rigid and soft constraining boundaries which appear to substantiate the speculation above concerning the effect of the surface field $-\nabla_1\Psi(z_1)$ upon the nature of the solution $\rho_{(1)}(z_1)$, and an extension to the case of the 'sticky' hard sphere surface has recently been made by Sullivan and Stell.[100]

In the case of liquid metals the intervention of quantum interference effects of the conduction electronic distribution at the surface may modify the distribution of ionic cores, and it would appear that, for certain systems at least, the electronic processes might well amplify any tendency towards the development of stable density oscillations in the ionic profile.[3] We shall discuss the case of liquid metal systems in Chapter 5.

2.6 Mechanical and thermodynamic stability of the BGY equation

The liquid–vapour transition profile develops subject to the constraints of constancy of the chemical potential and of the normal component of the pressure tensor across the transition zone. Appeal to the former constraint, constancy of the chemical potential, has yielded profiles which are characterized by chemical or thermodynamic equilibrium. It is straightforward to show that the single-particle Born–Green–Yvon[34] equation (2.5.6), or alternatively, the single-particle Kirkwood equation,[35] is equivalent to the virial equation for constant pressure $P_\perp$ normal to the interface, and as such ensures the mechanical or hydrostatic equilibrium of the surface. This mechanical constraint has been used recently by Pressing and Mayer[41] who expanded $P_\perp$ in powers of the spatial gradient and determined $\rho_{(1)}(z_1)$ near the critical temperature where the gradient is small. Here, however, we wish to determine the density profile at temperatures considerably below the critical and, following Harasima,[43] we rederive the BGY equation on the basis of explicit appeal to the condition for mechanical stability— incidentally obtaining an equation for non-equilibrium profiles which develop at intermediate stages of the iterative solution.

The virial equation for the pressure tensor has been derived by Irving and Kirkwood.[42] The isochores are taken to be in the xy plane; the normal and tangential components of the pressure tensor at z_1 are then given by (see 1.4.24)

$$P_\perp(z_1) = kT\rho_{(1)}(z_1) - \frac{1}{2}\int \nabla_1\Phi(r_{12})\rho_{(2)}(z_1,\mathbf{r}_{12})\frac{z_{12}^2}{r_{12}}\,d\mathbf{r}_{12} \qquad (2.6.1)$$

$$P_\parallel(z_1) = kT\rho_{(1)}(z_1) - \frac{1}{2}\int \nabla_1\Phi(r_{12})\rho_{(2)}(z_1,\mathbf{r}_{12})\frac{x_{12}^2}{r_{12}}\,d\mathbf{r}_{12}. \qquad (2.6.2)$$

Setting

$$\rho_{(2)}(z_1,\mathbf{r}_{12}) = \rho_{(1)}(z_1)\rho_{(1)}(z_2)g_{(2)}(z_1,\mathbf{r}_{12})$$

equations (2.6.1), (2.6.2) become

$$\frac{P_\perp(z_1)}{kT} = \rho_{(1)}(z_1)\left(1 - \frac{\pi}{kT}\int\!\!\int_{-\infty}^{\infty}\nabla_1\Phi(r_{12})\rho_{(1)}(z_2)g_{(2)}(z_1,\mathbf{r}_{12})z_{12}^2\,dz_{12}\,d\mathbf{r}_{12}\right) \quad (2.6.3)$$

$$\frac{P_\parallel(z_1)}{kT} = \rho_{(1)}(z_1)\left(1 - \frac{\pi}{kT}\int\!\!\int_{-\infty}^{\infty}\nabla_1\Phi(r_{12})\rho_{(1)}(z_2)g_{(2)}(z_1,\mathbf{r}_{12})(r_{12}^2 - z_{12}^2)\,dz_{12}\,d\mathbf{r}_{12}\right).$$
$$(2.6.4)$$

In homogeneous regions $g_{(2)}(z_1,\mathbf{r}_{12}) \to g_{(2)}(r_{12})$; $\rho_{(1)}(z_1) \to \rho$ whereupon the virial equations (2.6.3), (2.6.4) reduce to their uniform fluid form:

$$\frac{P_\perp}{kT} = \frac{P_\parallel}{kT} = \frac{P}{kT} = \rho\left(1 - \frac{2\pi\rho}{3kT}\int_0^{\infty}\nabla_1\Phi(r_{12})g_{(2)}(r_{12})r^3\,dr_{12}\right). \quad (2.6.5)$$

Differentiating (2.6.3) w.r.t. z_1 and simplifying (note $g_{(2)}(z_1,\mathbf{r}_{12}) = g_{(2)}(\mathbf{r}_{12}, z_1, z_2, z_2 - z_1)$, integrals involving $dg_{(2)}/dz_1$ vanish due to symmetry of $g_{(2)}$), we obtain

$$\frac{d\rho_{(1)}(z_1)}{dz_1} = \frac{2\pi}{kT}\int_{-\infty}^{\infty}\int_{z_{12}=-r_{12}}^{z_{12}=+r_{12}}\nabla_1\Phi(r_{12})g_{(2)}(z_1,\mathbf{r}_{12})\rho_{(1)}(z_1)\rho_{(1)}(z_2)z_{12}\,dr_{12}\,dz_{12}$$

$$+ \frac{1}{kT}\frac{dP_\perp(z_1)}{dz_1}. \quad (2.6.6)$$

Now, for hydrostatic equilibrium we set $dP_\perp(z_1)/dz_1 = 0$ whereupon equation (2.6.6) reduces to the single-particle BGY equation:

$$\frac{d}{dz_1}\ln\rho_{(1)}(z_1) = 2\pi\int_{-\infty}^{\infty}\rho_{(1)}(z_1 + z_{12})z_{12}\,dz_{12}\int_{|z_{12}|}^{\infty}\frac{\nabla_1\Phi(r_{12})}{kT}g_{(2)}(z_1,\mathbf{r}_{12})\,d\mathbf{r}_{12}$$
$$(2.6.7)$$

which may be integrated to yield

$$\rho_{(1)}(z_1) = \rho_L\exp\left\{-2\pi\int_{-\infty}^{z_1}dz_1'\int_{-\infty}^{\infty}dz_{12}\rho_{(1)}(z_1 + z_{12})\right.$$

$$\left.\times\int_{|z_{12}|}^{\infty}\frac{\nabla_1\Phi(r_{12})}{kT}g_{(2)}(z_1,\mathbf{r}_{12})\,d\mathbf{r}_{12}\right\}. \quad (2.6.8)$$

If, however, $dP_\perp(z_1)/dz_1 \neq 0$, corresponding to mechanical instability—a condition developing in the course of the iterative solution of (2.6.8) (see Section 2.8)—then we have the modulated distribution:

$$\rho_{(1)}(z_1) = \rho_L\exp\left\{-2\pi\int_{-\infty}^{z_1}dz_1'\int_{-\infty}^{\infty}dz_{12}\rho_{(1)}(z_1 + z_{12})z_{12}\right.$$

$$\left.\times\int_{|z_{12}|}^{\infty}\frac{\nabla_1\Phi(r_{12})}{kT}g_{(2)}(z_1,\mathbf{r}_{12})\,d\mathbf{r}_{12} + \frac{1}{kT}\int_{-\infty}^{z_1}\frac{1}{\rho_{(1)}(z_1')}\frac{dP_\perp(z_1')}{dz_1'}\,dz_1'\right\}.$$
$$(2.6.9)$$

It is found in the course of iterative solution of the BGY singlet equation that the intermediate profiles show a tendency to drift either towards the vapour phase or towards the liquid (evaporation or condensation), according to the sign of $(dP_{\perp}(z_1)/dz_1)$, in an attempt to equilibrate the inconsistent 'coexistence' densities ρ_L, ρ_V. It is invariably found that as the equilibrium profile is approached so the shift in the location of the profile diminishes, becoming zero (to within the specified numerical accuracy) for the converged solution when $(dP_{\perp}(z_1)/dz_1) = 0$.

Thermodynamic stability is also ensured, and is most easily seen in terms of the Kirkwood singlet equation, to which the BGY equation is formally equivalent. In specifying the quasi-thermodynamic local chemical potential $\mu(z_1)$, Hill[18] assumes mechanical stability of the free surface in forming his inhomogeneous expression (1.4.14):

$$\mu(z_1) = \text{constant} = kT \ln \rho_{(1)}(z_1)\Lambda^3 + \int_0^1 \int_V \Phi(r_{12})\rho_{(1)}(z_2)g_{(2)}(z_1, \mathbf{r}_{12}, \xi)\, d\mathbf{r}_{12}\, d\xi$$

$$(2.6.10)$$

which for thermodynamic stability may be equated to the bulk liquid chemical potential (1.4.12)

$$\mu = \mu_L = kT \ln \rho_L \Lambda^3 + \rho_L \int_0^1 \int_V \Phi(r_{12})g_{(2)}^L(r_{12})\, d\mathbf{r}_{12}\, d\xi \qquad (2.6.11)$$

where $\Lambda = h/(2\pi mkT)^{1/2}$. Combining equations (2.6.10) and (2.6.11) we regain Kirkwood's singlet equation

$$kT \ln \rho_{(1)}(z_1) = kT \ln \rho_L + \rho_L \int_0^1 \int_V \Phi(r_{12})g_{(2)}^L(r_{12}, \xi)\, d\mathbf{r}_{12}\, d\xi$$

$$- \int_0^1 \int_V \Phi(r_{12})\rho_{(1)}(z_2)g_{(2)}(z_1, \mathbf{r}_{12}, \xi)\, d\mathbf{r}_{12}\, d\xi. \qquad (2.6.12)$$

Provided Hill's quasi-thermodynamic specification of the point chemical potential $\mu(z_1)$ is correct, then the single-particle equation (2.6.12) (and the BGY equation) is an expression of both the thermodynamic and mechanical stability of the transition zone between liquid and vapour.

2.7 Surface correlations and the closure device

In the absence of an accurate and explicit expression for the two-particle distribution $\rho_{(2)}(z_1, \mathbf{r}_{12})$ arising in the various integral equations for the density profile a variety of largely *ad hoc* closures have been proposed, all of which are hierarchically inconsistent in the sense (1.3.7)

$$1 \neq \frac{1}{N-1} \int \rho_{(1)}(z_2)g_{(2)}(z_1, \mathbf{r}_{12})\, d\mathbf{r}_{12} \qquad (2.7.1)$$

(but which nevertheless represent a profile consistent with the closure prescription) and some of which are not invariant to particle exchange. Whilst these closures are physically unrealistic they are nevertheless mathematically convenient, and in much the same way as the Kirkwood superposition approximation in the theory of uniform fluids, the inconsistency may not prove to be as serious as *a priori* considerations might suggest, although this is a purely speculative suggestion. Otherwise we have few guidelines in our search for an adequate representation of the anisotropic two-particle distribution. However, recent developments seem to suggest that the difficulty in specifying $\rho_{(2)}(z_1, \mathbf{r}_{12})$ is a fundamental one involving the development of long range horizontal correlations within the transition zone (~ 1 mm in the earth's gravitational field) quite uncharacteristic of either bulk phase, but which are known to arise as a cooperative phenomenon associated with the spontaneous symmetry breaking of a phase transition[95]—an undeniable aspect of the interfacial region. We shall return to this important point in Section 2.10, and later in the present section.

As we have observed, the essential difficulty in the specification of $\rho_{(2)}(z_1, \mathbf{r}_{12})$ in the inhomogeneous region is the choice of a radial distribution function appropriate to the non-uniform density—clearly neither that of $\rho_{(1)}(z_1)$ nor $\rho_{(1)}(z_2)$, but some intermediate value which, moreover, exhibits sustained long range horizontal correlations. Clearly such a function cannot be contrived on the basis of any local density, mean field, or thermodynamic perturbation representation. On the other hand, long wavelength, sustained correlations characteristic of incipient phase change in fluid systems are unavailable. In fact the consequences of the inadequate representation of the interfacial pair distribution are largely suppressed; we are primarily concerned with integrals over the function rather than with the function itself. In the meantime we shall have to content ourselves with the relatively simple local density approximations and their derivatives.

Toxvaerd,[25,44] for example, solves the BGY equation iteratively for the square-well and Lennard–Jones systems having formed a linear interpolation between the isotropic bulk liquid and bulk vapour distributions:

$$g_{(2)}(z_1, \mathbf{r}_{12}) \sim \gamma g_{(2)}^{L}(r_{12}) + (1 - \gamma) g_{(2)}^{V}(r_{12}) \qquad (2.7.2)$$

where the weighting factor γ is a simple numeric which Toxvaerd relates to the 'effective local density'

$$\rho_{\text{eff}}(z_1, z_2) = \alpha\rho_{(1)}(z_1) + (1 - \alpha)\rho_{(1)}(z_2)$$
$$\gamma(z_1, z_2) = (\rho_{\text{eff}}(z_1, z_2) - \rho_V)/(\rho_L - \rho_V). \qquad (2.7.3)$$

Certainly (2.7.2) incorporates an angular dependence, at least for $\gamma \neq 1, 0$ although we have no guarantee of course that it is *the* angular dependence. Only for $\alpha = 0.5$ is the closure invariant to particle exchange. However, we note that for $\gamma = 1$ we regain Green's closure

$$g_{(2)}(z_1, \mathbf{r}_{12}) \sim g_{(2)}^{L}(r_{12}) \qquad (\gamma = 1) \qquad (2.7.4)$$

which is not angularly dependent, but is nonetheless invariant to particle exchange, whilst for $\alpha = 1$ we regain the approximation (2.4.15)

$$g_{(2)}(z_1, \mathbf{r}_{12}) \sim g_{(2)}(r_{12}, \rho_{(1)}(z_1)) \qquad (\alpha = 1) \qquad (2.7.5)$$

which is a refinement of Green's closure in as far as it is now density dependent, but is no longer invariant to particle exchange.

For the same square-well system as that considered in Section 2.4 (equation 2.4.12), at the same reduced temperature $(T^* = kT/\varepsilon = 1.00)$, Toxvaerd[25] solves the BGY equation on the basis of equations (2.7.2) and (2.7.3) for $\alpha = 0.44$ and $\alpha = 1$. These results are compared in Figure 2.7.1 with the profile calculated on the basis of constancy of the chemical potential $(\alpha = 1)$ and the excess free energy variational minimization of the tanh profile (2.4.16), the latter two being graphically indistinguishable. Also shown is the BGY profile determined on the basis of Green's closure (2.1.2) $g_{(2)}(z_1, \mathbf{r}_{12}) \sim g_{(2)}^{L}(r_{12})$ $(\gamma = 1)$ which, for all its shortcomings, is invariant to particle exchange. We see from Figure 2.7.1 that Green's closure seriously overestimates the gas density and consequently yields a much more spatially delocalized transition than do the other approximations. This is undoubtedly due to the density independence of the closure in retaining the bulk liquid distribution $g_{(2)}^{L}(r_{12})$ throughout, even in the vapour phase when the particle correlations are overestimated.

The BGY and quasi-thermodynamic profiles, both determined on the basis of (2.7.2) with $\alpha = 1$, are in good agreement. Toxvaerd, however, chooses to make his comparison of the quasi-thermodynamic curve with the BGY ($\alpha = 0.44$) profile. This he does on the grounds that only for this value of α does the Gibbs shift remain zero throughout the iterative process of numerical solution, although stable solutions are found for all $\alpha > 0$. We observe, parenthetically, that only for the *converged* profile must the Gibbs

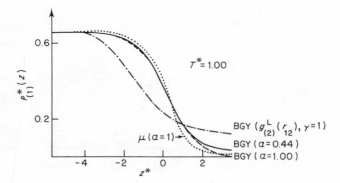

Figure 2.7.1 Born–Green–Yvon density transition profiles for a square-well fluid determined on the basis of the closures (2.7.2), (2.7.3) for a variety of parameters α.[25] The square-well profile based on constancy of the chemical potential across the transition zone is also shown for comparison

shift remain zero: partially converged, intermediate profiles, almost invariably shift between iterates, as we shall discuss in Section 2.8. Presumably the adoption of an exchange-invariant closure ($\alpha = 0.5$) would yield a profile bracketed by the BGY transitions $\alpha = 0.44$ and $\alpha = 1.0$. Jouanin[15] obtains a similar result for a square-well system on the basis of a rather more complicated and physically obscure linear interpolation between the bulk and vapour two-particle distributions.

Certainly it appears that the profiles are somewhat insensitive to the value of α: unfortunately we cannot conclude that invariance to particle exchange and angular dependence are relatively unimportant features of the choice of closure since they are inextricably related in (2.7.2): a different prescription could well yield a different conclusion.

In the case of a Lennard–Jones system at a variety of reduced temperatures ($T^* = 0.75$, 0.85, 1.00, and 1.20) Toxvaerd[25,45] again obtains structureless monotonic profiles (Figure 2.7.2) on the basis of the closure (2.7.2), but is able to obtain convergence only over the region $\alpha = 0.7$–0.8 depending upon the reduced temperature T^*. As Borstnik and Azman[37] point out, Toxvaerd's choice of closure is quite unphysical in as far as it is not symmetric with respect to particle exchange. Nevertheless, Toxvaerd obtains profiles graphically indistinguishable from those obtained by minimization of the interfacial free energy for a parametrically adjusted tanh profile, again on the basis of (2.7.2), but with $\alpha = 1.0$ (Figure 2.4.4).

In summary, it would appear that the quasi-thermodynamic and BGY profiles of Toxvaerd do not bear immediate comparison since the former adopt the closure (2.7.2) with $\alpha = 1$, whilst the BGY results use an expansion of $g_{(2)}[r, \rho(z)]$ in the interfacial gradient which is assumed to converge

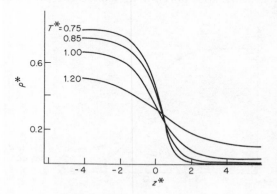

Figure 2.7.2 Born–Green–Yvon density transition profiles for a Lennard–Jones system determined on the basis of the closure (2.7.2) as a function of reduced temperature. Toxvaerd[25,45] obtained convergence only for $\alpha = 0.7$–0.8

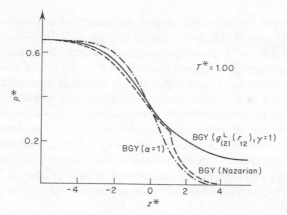

Figure 2.7.3 Born–Green–Yvon density transition
profiles for a square-well fluid using a variety of
closures

so rapidly that the zero and first-order terms are sufficient for an adequate representation.

Nazarian[46] emphasizes the importance of the statistical mechanical approach, and proposes two forms of two-particle closure, both of which are symmetrical with respect to particle exchange:

$$
\begin{aligned}
g_{(2)}(z_1, \mathbf{r}_{12}) &\sim g_{(2)}^{L}(r_{12}) \quad \text{if} \quad z_1 + z_2 \leqslant 0 \\
&\quad\ \ g_{(2)}^{V}(r_{12}) \quad \text{if} \quad z_1 + z_2 > 0
\end{aligned} \Bigg\}
\tag{2.7.6}
$$

which, in terms of the general closure expression (2.7.3), takes $\gamma = 1$ if the midpoint of particles 1 and 2 is on the liquid side of the dividing surface, otherwise it is zero. A second closure,

$$
g_{(2)}(z_1, \mathbf{r}_{12}) \sim g_{(2)}^{L}(r_{12}) + \left[\left(\frac{z_2}{z_{12}} \right) A(z_2) - \left(\frac{z_1}{z_{12}} \right) A(z_1) \right] \{ g_{(2)}^{V}(r_{12}) - g_{(2)}^{L}(r_{12}) \},
\tag{2.7.7}
$$

where $A(z)$ is the unit step function, assumes that the weighting factor γ is proportional to the fraction of the distance between points 1 and 2 on the liquid side of the dividing surface.

Toxvaerd has used the first of Nazarian's closures (2.7.6) to solve the BGY equation for the same square-well system as discussed above, and obtains the profile shown in Figure 2.7.3 which is compared with the simple Green closure $g_{(2)}(z_1, \mathbf{r}_{12}) \sim g_{(2)}^{L}(r_{12})$ ($\gamma = 1$) and equation (2.7.3) with $\alpha = 1$. Not surprisingly the three approximations are in good agreement in the liquid region, but the $\alpha = 1$ profile shows the first signs of its density dependence whilst the other two curves retain bulk liquid distributions until Nazarian's switches over rather dramatically from predominantly liquid-like to predominantly vapour-like correlations—the effect of the liquid persisting a short way beyond the dividing plane into the vapour.

For liquid argon at the triple point, Nazarian obtains strongly oscillatory profiles on the basis of the two closures (2.7.6), (2.7.7), Figure 2.7.4. Whilst the absolute value of the surface tension is largely independent of surface structure, its temperature dependence is not (1.4.7), and such a profile as this is most unlikely for liquid argon and is certainly inconsistent with what must be regarded as a reliable body of experimental evidence, although the profiles do bear comparison with other calculations and computer simulations. Nazarian's oscillations undoubtedly arise as a response to a closure which implies too strong a constraining field $-\nabla_1\Psi(z_1)$ at the surface, and this is clearly apparent from the nature of the closures (2.7.6), (2.7.7), and the discontinuity of slope in the square-well profile (Figure 2.7.3) where we relate the single-particle potential of mean force $\Psi(z_1)$ to the density profile by the Boltzmann form (2.5.7).

$$\rho_{(1)}(z_1) = \exp\left(\frac{-\Psi(z_1)}{kT}\right).$$

As we observed earlier, the kernel of the BGY equation is highly sensitive to the choice of input functions, and Borstnik and Azman[37] find it impossible to obtain convergence using Nazarian's algorithms in conjunction with a Lennard–Jones pair potential and Verlet's[47] molecular dynamic pair distribution. They did, however, obtain partially converged profiles using these input quantities, but using a modified Nazarian closure which was no longer symmetrical with respect to particle exchange. The profiles were again strongly oscillatory.

A subsequent attempt by Toxvaerd[48] to form a closure which remains invariant to renumbering of the particles is based upon a first order expansion of the pair distribution in terms of the density gradient, it being

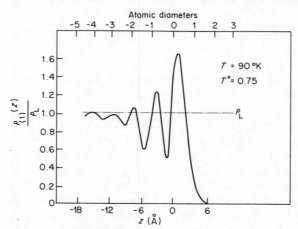

Figure 2.7.4 Nazarian's solution[46] of the single-particle Born–Green–Yvon equation for liquid argon at the triple point using the closure (2.7.7)

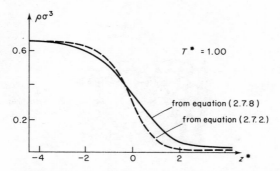

Figure 2.7.5 Solution of the Born–Green–Yvon equation (2.7.9) for the transition profile for a Lennard–Jones fluid at $T^* = \varepsilon/k = 1.00$ using a midpoint density closure[48] (2.7.8). Comparison is made for the same system using closure (2.7.2). See Reference 99, however

assumed that an expansion of the inhomogeneous radial distribution converges so rapidly that the zero and first order terms are sufficient to give the functional form of the density profile. This may be the case at critical temperatures, but in regions where the density varies by $\sim 10^3$ over an atomic diameter higher order terms will almost certainly be of importance. Thus, working to first order in an expansion in the vertical coordinates, the radial distribution appropriate to the density of the isochore plane is taken at the midpoint value:[82]

$$\rho_{(2)}(z_1, \mathbf{r}_{12}) \sim \rho_{(1)}(z_1)\rho_{(1)}(z_2)g_{(2)}[r_{12}, \rho_{(1)}(z_1 + \tfrac{1}{2}z_{12})] \qquad (2.7.8)$$

which is a symmetrical, angular-dependent form of (2.7.5). As we observed in Section 2.4, equation (2.7.8) is appropriate only for interfaces known to be structureless; indeed, the closure was initially applied to critical surfaces where the density gradient is small.[41] Nevertheless, insertion of (2.7.8) in the single-particle BGY equation yields

$$\rho_{(1)}(z_1') = \rho_L \exp\left\{ \int_{-\infty}^{z_1'} \int \nabla_1 \Phi(r_{12})\rho_{(1)}(z_2)g_{(2)}[r_{12}, \rho_{(1)}(z_1 + \tfrac{1}{2}z_{12})] \, d\mathbf{r}_{12} \, dz_1 \right\}$$

$$(2.7.9)$$

which Toxvaerd solves for a Lennard–Jones fluid at $T^* = kT/\varepsilon = 1.00$. The resulting monotonic profile is shown in Figure 2.7.5 where we have made a comparison with $\rho_{(1)}(z)$ determined on the basis of (2.7.2) (see Figure 2.7.2). The differences are seen to be slight. Profiles obtained on the basis of mechanical equilibrium—constancy of the normal component of the pressure tensor—yield identical profiles on the basis of the same closure device. Similar conclusions are found to hold for a square-well fluid. A subsequent comparison of the *ad hoc* closures (2.7.2), (2.7.8) by Salter and Davis[99] for Lennard–Jones fluids, however, suggests that improved approximation shows (2.7.8) to be superior.

54

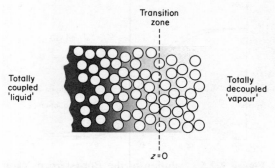

Figure 2.7.6 Schematic form for the anisotropic decoupling of particle interactions in the vicinity of the transition zone[49, 51]

Finally, in an analysis based on the BGY single-particle equation, Croxton and Ferrier[49,51] introduce a coupling operator $\xi(z)$ whose effect is to decouple anisotropically the pair interaction. Thus, any particle in the vicinity of the plane $z = 0$ is coupled anisotropically to its neighbours, even though the *distribution* of neighbours $g_{(2)}^L(r_{12})$ remains characteristic of the bulk liquid. This is shown in Figure 2.7.6 where the coupling varies from unity, corresponding to unimpaired (i.e. total) coupling, to virtually zero in the highly decoupled vapour phase. Such a device simulates a free surface somewhat in the Kirkwood manner, and is both angularly dependent and invariant to particle exchange. It does, of course, remain to determine the analytic variation of the coupling parameter $\xi(z)$. Clearly there will be three distinct regions corresponding to the two bulk phases, in which $\xi(z) = 1.0$ (liquid phase) and $\xi(z) = 0^+$ (vapour phase), and the transition region $0^+ < \xi(z) < 1$. An analytic form similar to, although not necessarily that of, the attractive branch of the reduced Lennard–Jones potential was chosen. There is no reason why, in principle, $\xi(z)$ should not be iteratively determined.

The coupling operator is introduced into the BGY equation in the following way

$$\rho_{(1)}(z_1) = \rho_L \exp\left\{\frac{-\rho_L}{kT}\int\limits_{-\infty}^{z_1}\int\!\!\int \{\Phi(r_{12})\nabla_1\xi(z_1) + \xi(z_1)\nabla_1\Phi(r_{12})\}g_{(2)}(r_{12})\frac{z_{12}}{r_{12}}\,\mathbf{dr}_{12}\,dz\right. \tag{2.7.10}$$

and for liquid argon at the triple point a small shoulder on an otherwise monotonic profile (Figure 2.7.7) is obtained.

Croxton and Ferrier[50] have also made a spherical harmonic expansion of the pair distribution in the vicinity of the transition zone, as follows:

$$\rho_{(2)}(z_1, \mathbf{r}_{12}) = \rho_L^2 g_{(2)}^L(r_{12})\left\{\sum_{l=0}^{\infty}\sum_{m=0}^{\pm l} A_{lm}(z_1, T)P_l^m(\cos\theta)\Phi(m\varphi)\right\}^2 \tag{2.7.11}$$

where it has been assumed that there is no radial distortion of the bulk liquid distribution $g^L_{(2)}(r_{12})$. Such an analysis has recently become fashionable in machine studies of the interfacial pair distribution (Section 10.10). The distribution (2.7.11) is, of course, invariant to particle exchange, and is symmetric about the z-axis (for nonhydrodynamic equilibrium systems of pairwise interacting spherical molecules), and therefore $m = 0$ throughout. Again, on symmetry grounds many of the $\{A_{l0}\} \equiv 0$, and Croxton and Ferrier retain only the $l = 0, 1$ harmonics in their angular description of the distribution in the transition zone:

$$\rho_{(2)}(z_1, \mathbf{r}_{12}) \sim \rho^2_L g^L_{(2)}(r_{12})\{A_{00}(z_1)P^0_0(\cos\theta) + A_{10}(z_1)P^0_1(\cos\theta)\}^2$$
$$= \rho^2_L g^L_{(2)}(r_{12})A^2_{00}(z_1)\{P^0_0 + \lambda(z_1)P^0_1\}^2 \qquad (2.7.12)$$

where $P^0_0 = 1$, $P^0_1 = \cos\theta$. Squaring the trial function increases its flexibility a little, and has been rewritten in terms of $\lambda(z_1) = A_{10}(z_1)/A_{00}(z_1)$ which has the significance of a *hybridization coefficient* between the spherically symmetric bulk modes and the surface angular modes of the appropriate symmetry. It is clear that $\lambda \rightarrow 0$ as $z_1 \rightarrow \pm\infty$; hybridization of bulk liquid and vapour modes with the specifically interphasal configurations will diminish as the bulk and surface correlations decouple far from the interphase, the pair distribution becoming spherically symmetric. The first unassociated harmonic, P^0_1, shows extensive hybridization with the spherically symmetric P^0_0 (bulk) modes only in the vicinity of the anisotropic liquid surface shown in Figure 2.7.8a for liquid argon at 84 °K.

The coefficients $A_{00}(z_1)$, $A_{10}(z_1)$ (and hence $\lambda(z_1)$) are determined so as to yield a minimum surface energy by standard variational techniques. λ is, of course, a function of temperature and its variation is shown schematically in Figure 2.7.8b.

Croxton and Ferrier[10] have estimated the surface tension of liquid argon

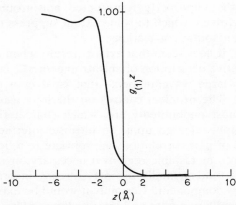

Figure 2.7.7 The density transition profile for liquid argon at the triple point using the anisotropic coupling scheme[50]

56

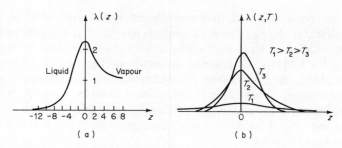

Figure 2.7.8 (a) The hybridization coefficient $\lambda(z) = A_{10}(z)/A_{00}(z)$ arising in the spherical harmonic representation of Croxton and Ferrier[50] in equation (2.7.12). (b) Schematic variation of $\lambda(z)$ with temperature

at the triple point on the basis of this model, and their results are given in Table 2.3.1. Whilst agreement with experiment is good, numerical coincidence is not an adequate criterion for the assessment of surface structure.

It is becoming increasingly apparent that the general agreement amongst the various determinations of density profile and surface tension based on a variety of closures is illusory. Indeed, it does seem that the specifications of $\rho_{(1)}(z_1, \mathbf{r}_{12})$ based on bulk phase distributions or their analytical continuations into metastable regions are fundamentally incorrect, and that our understanding of correlation within the transition zone has to be substantially revised. The difficulty arises in part from an observation of Wertheim[92] who pointed out the development of very long range horizontal correlations (~ 1 mm) at the surface when explicit account is taken of the earth's gravitational field. Clearly such long range correlation cannot derive from local density descriptions, and casts serious doubt upon all the mean field, local density, and thermodynamic perturbation theories described in this section. There is, perhaps, some consolation in the fact that generally integrations over the improperly represented anisotropic interfacial pair distributions are involved, which to some extent suppress the importance of these long range horizontal correlations.

More specifically, it does seem that extant closure schemes can provide no more than a qualitative description of anisotropic interfacial correlation only at intermediate to large wavenumbers—that is, over a correlation range typical of the bulk. The problem centres on the long wavelength capillary waves, as they almost undoubtedly are, which characterize the transition zone: are they simply released upon an intrinsic interfacial region determined on the basis of pair correlations appropriate to a 'local' or 'effective' density extended into metastable regions if necessary, or are the horizontal surface correlations fundamentally distinct—already incorporating the long wavelength capillary component? Certainly it would be extremely difficult to resolve the low k capillary modes and intermediate to large k fluid structure modes: they are inextricably coupled, and are shown qualitatively in Figure 2.7.9. More particularly, the two prescriptions for the *total* interfacial

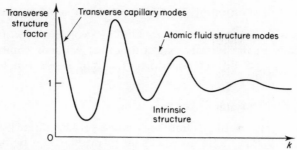

Figure 2.7.9 Schematic form of the transverse structure factor within the liquid–vapour transition zone. In addition to the conventional intermediate to large k structure characteristic of the bulk, a long wavelength divergence appears in the vicinity of $k = 0$ associated with high compressibility capillary wave states

description (capillary + intrinsic) make quite distinct k space demands upon the specification of the closure. The low k divergence in the transverse structure factor implies an interfacial tendency towards infinite transverse compressibility, ultimately deriving from the zero slope tie-line states rather than the metastable branch of the phase diagram which relates local pressure, density, and structure (Figure 2.7.10). Quite how interfacial structure may be related to tie-line states is another matter—Kalos *et al.*[93] have proposed a dynamical description in which a relatively sharp density discontinuity between gas and liquid fluctuates in location, providing a statistically defined surface in the temporal sense—however, the problem appears to be a fundamental one, and will undoubtedly cause some revision in our understanding of interfacial structure.

The role of capillary waves in the description of total surface structure will be considered in greater detail in Sections 2.11 and 10.11.

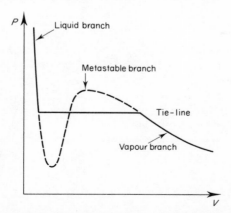

Figure 2.7.10 Typical subcritical isotherm for liquid–vapour transition

58

2.8 Numerical solution of the BGY equation: some further comments

The numerical solution of the Born–Green–Yvon equation for $\rho_{(1)}(z)$ consists essentially in the iterative substitution of a linear combination of the previous input and output functions, the kth input function being, for example

$$_{\text{IN}}\rho_{(1)}^{(k)}(z) = a_{\text{OUT}}\rho_{(1)}^{(k)}(z) + (1-a)_{\text{IN}}\rho_{(1)}^{(k-1)}(z). \qquad 0 \le a \le 1.0. \quad (2.8.1)$$

The iterative cycle continues until successive profiles differ by no more than some arbitrarily small amount, the density profile then being said to have converged. The fraction of the previous output a is known as the relaxation parameter and represents the amount of the new output combined with the previous input profile. a typically ranges in magnitude from 0.05–0.5; the largest value consistent with stable iteration is generally used—smaller and smaller relaxation parameters being necessary as the bulk fluid density increases. Invariably the combination of input and output density profiles is complicated by successive spatial shifts in the output profile of the order of an atomic diameter in the initial stages of iteration, decreasing to $\le 0.1\sigma$ after the iteration procedure has settled down (Figure 2.8.1). The movements are due in part to the settling down of the converging iterative procedure and arises as an artifact of the nonconstant normal pressure across the interfacial zone which develops prior to complete convergence of the BGY equation. An approximate discussion of this feature of the partially converged solution has been given by Osborn and Croxton[87] as follows.

Take $_{\text{IN}}\rho_{(1)}^{(i-1)}(z)$ to be the form of the density profile to be substituted into the r.h.s. of the BGY equation (2.6.8) where $_{\text{OUT}}\rho_{(1)}^{(i)}(z)$ is approximately $_{\text{IN}}\rho_{(1)}^{(i-1)}$ subject to a small displacement in the Gibbs surface Δz_{G} (Figure

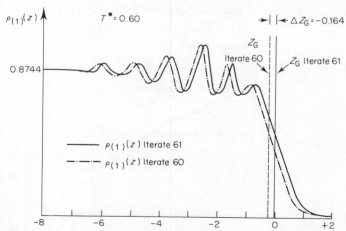

Figure 2.8.1 Successive profiles in the iterative solution of the BGY equation (after Osborn and Croxton[87])

2.8.1):

$$_{IN}\rho_{(1)}^{(i-1)}(z + \Delta z_G) \sim {}_{OUT}\rho_{(1)}^{(i)}(z) = \rho_L^{(i-1)} \exp\left\{-2\pi \int_{-\infty}^{z_1} dz_1' \int_{-\infty}^{\infty} dz_{12}{}_{IN}\right.$$

$$\left. \times \rho_{(1)}^{(i-1)}(z_1 + z_{12})z_{12} \int_{|z_{12}|}^{\infty} \frac{\nabla_1 \Phi(r_{12})}{kT} g_{(2)}(z_1, \mathbf{r}_{12}) \, d\mathbf{r}_{12}\right\}. \quad (2.8.2)$$

From (2.6.9), assuming the condition for mechanical stability has not yet been achieved, then $dP_\perp(z)/dz \neq 0$ arising from the pressure difference between liquid and vapour phases, and writing $\rho_{(1)}(z)$ for $_{IN}\rho_{(1)}^{(i-1)}(z)$ we have

$$\rho_{(1)}(z) = \rho_{(1)}(z + \Delta z_G) \exp\left\{\frac{1}{kT} \int_{-\infty}^{z} \frac{1}{\rho_{(1)}(z_1)} \frac{dP_\perp(z_1)}{dz_1} dz_1\right\}. \quad (2.8.3)$$

Expanding $\rho_{(1)}(z + \Delta z_G)$ for small Δz_G, to first order

$$\rho_{(1)}(z + \Delta z_G) \sim \rho_{(1)}(z) + \rho_{(1)}'(z)\Delta z_G$$

so that

$$\rho_{(1)}'(z)\Delta z_G \sim \rho_{(1)}(z)\left[\exp\left\{-\frac{1}{kT} \int_{-\infty}^{z} \frac{1}{\rho_{(1)}(z_1)} \frac{dP_\perp(z_1)}{dz_1} dz_1\right\} - 1\right].$$

$$(2.8.4)$$

Linearizing the exponential yields

$$\Delta z_G \sim \frac{\rho_{(1)}(z)}{\rho_{(1)}'(z)} \frac{1}{kT} \int_{-\infty}^{z} \frac{1}{\rho_{(1)}(z_1)} \frac{dP_\perp(z_1)}{dz_1} dz_1. \quad (2.8.5)$$

If we assume that the pressure difference $\Delta P = P_L - P_V$ is developed over the surface thickness w, and the approximate pressure gradient is $dP_\perp(z)/dz = -\Delta P/w$, we have at once

$$\Delta z_G \sim -\frac{1}{kT} \frac{\Delta P}{w} \left|\frac{\rho_{(1)}(z)}{\rho_{(1)}'(z)}\right| \int_{-w/2}^{w/2} \frac{dz_1}{\rho_{(1)}(z_1)}. \quad (2.8.6)$$

We see that without normal pressure equilibrium ($\Delta P = 0$) the Gibbs surface will move towards the phase with lower pressure, as it were to equilibrate the liquid–vapour coexistence by evaporation or condensation as the case may be. Clearly the correct thermodynamic specification of the liquid (or vapour) phase is an essential prerequisite, and at a given temperature is effectively set by the choice of ρ_L (or ρ_V). Equation (2.8.6) suggests that *the shift of the interface will vary within the interface itself.* This distortion of the new iterative profile cannot be corrected simply by relocating the Gibbs surface a distance $-\Delta z_G$, although such a procedure is invariably adopted,[45,88] and does have the advantage of not letting the profile wander across the numerically finite region of integration of the BGY equation (typically 20–40σ). In this case the new input function $_{IN}\rho_{(1)}^{(i)}(z)$ for a new location of the Gibbs surface z_G in $_{OUT}\rho^{(i)}_{(1)}(z)$ is, from (2.8.1):

$$_{IN}\rho^{(i)}_{(1)}(z) = a_{OUT}\rho^{(i)}_{(1)}(z - z_G) + (1-a)_{IN}\rho^{(i-1)}_{(1)}(z). \qquad (2.8.7)$$

An alternative relocation procedure proposed by Osborn and Croxton[87] determines that shift which minimizes the difference between successive input and output density profiles:

$$\text{Diff}\,|z_D| = \int |_{OUT}\rho^{(i)}_{(1)}(z + z_D) - _{IN}\rho^{(i-1)}_{(1)}(z)|\,dz \qquad (2.8.8)$$

the new profile being located at $z_{D_{min}}$ such that Diff is minimized. In this case the new input profile becomes

$$_{IN}\rho^{(i)}_{(1)}(z) = a_{OUT}\rho^{(i)}_{(1)}(z - z_{D_{min}}) + (1-a)_{IN}\rho^{(i-1)}_{(1)}(z). \qquad (2.8.9)$$

Apart from the fact that such a procedure is qualitatively more justifiable than the simple spatial shift which is generally adopted, *equation (2.8.9) may be shown to produce both oscillatory and monotonic solutions to the BGY equation which have otherwise been reported as being exclusively monotonic or numerically unstable on the basis of (2.8.7)*. A more careful reappraisal of our conclusions regarding the nature of the BGY profiles, and indeed, any iterative procedure involving the so-called 'marching phenomenon' would appear to be in order.

Following the introduction of the principal closures in Section 2.7, it is instructive to reconsider these solutions in the light of the preceding observations relating to the Gibbs shift. Consider first of all Toxvaerd's midpoint closure (2.7.8) for the anisotropic two-particle function

$$\rho_{(2)}(z_1, \mathbf{r}_{12}) \sim \rho_{(1)}(z_1)\rho_{(1)}(z_2)g_{(2)}[r_{12}\,|\,\rho_{(1)}(z_1 + \tfrac{1}{2}z_{12})]. \qquad (2.8.10)$$

The partially converged profiles using (2.8.10) and the linear combination (2.2.2)

$$g_{(2)}(r) = A\exp(-\Phi(r)/kT) + (1-A)g_{(2)}(r)_{WCA}$$

for the radial distribution function are shown in Figure 2.8.2 at a variety of reduced temperatures. The value of A is indicated. The profiles are found to develop a monotonic structure after a relatively small number of iterates. We should emphasize that the Gibbs surface is relocated on the basis of equation (2.8.9). Similar observations hold for Osborn's integral closure[87] which adopts for the local effective density

$$\begin{aligned}
\rho_{eff}(z_1, z_2) &\sim \frac{1}{z_{12}}\int_{z_1}^{z_1+z_{12}} \rho_{(1)}(z)\,dz & z_{12} \neq 0 \\
&\sim \rho_{(1)}(z) & z_{12} = 0.
\end{aligned} \qquad (2.8.11)$$

As we observed in Section 2.7, Toxvaerd was unable to obtain stable solutions to the BGY equation on the basis of the prescription (2.7.3)

$$\rho_{eff}(z_1, z_2) \sim \alpha\rho_{(1)}(z_1) + (1-\alpha)\rho_{(1)}(z_2) \qquad (2.8.12)$$

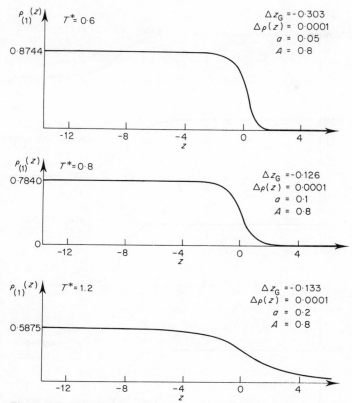

Figure 2.8.2 The partially converged profiles for liquid argon at a variety of reduced temperatures

unless $\alpha \sim 0.44$, for which $g_{(2)}(z_1, r_{12})$ is unphysical in as far as it is not invariant to renumbering of the particles. However, using the simple relocation (2.8.2), Toxvaerd obtained the monotonic profiles shown in Figure 2.7.2 for a Lennard–Jones system at a variety of reduced temperatures. The profiles resulting from setting $\alpha = 0.5$ in (2.8.12),[87] corresponding to invariance of the distribution with respect to particle exchange, are shown in Figure 2.8.3 at various reduced temperatures using the relocation (2.8.9), this notwithstanding Borstnik and Azman's inability to find stable profiles for $\alpha = 0.5$[37] Initially a monotonic profile is obtained which, after further iteration, develops stable density oscillations in complete contrast to Toxvaerd's earlier solutions. Undoubtedly the oscillatory profiles arise as a numerical response to an overconstraint implicit in (2.8.12) with $\alpha = 0.5$, and possibly in the use of the relocation procedure (2.8.8, 2.8.9). The two closure devices of Nazarian (2.7.6, 2.7.7) are both specified with respect to the Gibbs surface, and are not specifically dependent upon the actual density profiles. It follows that the precise location of the Gibbs surface is of great importance to the iterative development of the density profiles. Using the

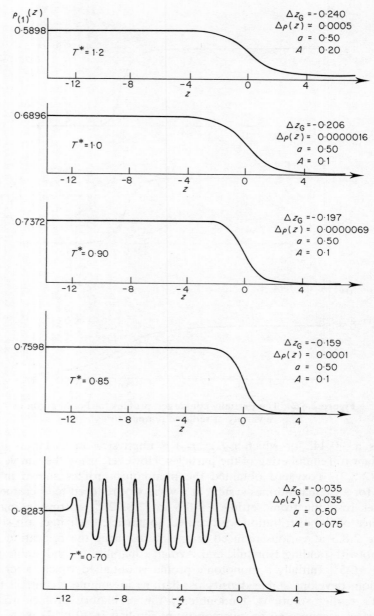

Figure 2.8.3 BGY density transition profiles for liquid argon at a variety of reduced temperatures on the basis of (2.8.12), using $\alpha = 0.5$

integral relocation (2.8.8, 2.8.9) Osborn and Croxton[87] find that no stable profiles appear to develop, although all solutions are characterized by an anomalous structured high-density surface layer which appears to grow at the expense of a broad depression (Figure 2.8.4) (subject of course to the boundary condition $\rho_{(1)}(-\infty) = \rho_L$), the net effect being to locate the Gibbs surface too far into the bulk liquid. It is evidently energetically advantageous for the particles to migrate towards the surface. Given the sensitivity of Nazarian's closures to the location of the dividing surface it is not altogether surprising that stable solutions cannot be found.

Unfortunately the introduction of a more sensitive relocation prescription (2.8.9) does little to resolve the dilemma regarding the nature of the density profile; indeed, it appears to sustain the controversy. Whilst the general consensus of opinion supports the existence of an essentially monotonic transition profile, at least for the liquid inert gases, a definitive numerical solution of the Born–Green–Yvon equation awaits an adequate assessment of the uncertainties embodied in the closure and relocation prescriptions. Moreover, the effect of a finite step width Δz between points at which the BGY equation is evaluated is found to have important consequences for the ultimate form of the transition profile.[87] It appears that practical limitations restrict the computations to $\Delta z > 0.04\sigma$, whilst the use of step widths $>0.08\sigma$ introduces numerical error, in some cases divergent numerical instability. A priori considerations suggest that the combination of a coarse step-length and an insensitive relocation procedure between iterates will have consequences for the density profile ranging from a spatial smearing of any intrinsic structure to divergent oscillations in the profile, either of which may be quite inappropriate, but both of which are initiated for numerical rather than physical reasons.

2.9 Surface correlations based on the Percus–Yevick equation

A recent determination by Perram and White,[52] based on an idea by Helfand, Frisch, and Lebowitz,[53] considers a *bulk* two-component system

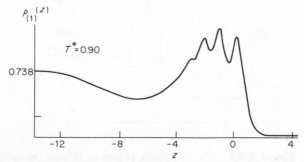

Figure 2.8.4 BGY density transition profile for liquid argon at $T^* = 0.90$ on the basis of Nazarian's closure

AB of collision diameters σ_A, σ_B and number densities ρ_A and ρ_B, respectively. Taking the limits $\sigma_B \to \infty$, $\rho_B \to 0$, one has effectively created an interface of type A particles interacting through a pair potential $\Phi_{AA}(r_{AA})$ whilst interacting with the type B 'particle' through $\Phi_{AB}(z)$. For the single-particle distribution about B we then have

$$\rho_{(1)}(z) = \rho_{L_A} \lim_{\substack{\sigma_B \to \infty \\ \rho_B \to 0}} g_{(2)}^{AB}(z). \tag{2.9.1}$$

If now we regard particle B as a vacuum bubble of large radius, the net force on particles A into the bulk of the fluid may be simulated by assigning a repulsive (i.e. surface constraining) field between the vacuum bubble and the surrounding particles. If the repulsive bubble field were of the form

$$\Phi_{AB}(z) = +\infty \qquad z > 0$$
$$= 0 \qquad \text{otherwise} \tag{2.9.2}$$

then it is obvious that such an interfacial density profile will exhibit oscillations. A softer potential would, of course, lead to a smoother profile and such a potential with its associated Boltzmann factor is shown schematically in Figure 2.9.1.

The correlations arising within the two-component system will be of the types AA, AB, and BB and so, generally, the total correlations $h_{\alpha\beta}(r)$ will be related to the direct correlations $c_{\alpha\beta}(r)$ through the Ornstein–Zernike equations[54]

$$h_{\alpha\beta}(r) = c_{\alpha\beta}(r) + \sum_{\gamma = \alpha,\beta} \rho_\gamma \int c_{\alpha\gamma}(|s|) h_{\gamma\beta}(|\mathbf{r}-\mathbf{s}|) \, \mathrm{d}\mathbf{s} \tag{2.9.3}$$

which Perram and White supplement by the Percus–Yevick approximation[55]

$$h_{\alpha\beta}(r) = -1 + c_{\alpha\beta}(r)/[1 - \exp(\Phi_{\alpha\beta}(r)/kT)] \tag{2.9.4}$$

so allowing the calculation of the various correlation functions.

The solution of (2.9.4), in particular for the correlation $g_{(2)}^{AB}$ in the limiting condition (2.9.1), enables us to determine the single-particle profile of type

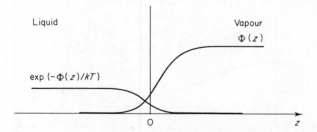

Figure 2.9.1 The exclusion potential Φ and its associated Boltzmann factor arising in the vacuum bubble model of the liquid surface

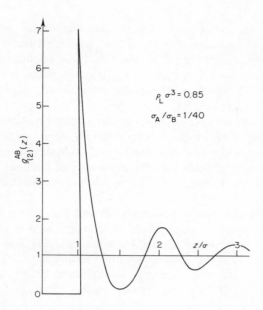

Figure 2.9.2 The distribution of hard spheres of species A about a hard sphere B: $\sigma_A/\sigma_B = 1/40$

A particles against a large type B particle. On the basis of the Andersen, Chandler, Weeks approach[56] the correlation $g_{(2)}^{AB}$ is determined for a variety of ratios $\sigma_A/\sigma_B = 1/40$, 1/25, 1/20, and 1/15 and is found to vary little with ratio. The curve $g_{(2)}^{AB}$ for $\sigma_A/\sigma_B = 1/40$ is shown in Figure 2.9.2 for hard spheres at a density $\rho_L\sigma_A^3 = 0.85$. On the basis of the ratio-dependence of the correlation, the curve obtained by extrapolating to $\sigma_A/\sigma_B = 0$, corresponding to a planar interface, seemed to differ very little from the $\sigma_A/\sigma_B = 1/40$ result.

Suppose now that the type A particles interact through the pair potential

$$\Phi_{AA}(r) = +\infty \qquad\qquad r < \sigma_{AA}$$
$$= -\varepsilon \left(\frac{\sigma_{AA}}{r}\right)^6 \qquad r > \sigma_{AA} \qquad (2.9.5)$$

and that the single-particle distribution is given as (2.9.1). Then the interaction energy of an A particle at z is

$$U(z) = \rho_L \int g_{(2)}^{AB}(z)\Phi_{AA}(r_{12})g_{(2)}^{AA}(r_{12})\,d\mathbf{r}_2 \qquad (2.9.6)$$

whereupon Φ_{AB}, the liquid–vapour exclusion bubble potential, may be defined as

$$\Phi_{AB}(z) = U(z) - U(\infty). \qquad (2.9.7)$$

Starting with an initial guess for $\Phi_{AB}(z)$ the correlation $g_{(2)}^{AB}$ may be determined from equations (2.9.3), (2.9.4), and used for the selfconsistent iterative solution through (2.9.6) for the interfacial profile which is shown in Figure 2.9.3 at $\varepsilon/kT = 1.5$ and density $\rho_L \sigma_{AA}^3 = 0.8$ but for a *curved* interface of radius 20 Å, which, it should be pointed out, does not represent a determination of $g_{(1)}(z)$.

As Croxton[57] has observed, however, the excess pressure associated with a curved interface of radius 20 Å is enormously large—of the same order of magnitude as the vapour pressure itself—and undoubtedly overestimates the hardness of the surface field, thereby initiating oscillations in the surrounding distribution. As it stands, the theory is more applicable to cavitation studies.

A more specific criticism of this approach has been given by Fischer[96] in a consideration of a hard sphere fluid in contact with a rigid wall in the Percus–Yevick and superposition (Green) approximations. In the former case the plane rigid wall is achieved within a binary hard sphere mixture AB by taking the limit $\rho_A \rightarrow 0$, $\sigma_A \rightarrow \infty$. Fischer[97] and Chapyak[98] have made a virial expansion for the density profile, based on the BGY hierarchy of equations:

$$\rho_{(1)}(z_1) = \rho_L[1 + \nu_1(z_1)\rho_L + \nu_2(z_1)\rho_L^2 + \cdots]. \tag{2.9.8}$$

For a hard sphere fluid against a rigid wall the exact virial coefficients may be evaluated in terms of cluster integrals with the results:

$$
\begin{aligned}
\nu_1(z_1) &= (2\pi\sigma^3/3)[1 - z^* + \tfrac{1}{3}z^{*3}] && z_1 < \sigma \\
\nu_1(z_1) &= 0 && z_1 \geqslant \sigma
\end{aligned} \tag{2.9.9}
$$

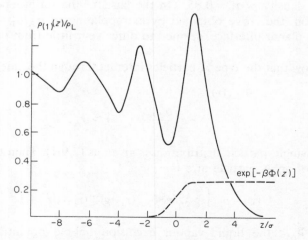

Figure 2.9.3 The interfacial profile for a system of type A particles interacting through a Sutherland potential (2.9.5) in the Percus–Yevick approximation

and

$$v_2(z_1) = \frac{10}{36}\pi^2\sigma^6\left[-\frac{16}{5}+\frac{96}{25}z_1^* - \frac{8}{5}z_1^{*4} - \frac{1}{50}z_1^{*6}\right] \qquad \sigma \leqslant z_1 < 2\sigma$$

$$v_2(z_1) = 0 \qquad\qquad\qquad\qquad\qquad\qquad\qquad\qquad z_1 \geqslant 2\sigma.$$

$$(2.9.10)$$

The Percus–Yevick (PY) and Green (G) closures, on the other hand, yield

$$v_2^{PY}(z_1) = \frac{10}{36}\pi^2\sigma^6\left[\frac{7}{10} - 3z_1^* + \frac{9}{10}z_1^{*2} + \frac{16}{10}z_1^{*3} - \frac{6}{10}z_1^{*4} + \frac{1}{50}z_1^{*6}\right]$$

$$v_2^{G}(z_1) = \frac{10}{36}\pi^2\sigma^6\left[1 - \frac{219}{50}z_1^* + \frac{27}{10}z_1^{*2} + \frac{13}{10}z_1^{*3} - \frac{6}{5}z_1^{*4} + \frac{1}{5}z_1^{*6}\right]$$

$$(2.9.11)$$

both for $0 \leqslant z_1 \leqslant \sigma$. For a hard sphere fluid both approximations yield the exact v_1, and the exact result (2.9.10) for $z_1 \geqslant \sigma$. Moreover, at the wall ($z_1 = 0$) the Green closure yields the exact v_2 whilst $v_2^{PY}(0) = 0.7v_2(0)$.

As Fisher has shown, at the ideal wall we have

$$\rho_{(1)}(0)/\rho_L = p/\rho_L kT \qquad (2.9.12)$$

where p denotes the bulk fluid pressure. Adopting the Carnahan–Starling hard sphere equation of state we have then

$$\rho_{(1)}(0)/\rho_L = (1 + \eta + \eta^2 - \eta^3)/(1 - \eta)^3 \qquad (2.9.13)$$

where $\eta = \rho_L\pi\sigma^3/6$. On the other hand, the Percus–Yevick approximation yields

$$\rho_{(1)}(0)/\rho_L = (1 + 2\eta)/(1 - \eta)^2 \qquad (2.9.14)$$

which is substantially lower than that required by (2.9.13), whilst the Green closure appears to give the exact value (2.9.12) to within numerical accuracy.

Numerical solution of the Ornstein–Zernike equation (2.9.3) in conjunction with (2.9.4) in the limit $\sigma_B \to \infty$ is particularly difficult as it stands. Henderson et al.[101] have accordingly reformulated the approach, taking the limit $\sigma_B \to \infty$, and obtain results which are qualitatively similar to those of Perram and White.

An extension of this analysis to the mean spherical approximation for a system of hard spheres interacting with an exponentially attractive wall has been considered by Waisman et al.[102] The results are in excellent agreement with the Monte Carlo simulations of Liu et al.[103]

2.10   Direct correlation at the liquid surface

On the basis of the binary representation developed in Section 2.9, we may take the limit $\sigma_B \to \infty$ and write for the wall–molecule Ornstein–Zernike equation

$$h(z) = c(z) + \rho \int c(|\mathbf{r} - \mathbf{r}'|) h(z') \, d\mathbf{r}'$$

where $c(z)$ is the direct correlation function for the wall–molecule system. $c(r)$ is the bulk direct correlation for a pair of molecules at separation r, and ρ is the bulk fluid density. The single particle density profile is simply related to $h(z)$ as $\rho(z) = \rho[1 + h(z)]$. The above equation has been considered by a number of investigators in the hard-sphere Percus–Yevick approximation,[101,104] in the mean-spherical approximation in which an exponential attraction develops between the wall and hard sphere molecules,[102] and for more general interactions.[105] Perram and Smith,[106] working in the PY approximation, have extended the treatment to a system in which all interactions are of the Baxter[107] sticky hard sphere type. In each case the single-particle distribution is against an impenetrable wall.

More pertinent for present purposes is the development of Sullivan and Stell[100] in which they remove the restriction of an impenetrable wall, allowing the coupling and exchange of particles between two fluid phases of densities ρ_L, ρ_V across an interfacial plane. To do this they write the two coupled Ornstein–Zernike equations

$$\left. \begin{aligned} h_L(z) &= c_L(z) + \rho_L \int c_L(|\mathbf{r} - \mathbf{r}'|) h_L(z') \, d\mathbf{r}' \\ h_V(z) &= c_V(z) + \rho_V \int c_V(|\mathbf{r} - \mathbf{r}'|) h_V(z') \, d\mathbf{r}' \end{aligned} \right\} \tag{2.10.1}$$

where the continuous density profile is now given as $\rho_{(1)}(z) = \rho_L[1 + h_L(z)] = \rho_V[1 + h_V(z)]$. The solution of these coupled equations will yield the density profile; a knowledge of c_L, c_V is a prerequisite, however. Sullivan and Stell make the decomposition

$$c_L(z) = c_L^{(0)}(z) - \phi_L(z)$$
$$c_L(z) = c_V^{(0)}(z) - \phi_V(z)$$

where

$$\begin{array}{lll} c_L^{(0)} = 0 & \phi_V(z) = 0 & z > 0 \\ c_V^{(0)}(z) = 0 & \phi_L(z) = 0 & z < 0 \end{array}$$

Fourier transformation of (2.10.1), and use of Baxter's factorization of uniform phase direct correlation functions gives

$$1 - \rho_L \tilde{c}_L(k) = \tilde{Q}_L(k) \tilde{Q}_L(-k)$$
$$1 - \rho_V \tilde{c}_V(k) = \tilde{Q}_V(k) \tilde{Q}_V(-k)$$

enabling us to rewrite (2.10.1)

$$\tilde{Q}_V(-k)\tilde{Q}_L(k)[\Delta\rho(k)-\Delta\rho\tilde{\theta}_V(k)] = \rho_L \frac{\tilde{Q}_V(-k)}{\tilde{Q}_L(-k)}[\tilde{c}_L^{(0)}-\tilde{\phi}_L(k)]$$

$$\tilde{Q}_V(-k)\tilde{Q}_L(k)[\Delta\rho(k)+\Delta\rho\tilde{\theta}_L(k)] = \rho_V \frac{\tilde{Q}_L(k)}{\tilde{Q}_V(k)}[\tilde{c}_V^{(0)}(k)-\tilde{\phi}_V(k)]$$

where

$$\Delta\rho(z) = \rho_{(1)}(z) - \rho_L\theta(z) - \rho_V\theta(-z), \qquad \Delta\rho = \rho_L - \rho_V$$

and the step function

$$\theta(z) = 1 \qquad z \geqslant 0$$
$$= 0 \qquad z < 0$$

where $\tilde{\theta}(k)$ is the Fourier transformation of $\theta(z)$.

Sullivan and Stell show that the unknown functions $\tilde{c}_L^{(0)}(k)$, $\tilde{c}_V^{(0)}(k)$ may be eliminated, yielding

$$\Delta\tilde{\rho} = \frac{\Delta\rho}{ik} + \frac{[ik(\tilde{\psi}_L(k)+\tilde{\psi}_V(k)) - \Delta\rho\tilde{Q}_L(0)\tilde{Q}_V(0)]}{ik\tilde{Q}_L(k)\tilde{Q}_V(k)} \qquad (2.10.2a)$$

where now

$$\tilde{\psi}_L(k) = \frac{-\rho_L}{2\pi} \int_{-\infty}^{\infty} e^{-ikz} \frac{\tilde{Q}_V(-k)}{\tilde{Q}_L(-k)} \tilde{\phi}_L(k)\, dk \qquad (2.10.2b)$$

with a corresponding exchange of subscripts for $\tilde{\psi}_V(k)$.

Presuming the bulk correlations $c_L(r)$, $c_V(r)$ to be known, it only remains to specify ϕ_L, ϕ_V for the density profile to be determined by inversion and integration.

The simplest analytical form for $\phi(z)$ which ensures continuity of both $\rho_{(1)}(z)$ and its first derivative are the exponential sums

$$\phi_L(z) = a_L e^{-\alpha_L z} + b_L e^{-\beta_L z}$$

$$\phi_V(z) = a_V e^{\alpha_V z} + b_V e^{\beta_L z}$$

(where a, α, b, β are constants determined for a specific model) which, inserted in (2.10.2) yield the profiles shown in Figure 2.10.1(a) for Baxter's sticky hard sphere interaction. In contrast, if we set $\phi_L = \phi_V = 0$ we obtain the expected discontinuity at the surface (Figure 2.10.1b) together with a strongly oscillatory structure, illustrating the crucial role of $\phi_L, \phi_V \neq 0$.

This approach draws together nicely the free and constrained surface analyses, although unless further conditions are able to be placed on the specification of the ϕ-function, the treatment cannot yet participate in the controversy surrounding the nature, structured or otherwise, of the free liquid surface.

A development in terms of an anisotropic particle–particle correlation

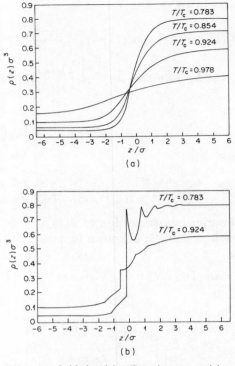

Figure 2.10.1 (a) Density transition profiles for sticky-sphere molecules of diameter σ using a 2-exponential ϕ function. (b) Density transition profiles for sticky-sphere molecules of diameter σ for
$$\phi = 0$$

$c(\mathbf{r}_1, \mathbf{r}_2)$ rather than the wall–particle correlation $c(z)$ may also be given. A connection between the two functions has been provided by Sullivan and Stell.[100]

An analysis of the graphical contributions to the total correlation in a uniform fluid[55] leads us to express the total correlation as a density expansion in terms of the propogation of direct correlation between particles 1 and 2 through all possible chains of connection. Thus, in a planar region of density inhomogeneity along the z-axis we may write for particles 1,2

$$\rho_{(2)}(z_1, \mathbf{r}_{12}) = \rho_{(1)}(z_1)\rho_{(1)}(z_2)g_{(2)}(z_1, \mathbf{r}_{12}) = \rho_{(1)}(z_1)\rho_{(1)}(z_2)[h(z_1, \mathbf{r}_{12}) + 1]$$

$$= \rho_{(1)}(z_1)\rho_{(1)}(z_2)\left[1 + c(z_1, \mathbf{r}_{12}) + \int \rho_{(1)}(z_3)c(z_1, \mathbf{r}_{13})c(z_2, \mathbf{r}_{23})\, d\mathbf{3}\right.$$

$$\left. + \int\int \rho_{(1)}(z_3)\rho_{(1)}(z_4)c(z_1, \mathbf{r}_{14})c(z_2, \mathbf{r}_{23})c(z_3, \mathbf{r}_{34})\, d\mathbf{3}\, d\mathbf{4} + \cdots\right] \quad (2.10.3)$$

whilst for particles 2,3

$$\rho_{(2)}(z_2, \mathbf{r}_{23}) = \rho_{(1)}(z_2)\rho_{(1)}(z_3)g_{(2)}(z_2, \mathbf{r}_{23}) = \rho_{(1)}(z_2)\rho_{(1)}(z_3)[h(z_2, \mathbf{r}_{23}) + 1]$$

$$= \rho_{(1)}(z_2)\rho_{(1)}(z_3)\left[1 + c(z_2, \mathbf{r}_{23}) + \int \rho_{(1)}(z_4)c(z_2, \mathbf{r}_{23})c(z_3, \mathbf{r}_{34})\, \mathrm{d}4 \right.$$

$$\left. + \iint \rho_{(1)}(z_4)\rho_{(1)}(z_5)c(z_2, \mathbf{r}_{25})c(z_3, \mathbf{r}_{34})c(z_4, \mathbf{r}_{45})\, \mathrm{d}4\, \mathrm{d}5 + \cdots \right] \quad (2.10.4)$$

If we multiply (2.10.4) by $\rho_{(1)}(z_1)c(z_1, \mathbf{r}_{13})$ and integrate over all positions of particle 3, we obtain, after eliminating the integral terms in (2.10.3),

$$h(z_1, \mathbf{r}_{12}) = c(z_1, \mathbf{r}_{12}) + \int \rho_{(1)}(z_3)h(z_2, \mathbf{r}_{23})c(z_1, \mathbf{r}_{13})\, \mathrm{d}3 \quad (2.10.5)$$

which represents the generalization of the Ornstein–Zernike equation to a planar region of inhomogeneity along the z-axis. The single-particle distribution may now be obtained by inserting (2.10.5) in the hierarchical relation (1.3.6)

$$g_{(1)}(z_1) = \frac{1}{N-1} \int \rho_{(1)}(z_2)g_{(2)}(z_1, \mathbf{r}_{12})\, \mathrm{d}2 \quad (2.10.6)$$

where, of course,

$$g_{(2)}(z_1, \mathbf{r}_{12}) = h(z_1, \mathbf{r}_{12}) + 1.$$

No attempt appears to have been made to use equation (2.10.5) in conjunction with (2.10.6) as a means of determining the single-particle density distribution. Of course, the generalized Ornstein–Zernike equation tells us nothing new—it is merely a defining relation for $c(z_1, \mathbf{r}_{12})$ and the solution of the structural problem is now transferred to the determination of the anisotropic direct correlation. There is reason to suppose, however, that the specification of an approximate closure for $c(z_1, \mathbf{r}_{12})$ is likely to be more efficacious than the models discussed in the preceding sections for $g_{(2)}(z_1, \mathbf{r}_{12})$ simply because of the short range nature of the direct correlation ($\sim\sigma$) in comparison with that of the radial distribution function ($\sim 4\sigma$). At temperatures well below the critical point the spatial delocalization of the liquid surface is of the order of the range of $g_{(2)}(z_1, \mathbf{r}_{12})$, which undoubtedly makes its specification in terms of a simple closure device considerably less wieldy than for the much more localized function $c(z_1, \mathbf{r}_{12})$. Of course, before (2.10.5) can be solved for the anisotropic total correlation an approximate specification of the direct correlation has to be made, which at first sight would seem to offset some of the advantages of the Ornstein–Zernike approach. It has to be remembered, however, that the specification of $g_{(2)}(z_1, \mathbf{r}_{12})$ in terms of a closure device involves approximate (bulk) radial distributions determined on the basis of the various integral equation methods of uniform liquid theory.

Two expressions which immediately suggest themselves for $c(z_1, \mathbf{r}_{12})$ are, from the theory of uniform fluids, the Percus–Yevick and hypernetted chain approximations, suitably adapted to the inhomogeneous surface region:

$$c_{PY}(z_1, \mathbf{r}_{12}) = h(z_1, \mathbf{r}_{12}) - \left\{ g_{(2)}(z_1, \mathbf{r}_{12}) \exp\left(\frac{\Phi(r_{12})}{kT}\right) - 1 \right\} \quad (2.10.7)$$

and

$$c_{HNC}(z_1, \mathbf{r}_{12}) = g_{(2)}(z_1, \mathbf{r}_{12}) - 1 - \ln g_{(2)}(z_1, \mathbf{r}_{12}) - \frac{\Phi(r_{12})}{kT}. \quad (2.10.8)$$

Local closures for c_{PY} or c_{HNC} of the form (2.7.2) are likely to yield significantly better results than their total correlation (or radial distribution) counterparts.

Of course, the two formulations (2.10.5), (2.10.6) and those based on $g_{(2)}(z_1, \mathbf{r}_{12})$ should be equivalent; a general proof requires a knowledge of the interrelationships between the inhomogeneous direct and total correlations and the pair potential. However, neglecting the angular dependence of c and g, it is clear that the two equations become identical in the low density limit as $c \to \exp(-\Phi/kT) - 1$ and $g \to \exp(-\Phi/kT)$.

Whilst the direct correlation approach has not as yet been used for the determination of the density profile, alternative statistical mechanical expressions for the surface tension have been proposed[58–62] in which the fluid structure is expressed in terms of $c(z_1, \mathbf{r}_{12})$ rather than $g_{(2)}(z_1, \mathbf{r}_{12})$.

Lovett et al.,[61] for example, consider a small local curvature of radius R to develop in the planar interface, whereupon the single-particle distribution $\rho_{(1)}(z_1)$ becomes (Figure 2.10.2)

$$\rho_{(1)}(x_1, y_1, z_1) = \rho_{(1)}\{[x_1^2 + y_1^2 + (z_1 + R)^2]^{1/2} - R\}. \quad (2.10.9)$$

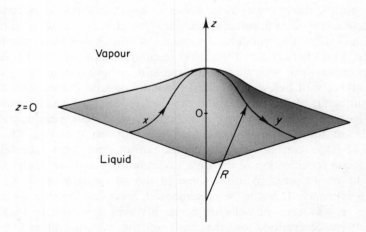

Figure 2.10.2 Geometry of a surface capillary wave

The maintenance of a curved interface requires a pressure difference between the two bulk phases

$$\Delta p = \frac{2\gamma}{R} + O\left(\frac{1}{R^2}\right) \tag{2.10.10}$$

where for small curvatures γ is the *planar* surface tension. Such a curvature could, in principle, be achieved instead by some local external field whose potential we designate $\phi(\mathbf{r})$, and which may be determined as a functional expansion in the singlet density about the unperturbed distribution:

$$-\beta\phi(\mathbf{r}_1) = \int \{\delta[-\beta\phi(\mathbf{r}_1)/\delta\rho(\mathbf{r}_2)]\}_{\text{planar}} \Delta\rho(\mathbf{r}_2)\, d\mathbf{r}_2 + O(\Delta\rho)^2 \tag{2.10.11}$$

where $\beta = (kT)^{-1}$. Percus[63] has shown

$$\delta[-\beta\phi(\mathbf{r}_1)]/\delta\rho(\mathbf{r}_2) = [\delta(\mathbf{r}_1 - \mathbf{r}_2)/\rho(\mathbf{r}_1)] - c(\mathbf{r}_1, \mathbf{r}_2) \tag{2.10.12}$$

where $c(\mathbf{r}_1, \mathbf{r}_2)$ is the anisotropic direct correlation function and

$$\Delta\rho(\mathbf{r}_1) = \rho_{(1)}\{[x_1^2 + y_1^2 + (z_1 + R)^2]^{1/2} - R\} - \rho_{(1)}(z_1)$$

$$= \frac{1}{2}\left(\frac{x_1^2 + y_1^2}{R}\right)\left(\frac{d\rho_{(1)}(z_1)}{dz_1}\right) + O(R^{-2}) \tag{2.10.13}$$

whereupon we may rewrite (2.10.11)

$$-\beta\phi(\mathbf{r}_1) = \int\left\{\left(\frac{\delta(\mathbf{r}_1 - \mathbf{r}_2)}{\rho_{(1)}(\mathbf{r}_1)}\right) - c(\mathbf{r}_1 - \mathbf{r}_2)\right\}\left(\frac{x_2^2 + y_2^2}{2R}\right)\left(\frac{d\rho_{(1)}(z_2)}{dz_2}\right) d\mathbf{r}_2 + O(R^{-1})$$

$$\tag{2.10.14}$$

where $c(\mathbf{r}_1, \mathbf{r}_2)$ is the direct correlation in the planar interface. On the z axis ($x = y = 0$) the term in (2.10.14) involving the δ-function drops out leaving

$$\beta\phi(\mathbf{r}_1) = \int c(\mathbf{r}_1, \mathbf{r}_2)\left(\frac{x_2^2 + y_2^2}{2R}\right)\left(\frac{d\rho_{(1)}(z_2)}{dz_2}\right) d\mathbf{r}_2 + O(R^{-2}). \tag{2.10.15}$$

The z component of the associated force $(\partial\varphi(z)/\partial z)$ integrated along the z axis is equivalent to the total pressure difference $\Delta p'$:

$$\Delta p' = \int_{-\infty}^{\infty} \rho_{(1)}(z_1)\frac{\partial}{\partial z_1}[\varphi(z_1)]\, dz_1$$

$$= \frac{kT}{2R}\int_{-\infty}^{\infty} \rho_{(1)}(z_1)\frac{\partial}{\partial z_1}\int c(\mathbf{r}_1, \mathbf{r}_2)\left(\frac{d\rho_{(1)}(z_2)}{dz_2}\right)(x_2^2 + y_2^2)\, d\mathbf{r}_2\, dz_1 + O(R^{-2}). $$

$$\tag{2.10.16}$$

And since, for mechanical stability $\Delta p = -\Delta p'$, we have from equations (2.10.10) and (2.10.16)

$$2\gamma = -\frac{kT}{2}\int_{-\infty}^{\infty} \rho_{(1)}(z_1)\frac{\partial}{\partial z_1}\int c(\mathbf{r}_1, \mathbf{r}_2)\left(\frac{d\rho_{(1)}(z_2)}{dz_2}\right)(x_2^2 + y_2^2)\, d\mathbf{r}_2\, dz_1.$$

$$\tag{2.10.17}$$

Assuming the direct correlation function has a finite range,

$$\int c(\mathbf{r}_1, \mathbf{r}_2)\left(\frac{d\rho_{(1)}(z_2)}{dz_2}\right)(x_2^2 + y_2^2)\, d\mathbf{r}_2$$

is zero at the limits of integration on z, and integration by parts gives

$$\gamma = \frac{1}{4}kT \int_{-\infty}^{\infty} dz_1 \int d\mathbf{r}_2 \frac{d\rho_{(1)}(z_1)}{dz_1} \frac{d\rho_{(1)}(z_2)}{dz_2} c(z_1, \mathbf{r}_{12})(x_{12}^2 + y_{12}^2)$$

(2.10.18)

or, in terms of relative coordinates,

$$\gamma = \frac{1}{4}kT \int_{-\infty}^{\infty} dz_1 \iiint dz_2\, dx_{12}\, dy_{12} \frac{d\rho_{(1)}(z_1)}{dz_1} \frac{d\rho_{(1)}(z_2)}{dz_2} c(z_1, \mathbf{r}_{12})(x_{12}^2 + y_{12}^2).$$

(2.10.19)

Transforming to cylindrical polar coordinates and performing the angular integration this expression reduces to a three-fold integral[62]

$$\gamma = \frac{\pi kT}{2} \int_{-\infty}^{\infty} dz_1 \frac{d\rho_{(1)}(z_1)}{dz_1} \int_{-\infty}^{\infty} dz_2 \frac{d\rho_{(1)}(z_2)}{dz_2} \int_{|z_{12}|}^{\infty} c(z_1, \mathbf{r}_{12})r_{12}(r_{12}^2 - z_{12}^2)\, dr_{12}.$$

(2.10.20)

Lekner and Henderson[62] make a comparison between the direct correlation expression of the surface tension (2.10.20) and its radial distribution function counterpart. Integrating (2.10.20) by parts twice gives

$$\gamma = \frac{\pi kT}{2} \int_{-\infty}^{\infty} dz_1 \rho_{(1)}(z_1) \int_{-\infty}^{\infty} dz_2 \rho_{(1)}(z_2) \frac{\partial^2}{\partial z_1 \partial z_2} \int_0^{\infty} c(z_1, \mathbf{r}_{12})(r_{12}^2 - z_{12}^2)r_{12}\, dr_{12}$$

(2.10.21)

whilst (1.4.24) may be written in terms of polar coordinates

$$\gamma = \frac{\pi}{2k} \int_{-\infty}^{\infty} dz_1 \rho_{(1)}(z_1) \int_{-\infty}^{\infty} dz_2 \rho_{(1)}(z_2) \int_{|z_{12}|}^{\infty} g_{(2)}(z_1, \mathbf{r}_{12})$$

$$\times \frac{d\Phi(r_{12})}{dr_{12}} (r_{12}^2 - 3z_{12}^2)\, dr_{12}. \quad (2.10.22)$$

Both theories are claimed to be general, and whilst Lovett et al. maintain they are 'complementary', and March and Tosi[91] that they are alternative but approximate representations, it remains to be shown that they are generally equivalent. For central pairwise interactions the two theories are evidently equivalent if

$$k^2 T \frac{\partial^2}{\partial z_1 \partial z_2} \int_{|z_{12}|}^{\infty} c(z_1, \mathbf{r}_{12})(r_{12}^2 - z_{12}^2)r_{12}\, dr_{12}$$

$$= \int_{|z_{12}|}^{\infty} g_{(2)}(z_1, \mathbf{r}_{12}) \frac{d\Phi(r_{12})}{dr_{12}} (r_{12}^2 - 3z_{12}^2)\, dr_{12}. \quad (2.10.23)$$

Using the identity

$$\frac{\partial^2}{\partial z_1 \partial z_2} f(|z_{12}|) = -2\delta(z_{12})f'(|z_{12}|) - f''(|z_{12}|) \qquad (2.10.24)$$

the l.h.s. of (2.10.23) reduces to

$$2k^2 T \left\{ \int_{|z_{12}|}^{\infty} c(r_{12})r_{12}\, dr_{12} - z_{12}^2 c(|z_{12}|) \right\} \qquad (2.10.25)$$

where the angular dependence in $c(z_1, \mathbf{r}_{12})$ has been neglected. The r.h.s. of (2.10.23) becomes identical to (2.10.25) when we take the low density limiting expressions $g_{(2)} = \exp(-\Phi/kT)$ and $c = \exp(-\Phi/kT) - 1$. Unfortunately this conclusion cannot be extended to the low temperature, high density systems of interest here: a more general demonstration would require a precise knowledge of the inhomogeneous functions $c(z_1, \mathbf{r}_{12})$ and $g_{(2)}(z_1, \mathbf{r}_{12})$. Approximate closures of the type (2.7.2) would be quite inappropriate in any attempt to prove the consistency of (2.10.23) except, of course, for low densities and small interfacial density gradients.

Although analytic equivalence of the two expressions at subcritical temperatures has never been demonstrated for the general pairwise interacting fluid, doubts have nevertheless been raised concerning the propriety of mean field derivations, such as that of the fluctuation theory development, in regions of highly density gradient. The asymptotic equivalence as $T \to T_c$ demonstrated above confirms this suspicion. Yang et al.,[90] however, assert that the two approaches *must* be equivalent since the direct correlation determines the total correlation (i.e. the radial distribution function). A recent demonstration of their analytic equivalence has been provided at all temperatures by Leng et al.,[89] admittedly in a mean field approximation, for the penetrable sphere model. These authors go on to assert that since the model is not atypical of fluids in general (apart from its tractability) doubts concerning the equivalence of the two approaches may finally be banished. Naturally, attention focuses on the specification of the anisotropic direct correlation function, which for the penetrable sphere model is

$$c(\mathbf{r}_1, \mathbf{r}_2) = \frac{1}{v_0} \int f(r_{13})f(r_{23})\rho_{(1)}(-z_3)\, d\mathbf{r}_3 \qquad (2.10.26)$$

where $f(r)$ is the negative unit step function. v_0 is the volume of the penetrable sphere. Assessment of the model undoubtedly centres on the adequacy of this mean field specification under conditions of pronounced density gradient. Indeed, as the authors observe, it is surprising at first sight that the equivalence should be demonstrable on the basis of a mean field approximation since different approximations generally yield different results. However, it is pointed out that for a homogeneous fluid the two approaches yield the same pressure, and so it is not surprising that they should also yield the same surface tension. One has to decide whether the coincidence of the results is an artifact of the approximation, albeit a remarkable one, before banishing our doubts completely, as Leng *et al.* encourage us to do.

2.11 Surface capillary waves and spatial delocalization of the interface

Direct experimental investigation of the surface structure of simple fluids and liquid metals by high and low energy electron diffraction, neutron, X-ray studies and optical scattering have been attempted, but no convincing structural determinations of anything other than the gross features, such as surface thickness for example, have been reported as yet.

Recent optical scattering experiments in the vicinity of the critical point appear to confirm that the density transition profile is a monotonic decreasing function from the liquid to the vapour phase. It has to be realized, however, that such optical measurements provide an assessment only of the dielectric profile, and whilst for critical systems this may reasonably be assumed proportional to the local density, at lower temperatures a more subtle interpretation is required. Incident light polarized at 45° with respect to a discontinuously sharp plane of incidence is scattered with an ellipticity[64]

$$\mathcal{E} = \frac{k}{\sqrt{2}} \frac{\varepsilon - 1}{\rho_L^2} \int_{-\infty}^{\infty} \rho_{(1)}(z_1)[\rho_L - \rho_{(1)}(z_1)] \, dz_1 \qquad (2.11.1)$$

where k is the optical wavenumber and ε is the dielectric constant. This expression assumes that the wavelength of the light is large compared with the interfacial thickness, and that the Clausius–Mossotti relation holds. The assumption that the dielectric tensor remains isotropic throughout the transition zone has recently been reformulated by Buff and Lovett[59,65] in such a way that it is valid at low temperatures. Nevertheless, it is clear that $\mathcal{E}$ in (2.11.1) measures a surface thickness defined by[66]

$$l = \int_{-\infty}^{\infty} \left(\frac{\rho_{(1)}(z_1)(\rho_L - \rho_{(1)}(z_1))]}{\rho_L^2} \right) dz_1. \qquad (2.11.2)$$

The optical reflectivity of normally incident light has been measured over a range of wavelengths[67] but, of course, such an experiment is sensitive only to the long wavelength Fourier components of the density profile, the more detailed structure which may be present being completely washed out. In these measurements the profile is characterized by the length[68]

$$l_0 = (\rho_L - \rho_V) \left| \frac{d\rho_{(1)}(z_1)}{dz_1} \right|^{-1} \qquad (2.11.3)$$

where the density gradient is evaluated on the Gibbs dividing surface. This definition has been criticized[69] since it is determined solely by the density at the Gibbs surface, and is therefore insensitive to the wings of the profile. Lekner and Henderson[69] compare a number of possible definitions of surface thickness which have been proposed previously, and introduce a new measurement t, the '10–90' thickness—the distance over which the density rises from $\rho_V + \frac{1}{10}(\rho_L - \rho_V)$ to $\rho_V + \frac{9}{10}(\rho_L - \rho_V)$.

The ratio of the interfacial thickness to the atomic diameter t/σ is determined for a number of profiles, most of which are of variationally optimized Fermi or exponential form, and find $1.3 < t/\sigma < 3.8$ for liquid argon in the range 84–90 °K. Although the results must be regarded as somewhat inconclusive, and certainly do not advance our understanding of interfacial structure, some interesting points emerge, particularly concerning the contribution of surface capillary waves to the spatial delocalization of the transitional zone.

This recent and controversial aspect of the interfacial structure has acquired added importance with Wertheim's[92] demonstration of the existence of long range horizontal correlations within the transition zone, and their identification by Kalos et al.[93] as surface capillary waves.

Certainly the existence of surface capillary waves will effectively broaden the interfacial zone, but may we regard the mean field, van der Waals, and the capillary wave models as being essentially equivalent? Or should we regard the van der Waals profile as being the 'intrinsic' profile which is subsequently subjected to capillary undulation? Whilst not pre-empting the discussion, we shall show that each of these latter procedures yield profiles and surface tensions which depend upon surface area and the strength of the gravitational field. In other words, characteristics which are not intrinsic to the liquid–vapour coexistence: such a situation is quite at variance to what we have generally taken it to be.

Without assigning any specific form to the intrinsic profile we now consider the effect of releasing capillary waves upon the interface.

A long wavelength thermal oscillation $\zeta(x, y)$ about an equilibrium plane $\zeta = 0$ of the form shown in Figure 2.11.1 will develop with the probability $\exp(-\beta A)$, where A is the Helmholtz free energy of that particular configuration. The free energy at the liquid surface may be resolved into two components

$$A = A_0 + W \tag{2.11.4}$$

where A_0 is the 'intrinsic' mean field energy of the liquid surface unperturbed by capillary waves, and W represents the reversible work associated with the creation of undulations. This representation assumes, of course, that the undulations are of sufficiently small wavelength so as not to modify the intrinsic structure. Thus there are two different characteristic interfacial widths: the intrinsic width over which the density of a planar interface (almost) changes from one bulk value to the other, and the second, which is the r.m.s. capillary wave dispersion $\langle \bar{\zeta}^2 - \langle \bar{\zeta} \rangle^2 \rangle^{1/2}$ of the planar interface arising from fluctuations about the plane $\zeta = 0$. We now consider the development of the term W in (2.11.4) which arises from the capillary waves.

The reversible work W done to create a convoluted dividing surface

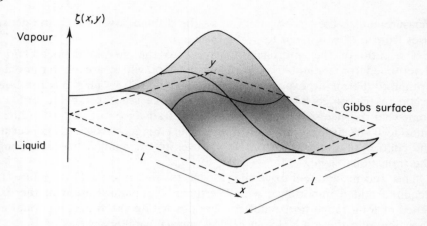

Figure 2.11.1 Geometry of a long wavelength thermal oscillation $\zeta(x, y)$ about the equilibrium plane $\zeta = 0$

$\zeta(x, y)$ with respect to a fixed xy plane (Figure 2.11.1) will involve two terms—a gravitational component and a component arising from the isothermal increase of the area $\zeta(x, y)$ with respect to that of the equilibrium xy plane. This latter term will, of course, be proportional to the intrinsic surface tension γ_0. We may therefore write

$$W = (\rho_L - \rho_V) \int_l \int_l \left[\int \zeta(x, y) gz \, dz \right] dx \, dy$$

$$+ \gamma \int_l \int_l \left[1 + \left(\frac{\partial \zeta}{\partial x} \right)^2 + \left(\frac{\partial \zeta}{\partial y} \right)^2 \right]^{1/2} dx \, dy \quad (2.11.5)$$

where g is the acceleration due to gravity. The surface undulation $\zeta(x, y)$ may be expressed as a sum of decoupled surface waves as for a harmonic analysis of a vibrating membrane:

$$\zeta(x, y) = \sum_{\mathbf{k}} A(\mathbf{k}) \exp(i\mathbf{k}.\mathbf{s}) \qquad \mathbf{s} = x\mathbf{i} + y\mathbf{j}. \quad (2.11.6)$$

Expanding $(1 + (\partial \zeta/\partial x)^2 + (\partial \zeta/\partial y)^2)^{1/2}$ to second order and substituting (2.11.6) in (2.11.5) we obtain

$$W = l^2 \gamma_0 + \sum_{\mathbf{k}, \mathbf{k}'} A(\mathbf{k}) A(\mathbf{k}') \int_0^l \int_0^l \exp[i(\mathbf{k} + \mathbf{k}') \cdot \mathbf{s}$$

$$\times \{\tfrac{1}{2}(\rho_L - \rho_V)g - \tfrac{1}{2}\gamma(k_x k_x' + k_y k_y')\} \, dx \, dy. \quad (2.11.7)$$

With periodic boundary conditions $\zeta(0, y) = \zeta(l, y)$ and $\zeta(x, 0) = \zeta(x, l)$, $\mathbf{k}$ may take the values

$$\mathbf{k} + \frac{2\pi}{l} (n_x \mathbf{i} + n_y \mathbf{j}) \qquad n_x, n_y = 0, \pm 1, \pm 2, \ldots \quad (2.11.8)$$

whereupon

$$W = \gamma_0 l^2 + \sum_{\mathbf{k}} |A(\mathbf{k})|^2 [\tfrac{1}{2}(\rho_L - \rho_V)g + \tfrac{1}{2}\gamma k^2] l^2. \tag{2.11.9}$$

We may now calculate $\bar{\zeta}$ and $\bar{\zeta}^2$ for a given set of amplitudes $A(\mathbf{k})$, thus

$$\bar{\zeta} = \frac{1}{l^2} \int_0^l \int_0^l \zeta \, dx \, dy = A(0)$$

$$\bar{\zeta}^2 = \frac{1}{l^2} \int_0^l \int_0^l \zeta^2 \, dx \, dy = \sum_{\mathbf{k}} A(\mathbf{k})^2. \tag{2.11.10}$$

The probability of an oscillation developing involving an amount of reversible work W is, of course, $\exp(-\beta W)$. Computing the thermal average of $\bar{\zeta}^2$, we find for the mean square capillary dispersion

$$\langle \bar{\zeta}^2 - \langle \bar{\zeta} \rangle^2 \rangle = \sum_{k>0} \{\beta l^2 [\tfrac{1}{2}(\rho_L - \rho_V)g + \tfrac{1}{2}\gamma_0 k^2]\}^{-1}. \tag{2.11.11}$$

The sum over $\mathbf{k}$ may be replaced by an integral since $\Delta n_x \Delta n_y = 1 = (l/2\pi)^2 \Delta k_x \Delta k_y$ whereupon (2.11.11) becomes

$$\langle \bar{\zeta}^2 - \langle \bar{\zeta} \rangle^2 \rangle = \frac{1}{(2\pi)^2 \beta \gamma_0} \iint_{k_{min}<k<k_{max}} (\xi^2 + k^2)^{-1} \, d^2 k$$

$$= \frac{1}{4\pi\beta\gamma_0} \ln \left\{ \frac{\xi^2 + k_{max}^2}{\xi^2 + k_{min}^2} \right\} \tag{2.11.12}$$

where $\xi = [(\rho_L - \rho_V)g/2\gamma_0]^{-1/2}$ and represents the intrinsic correlation length of the transition zone. We may take $k_{min} = 2\pi/l$ as an obvious lower limit; however, no unique specification for k_{max} may be made: here we take $k_{max} = 2\pi/\sigma$ for convenience, whilst Davis[76] sets $k_{max} = 2\pi/l_{min}$, where l_{min} is a few molecular diameters diverging as $(T_c - T)^{-1/2}$ as the critical point is approached.

The broadening of the intrinsic interface by capillary waves is demonstrated most clearly by taking a step function density as the intrinsic profile. This yields the broadened profile[93,94]

$$\rho_{(1)}^{(z)} = \tfrac{1}{2}\rho_L \, \text{erfc} \left(\frac{z}{(2\langle \bar{\zeta}^2 - \langle \bar{\zeta} \rangle^2 \rangle)^{1/2}} \right)$$

which we see from (2.11.11) depends explicitly upon the area of the interface and g. In the absence of gravity $\xi = 0$ and $\langle \bar{\zeta}^2 - \langle \bar{\zeta} \rangle^2 \rangle = (2\pi\beta\gamma)^{-1} \ln(l/\sigma)$, the mean square dispersion diverging logarithmically with area.

If, in (2.11.12) we set $k_{min} = 0$ (i.e. $l = \infty$), then we find for the dispersion $\langle \bar{\zeta}^2 - \langle \bar{\zeta} \rangle^2 \rangle = (4\pi\beta\gamma)^{-1} \ln(1 + \xi^{-1} k_{max})$, which again shows that the mean square dispersion of the interface diverges as $(-\ln g)^{-1/2}$. In both cases, then, the interfacial thickness becomes *infinite* in zero gravitational field; the capillary wave representation evidently does not coincide with the finite width mean field profiles determined on the basis of the integral equations, which are, of course, zero gravity solutions. (In the latter case, however, it must be said that the use of closures for $\rho_{(2)}(z_1, \mathbf{r}_{12})$ in the sense of Wertheim (Section 2.10) does not allow a final conclusion to be drawn.) Moreover, the surface tension ceases to be an intrinsic property of the liquid surface, but instead becomes dependent upon area and gravity. Evidently capillary undulation needs to be incorporated more subtly before we have a satisfactory representation of the interfacial zone. Weeks' treatment[94] is essentially a capillary fluctuation treatment, and he too arrives at an interfacial divergence: $d\rho_{(1)}(z)/dz \rightarrow 0$ as $(-\ln g)^{-1/2}$ as $g \rightarrow 0$, and attributes the divergence to the long range horizontal correlations of Wertheim.

Closely related is the recent essentially *dynamical* description of the interfacial profile and the associated specification of the anisotropic pair distribution of Kalos *et al.*[93] Suppose, for simplicity, we assume that the two-phase system is sharply resolved by a plane of density discontinuity between liquid and vapour located at Z with pure vapour for $z < Z$ and pure liquid for $z > Z$ (Figure 2.11.2a). It follows that for given location Z of the intrinsic Gibbs surface,

$$\rho_{(1)}(r)_Z = \rho_L H(z-Z) + \rho_V H(Z-z)$$
$$\rho_{(2)}(r_1, r_2)_Z = \rho_L^2 g_{(2)}^L(r_1 - r_2) H(z_1 - Z) H(z_2 - Z)$$
$$+ \rho_L \rho_V \{1 - H(z_1 - Z)H(z_2 - Z) - H(Z - z_1)H(Z - z_2)\}$$
$$+ \rho_V^2 g_{(2)}^V(r_1 - r_2)H(Z - z_1)H(Z - z_2) \quad (2.11.13)$$

where $H(x)$ is the Heaviside step function. If, now, the location of Z is distributed with probability density $f(Z)$ (Figure 2.11.2b), then the familiar delocalized surface is realized in terms of the dynamical fluctuation of the gas–liquid discontinuity past any given spatial point—however, the point may be regarded as being in either the gas or liquid phase at any instant. Such an instantaneous representation, determined on the basis of an energy criterion, is shown in Figure 2.11.3: the intrinsic Gibbs surface is seen to be strongly convoluted in this molecular dynamics simulation discussed by Kalos *et al.* (Section 10.11).

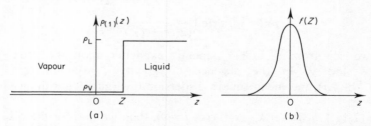

Figure 2.11.2 (a) Discontinuous density transition profile. (b) Spatial distribution of probability for the Gibbs surface

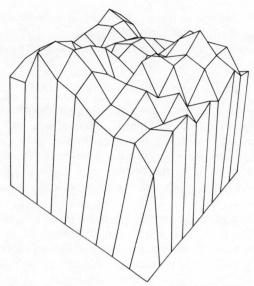

Figure 2.11.3 Typical surface configuration of atoms resolved into liquid and vapour phases for the criterion $E_0 = 0$

So, integrating over the probability distribution $f(Z)$, we have

$$\rho_{(1)}(r) = \int \rho_{(1)}(r)_Z f(Z)\, dZ$$

$$= \rho_L F(z) + \rho_V[1 - F(z)]$$

$$\rho_{(2)}(r_1, r_2) = \rho_L^2 g_{(2)}^L(r_1 - r_2) F(\text{Min}\,(z_1, z_2)) \qquad (2.11.13a)$$

$$+ \rho_L \rho_V[F(\text{Max}\,(z_1, z_2)) - F(\text{Min}\,(z_1, z_2))]$$

$$+ \rho_V^2 g_{(2)}^V(r_1 - r_2)[1 - F(\text{Max}\,(z_1, z_2))]$$

where $F(Z) = \int_{-\infty}^{Z} f(Z')\, dZ'$. Eliminating $F(Z)$ yields the closure between $\rho_{(1)}$ and $\rho_{(2)}$:

$$\rho_{(2)}(r_1, r_2) = \left(\frac{\rho_L^2}{\rho_L - \rho_V}\right)[\rho_{(1)}(\text{Min}\,(z_1, z_2)) - \rho_V]g_{(2)}^L(r_1 - r_2)$$

$$+ \left(\frac{\rho_L \rho_V}{\rho_L - \rho_V}\right)[\rho_{(1)}(\text{Max}\,(z_1, z_2)) - \rho(\text{Min}\,(z_1, z_2))]$$

$$+ \left(\frac{\rho_V^2}{\rho_L - \rho_V}\right)[\rho_L - \rho_{(1)}(\text{Max}\,(z_1, z_2))]g_{(2)}^V(r_1 - r_2). \qquad (2.11.14)$$

Provided $\rho_V \ll \rho_L$, the above expression simplifies to yield

$$\rho_{(2)}(r_1, r_2) \simeq \rho_L \rho_{(1)}(\text{Min}\,(z_1, z_2))g_{(2)}^L(r_1 - r_2) \qquad (2.11.15)$$

which, we recall, is based on a *sharp* interfacial density discontinuity, spatially delocalized with probability $f(Z)$.

The distinction between (2.11.15) and the conventional local density approximation

$$\rho_{(2)}(r_1, r_2) = \rho_{(1)}(r_1)\rho_{(1)}(r_2)g_{(2)}[r_1 - r_2 \mid \bar{\rho}] \qquad (2.11.16)$$

centres primarily on the specification of the pair distribution. In the latter case the continuum of interfacial densities between ρ_L and ρ_V is realized by passing along the metastable branch B'B (Figure 2.11.4) (the density inhomogeneity being stabilized by the bulk field) and determining the corresponding pair distribution. In the former case (2.11.15) the distribution is realized as a weighted combination of the states A, A' along the tie-line A'A.

As far as vertical correlations (along the z-axis) are concerned, neither approximation shows any major qualitative differences or deficiencies. For horizontal correlations, however, markedly different distributions are obtained. Thus, for correlations parallel to the surface at depth z, from (2.11.14)

$$\rho_{(2)}^{\parallel}(z_1, \mathbf{r}_{12}) \sim \frac{\rho_{(1)}(z_1) - \rho_V}{\rho_L - \rho_V} \rho_L^2 g_{(2)}^L(\mathbf{r}_{12}) + \frac{\rho_L - \rho_{(1)}(z_1)}{\rho_L - \rho_V} \rho_V^2 g_{(2)}^V(\mathbf{r}_{12})$$

$$(2.11.14a)$$

where $\mathbf{r}_{12} = [(x_1 - x_2)^2 + (y_1 - y_2)^2]^{1/2}$ whilst, according to (2.11.16),

$$\rho_{(2)}^{\parallel}(z_1, \mathbf{r}_{12}) \sim (\rho_{(1)}(z_1))^2 g_{(2)}(\mathbf{r}_{12} \mid \rho_{(1)}(z_1)). \qquad (2.11.16a)$$

Fourier inversions of these quantities are quite similar for $\mathbf{k} \neq 0$; however in

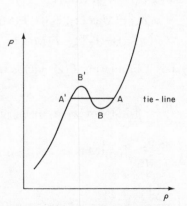

Figure 2.11.4 Schematic form of the (p, ρ) isotherm showing the connection of coexisting fluid states via the tie-line A'A, and the metastable branch B'B

their long wavelength behaviour ($\mathbf{k} \to 0$) they differ markedly, being regular for (2.11.16a) but singular for (2.11.14a). Of course, such a pronounced singularity is not to be expected for a delocalized intrinsic interface, but is nevertheless indicative of the development of long-range horizontal correlations: a somewhat more realistic representation showing the development of long wavelength horizontal modes under certain limiting conditions has been proposed recently by Weeks.[94] A qualitatively similar small k divergence was predicted by Croxton[95] along the solid–fluid tie-line, and was associated there, as it undoubtedly is here, with a region of infinite compressibility which is, of course, forfeited in passing along the metastable branch B'B (Figure 2.11.4) in adopting the closure (2.11.16a).

We shall consider machine investigations of the long wavelength behaviour in Section 10.11.

For a surface of infinite extent ($l = \infty$), equation (2.11.12) reduces to that of Buff et al.[75] It is clear from (2.11.12) that the principal contribution to capillary surface dispersion comes from long wavelengths—just those suppressed in computer simulations of the liquid surface and missing in a mean field analysis where excitations of wavelength $\sim$ correlation length only are included. Indeed, Davis observed that for the simulation of Lee et al.[77] ($T = 84\,°\text{K}$, $\rho_L - \rho_V = 1.39\,\text{g/cm}^3$, $\gamma = 16.5\,\text{dyn/cm}$) the capillary wave dispersion of the surface for a fluid of infinite extent ($l = \infty$) is 4.2 Å, whilst for the above simulation ($l = 22.2\,\text{Å}$ 'extended' by periodic boundary conditions) the dispersion would be only 0.74 Å. Since the reported profile width was 13.5 Å it is quite clear that suppression of long wavelength capillary waves is responsible for considerable spatial confinement of the surface delocalization. The effect of increasing the cross-sectional area of the simulation should be accompanied by an enhancement of capillary dispersion, as indeed was found to be the case by Chapela et al.[78] Davis concludes that the one-dimensional mean field theory and the capillary wave theory describe different density dispersion aspects of fluid interfaces, and concurs with Webb's interpretation of the existing light scattering and reflection data for interfaces as implying 'that when the interface undergoes its equilibrium vibrations, it does so as a membrane that has first been endowed with an intrinsic profile.'

In Davis' mean field analysis it is, of course, assumed that the local free energy is determined by the local density alone: the interface is effectively subdivided into elementary strata, each of which constitutes an equilibrium bulk fluid of the local density $\rho_{(1)}(z)$. Certainly *bulk* excitations of wavelength of the order of the bulk correlation length appropriate to the local density are included in such an analysis of the surface excitation spectrum, but long wavelength surface-specific excitations which are known to occur in the interfacial region are not included, and it is these which Davis attempts to incorporate. It would appear that in his analysis a proper decoupling of the bulk-specific and surface-specific modes has not been fully achieved in the vicinity of λ_{min} and there may be some overlap in the

84

excitations accounted for in both the mean field and capillary wave contributions to the excess Helmholtz free energy. In principle, a variational redistribution of the excitations between the bulk and surface capillary components so as to minimize the excess free energy γ may be of use, but it is difficult to see how this could be achieved at present.

References

1. H. S. Green, *Hand. Phys.* **10,** 79 (1960).
2. T. Osborn and C. A. Croxton, *Mol. Phys.* (1980) (in press).
3. C. A. Croxton, In C. A. Croxton (Ed.), *Progress in Liquid Physics*, Wiley, London (1978), Ch. 2.
4. J. G. Kirkwood and F. P. Buff, *J. Chem. Phys.*, **17,** 338 (1949).
5. R. H. Fowler, *Proc. Roy. Soc.* **A159,** 229 (1937).
6. A. Eisenstein and N. S. Gingrich, *Phys. Rev.*, **62,** 261 (1942).
7. T. L. Hill, *J. Chem. Phys.*, **19,** 261 (1951).
   T. L. Hill, *J. Chem. Phys.*, **19,** 1203 (1951).
   T. L. Hill, *J. Chem. Phys.*, **20,** 1510 (1952).
8. I. W. Plesner and O. Platz, *J. Chem. Phys.* **48,** 5361 (1968).
9. P. D. Shoemaker, G. W. Paul, and L. E. Marc de Chazal, *J. Chem. Phys.*, **52,** 491 (1970).
10. C. A. Croxton and R. P. Ferrier, *Phys. Lett.*, A, **35,** 330 (1971).
    C. A. Croxton and R. P. Ferrier, *J. Phys. C.*, **4,** 2447 (1971).
11. See, for example, F. Reif, *Statistical and Thermal Physics*, McGraw-Hill, New York (1965), p. 314.
12. F. P. Buff, *Phys. Rev.* **82,** 773(T) (1951).
13. A. Harasima, *Proc. Int. Conf. Theor. Phys.*, Science Council of Japan (1954).
    A. Harasima. In I. Prigogine (Ed.), *Advances in Chemical Physics*, Vol. 1, New York, Interscience (1958).
14. M. V. Berry, R. F. Durrans, and R. Evans, *J. Phys. A*, **5,** 166 (1972).
15. C. Jouanin, *C.R. Acad. Sci. Paris*, B, **268,** 1597 (1969).
16. D. Fitts, *Physica*, **42,** 205 (1969).
17. S. Toxvaerd, *J. Chem. Phys.*, **55,** 3116 (1971).
18. T. L. Hill, *J. Chem. Phys.*, **30,** 1521 (1959).
19. S. Ono and S. Kondo, *Hand. Phys.* **10,** 219 (1960).
20. T. L. Hill, *J. Chem. Phys.*, **20,** 141 (1952).
21. C. A. Croxton, *Adv. Phys.*, **22,** 385 (1973).
22. L. Tonks, *Phys. Rev.*, **50,** 955 (1936).
23. H. C. Longuet-Higgins and B. Widom, *Mol. Phys.*, **8,** 549 (1964).
    E. A. Guggenheim, *Mol. Phys.*, **9,** 43 (1965).
    H. Reiss, H. L. Frisch, and J. L. Lebowitz, *J. Chem. Phys.*, **31,** 361 (1959).
    E. Helfand, H. L. Frisch, and J. L. Lebowitz, *J. Chem. Phys.*, **34,** 1037 (1961).
24. F. P. Buff and F. H. Stillinger, *J. Chem. Phys.*, **39,** 1911 (1963).
    I. W. Plesner and I. Michaeli, *J. Chem. Phys.*, **60,** 3016 (1974).
25. S. Toxvaerd, *J. Chem. Phys.*, **57,** 4092 (1972). (Note that the z scale in Figure 3 of Toxvaerd's paper is incorrectly labelled.)
26. J. A. Barker and D. Henderson, *J. Chem. Phys.*, **47,** 2856 (1967).
27. R. W. Zwanzig, *J. Chem. Phys.*, **22,** 1420 (1954).
28. J. Lekner, *Prog. Theor. Phys.*, **45,** 42 (1971).
29. L. Verlet and J. J. Weiss, *Phys. Rev.*, A, **5,** 939 (1972).
30. W. R. Smith, D. Henderson, and J. A. Barker, *J. Chem. Phys.*, **55,** 4027 (1971).

31. S. Toxvaerd, *J. Chem. Phys.*, **55**, 3116 (1971).
32. S. Fisk and B. Widom, *J. Chem. Phys.*, **50**, 3219 (1969). Experimental support for a tanh profile in the vicinity of the critical point is given by J. Zollweg, G. Hawkins, and G. B. Benedek, *Phys. Rev. Lett.*, **27**, 1182 (1971).
33. C. A. Croxton, *Liquid State Physics—A Statistical Mechanical Introduction*, Cambridge University Press, London and New York (1974), Ch. 4.
    S. Ono and S. Kondo, *Hand. Phys.* **10**, 179 (1960).
34. M. Born and H. S. Green, *Proc. Roy. Soc.* A, **188**, 10 (1946).
    J. Yvon, *La théorie Statistique des fluids et l'equation d'état*, Hermann, Paris (1935).
35. J. Kirkwood, *J. Chem. Phys.*, **3**, 300 (1935).
36. T. R. Osborn and C. A. Croxton, *Mol. Phys.* (in press, 1980).
37. B. Borstnik and A. Azman, *Mol. Phys.*, **29**, 1165 (1975).
38. B. U. Felderhof, *Phys. Rev.*, A, **1**, 1185 (1970).
39. I. Z. Fisher and B. V. Bokut', *Zh. Fiz. Khim*, **31**, 200 (1957).
    I. Z. Fisher, *Statistical Theory of Liquids*, University of Chicago Press, Chicago (1964), p. 170.
40. J. D. Bernal, *Proc. Roy. Soc.*, A, **280**, 299 (1964).
41. J. Pressing and J. E. Mayer, *J. Chem. Phys.*, **59**, 2711 (1973).
42. J. H. Irving and J. G. Kirkwood, *J. Chem. Phys.*, **18**, 817 (1950).
43. A. Harasima, *Adv. Chem. Phys.*, **1**, 203 (1958).
44. S. Toxvaerd, *Prog. Surf. Sci.*, **3**, 189 (1972).
45. S. Toxvaerd, *Mol. Phys.*, **26**, 91 (1973).
46. G. M. Nazarian, *J. Chem. Phys.*, **56**, 1408 (1972).
47. L. Verlet, *Phys. Rev.*, **165**, 201 (1968).
48. S. Toxvaerd, *J. Chem. Phys.*, **64**, 2863 (1976).
49. C. A. Croxton and R. P. Ferrier, *J. Phys. C.*, **4**, 1909 (1971).
    C. A. Croxton and R. P. Ferrier, *Phil. Mag.*, **24**, 489 (1971).
50. C. A. Croxton and R. P. Ferrier, *J. Phys. C.*, **4**, 1921 (1971).
    C. A. Croxton and R. P. Ferrier, *Phil. Mag.*, **24**, 493 (1971).
51. C. A. Croxton, *Adv. Phys.*, **22**, 385 (1973).
52. J. W. Perram and L. R. White, *Faraday General Discussion, Physical Adsorption in Condensed Phases*, April, 1975.
53. E. Helfand, H. L. Frisch, and J. L. Lebowitz, *J. Chem. Phys.*, **34**, 1037 (1961).
54. H. Reiss, H. L. Frisch, and J. L. Lebowitz, *J. Chem. Phys.*, **31**, 369 (1959).
55. See for example, C. A. Croxton, *Liquid State Physics—A Statistical Mechanical Introduction*, Cambridge University Press, London and New York (1974).
    C. A. Croxton, *Introduction to Liquid State Physics*, Wiley, London (1975).
56. H. C. Andersen, D. Chandler, and J. D. Weeks, *J. Chem. Phys.*, **56**, 3812 (1972).
57. C. A. Croxton, *Faraday General Discussion, Physical Adsorption in Condensed Phases*, April 1975.
58. J. Yvon, *Proc. IUPAP symposium on Thermodynamics*, Brussels (1948), p. 9.
59. F. P. Buff and R. Lovett. In H. L. Frisch and Z. W. Salsburg (Eds.), *Simple Dense Fluids*, Academic Press, New York (1968), p. 17.
60. D. G. Triezenberg and R. Zwanzig, *Phys. Rev. Lett.*, **28**, 1183 (1972).
61. R. Lovett, P. W. De Haven, J. J. Vieceli, and F. P. Buff, *J. Chem. Phys.*, **58**, 1880 (1973).
62. J. Lekner and J. R. Henderson, *Mol. Phys.*, **34**, 333 (1977); *Physica* **94A**, 545 (1978).
63. J. K. Percus. In H. L. Frisch and J. L. Lebowitz (Eds), *The Equilibrium Theory of Classical Fluids*, Benjamin, New York (1964), pp. 11–33.
64. P. Drude, *Theory of Optics*, Longmans, Green and Co., New York (1907), p. 292.

86

65. F. P. Buff and R. A. Lovett, *1966 Saline Water Conversion Report*, U.S. Government Printing Office, Washington D.C., p. 26.
66. C. Ebner and W. F. Saam, *Phys. Rev. B*, **12**, 923 (1975).
67. E. S. Wu and W. W. Webb, *Phys. Rev. A*, **8**, 2065 (1973).
68. J. S. Huang and W. W. Webb, *J. Chem. Phys.*, **50**, 3677 (1969).
69. J. Lekner and J. R. Henderson, *Mol. Phys.*, **36**, 781 (1978).
70. F. F. Abraham, *J. Chem. Phys.*, **63**, 157 (1975).
71. V. Buongiorno and H. T. Davis, *Phys. Rev.*, A, **12**, 2213 (1975).
72. L. P. Kadanoff, W. Götze, D. Hamblen, R. Hecht, E. A. S. Lewis, V. V. Palciauskas, M. Rayl, J. Swift, D. Aspnes, and J. Kane, *Rev. Mod. Phys.*, **39**, 395 (1967).
73. Y. Singh and F. F. Abraham, *J. Chem. Phys.*, **67**, 537 (1977).
74. C. A. Croxton, *J. Phys. C.*, **12**, 2239 (1979).
75. F. P. Buff, R. A. Lovett, and F. H. Stillinger, Jr., *Phys. Rev. Lett.* **15**, 621 (1965).
76. H. T. Davis, *J. Chem. Phys.*, **67**, 3636 (1977).
77. J. K. Lee, J. A. Barker, and G. M. Pound, *J. Chem. Phys.*, **60**, 1976 (1974).
78. G. A. Chapela, G. Saville, S. M. Thompson, and J. S. Rowlinson, *J. Chem. Soc. Faraday Trans. II*, **73**, 1133 (1977).
79. Y. Singh and F. Abraham, *J. Chem. Phys.*, **67**, 5960 (1977).
80. C. E. Upstill and R. Evans, *J. Phys. C.*, **10**, 2791 (1977).
81. I. W. Plesner, O. Platz, and S. E. Christiansen, **48**, 5364 (1968).
82. This approximation has also been used recently by S. Salter and H. T. Davis, *J. Chem. Phys.*, **63**, 3295 (1978) in their discussion of interphasal properties of simple liquids.
83. J. W. Cahn and J. E. Hilliard, *J. Chem. Phys.*, **28**, 258 (1958); **31**, 688 (1959).
84. C. A. Croxton, *J. Phys. C.*, **12**, 2239 (1979).
85. J. Barker and D. Henderson, *J. Chem. Phys*, **47**, 2856 (1967)
    J. D. Weeks, D. Chandler, and H. C. Anderson, *J. Chem. Phys.*, **54**, 5237 (1971).
86. H. S. Green, *The Molecular Theory of Fluids*, North-Holland, Amsterdam (1952), Section 6.1.
87. T. R. Osborn and C. A. Croxton, *Mol. Phys.* (in press, 1980).
88. K. U. Co, J. J. Kozak, and K. D. Luks, *J. Chem. Phys.*, **66**, 1002 (1977).
89. C. A. Leng, J. S. Rowlinson, and S. M. Thompson, *Proc. Roy. Soc.*, A, **358**, 267 (1977).
90. A. J. M. Yang, P. D. Fleming, and J. H. Gibbs, *J. Chem. Phys.*, **64**, 3722 (1976).
91. N. H. March and M. P. Tosi, *Atomic Dynamics in Liquids*, Macmillan, London (1976), p. 260.
92. M. S. Wertheim, *J. Chem. Phys.*, **65**, 2377 (1976).
93. M. H. Kalos, J. K. Percus, and M. Rao, *J. Stat. Phys.*, **17**, 111 (1977).
94. J. D. Weeks, *J. Chem. Phys.*, **67**, 3106 (1977).
95. C. A. Croxton, *J. Phys. C.*, **7**, 3723 (1974).
96. J. Fischer, *Mol. Phys.*, **33**, 75 (1977).
97. I. Z. Fisher and B. V. Bokut', *Zh. Fiz. Khim.*, **30**, 2747 (1956).
98. E. J. Chapyak, *J. Chem. Phys.*, **57**, 4512 (1972).
99. S. J. Salter and H. T. Davis, *J. Chem. Phys.*, **63**, 3295 (1975).
100. D. E. Sullivan and G. Stell, *J. Chem. Phys.*, **67**, 2567 (1977).
101. D. Henderson, F. Abraham, and J. Barker, *Mol. Phys.*, **31**, 1291 (1976).
102. E. Waisman, D. Henderson, and J. Lebowitz, *Mol. Phys.*, **32**, 1373 (1976).
103. K. S. Liu, M. H. Kalos, and G. V. Chester, *Phys. Rev. A.*, **10**, 303 (1975).
104. J. K. Percus, *Lecture Notes on Non-Uniform Fluids* (Enseignement du 3 ème

Cycle de Physique en Suisse Romande, Eté 1975); *J. Stat. Phys.*, **15,** 423 (1976).
105. L. Blum and G. Stell., *J. Stat. Phys.*, **15,** 439 (1976).
106. J. W. Perram and E. R. Smith, *Chem. Phys. Lett.*, **35,** 138 (1975); *Chem. Phys. Lett.*, **39,** 328 (1976); *Phys. Lett.* A, **59,** 11 (1976).
107. R. J. Baxter, *J. Chem. Phys.*, **49,** 2770 (1968).
108. F. B. Sprow and J. M. Prausnitz, *Trans. Faraday Soc.*, **62,** 1097 (1966).

CHAPTER 3

Molecular Liquid Surfaces

3.1 Introduction

Almost all rigorous theoretical work on interfacial phenomena has been for fluids with spherically symmetric molecular systems in which the central interaction is solely a function of the intermolecular separation of their centres of mass. It will be shown in this chapter that the same techniques developed in the preceding chapters bear straightforward extension to nonspherical systems—techniques which are also generally applicable to nonpolar polyatomics and their mixtures. These approaches prove inadequate, however, for strongly multi-polar or associating molecular liquids. Consequently we shall defer their discussion to later chapters.

The specification of the internal set of Euler angles $\mathbf{e}_i = \{\theta_i \varphi_i \chi_i\}$ describing the orientation of molecule i relative to an external space frame distinguishes the system from its spherically symmetric counterpart (Figure 3.1.1). If the molecule is axially symmetric then, of course, the variable χ is redundant and integrations over this variable simply introduce the factor $2\pi = \int_0^{2\pi} d\chi$, generally designated Ω. The system Hamiltonian is of the form

$$\mathcal{H} = \sum_{i=1}^{N} \frac{\mathbf{p}_i^2}{2m_i} + \Phi_N(\mathbf{r}^N; \mathbf{I}^N) + \mathcal{H}_{\text{int}} \qquad (3.1.1)$$

where $\mathbf{p}_i$ and m_i are the centre of mass momentum and the mass of the ith particle, whilst Φ_N represents the total N-body intermolecular interaction energy depending both upon molecular positions $\mathbf{r}^N$ and the internal states $\mathbf{I}^N$. Assuming that the total potential is pairwise additive, we have

$$\Phi_N(\mathbf{r}^N; \mathbf{I}^N) = \sum_{i>j}\sum \Phi_{ij}(\mathbf{r}_i, \mathbf{r}_j; \mathbf{I}_i, I_j)$$

where Φ_{ij}, the pair potential, depends upon the positions $\{\mathbf{r}_i, \mathbf{r}_j\}$ and internal states $\{\mathbf{I}_i, \mathbf{I}_j\}$ of the two molecules. $\mathcal{H}_{\text{int}}$ is the internal energy Hamiltonian which includes contributions from intramolecular interactions, rotations, and vibrations from electronic and nuclear interactions and motions. Of these internal degrees of freedom some are generally classical (e.g. unhindered rotations and molecular rotation about the centre of mass, although low temperature, low mass systems such as H_2 may show departures from classical behaviour) whilst others may be quantum mechanical.

Separability of the Hamiltonian (3.1.1) and integration over the momen-

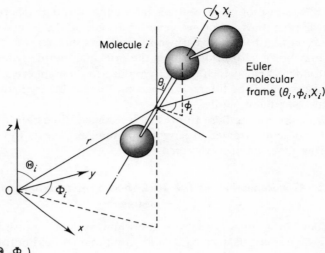

Figure 3.1.1 Relation between the internal rigid molecular frame $(\theta_i, \varphi_i, \chi_i)$ and an external frame (r_i, Θ_i, Φ_i)

tum variables enables us to write

$$Z_N = \frac{1}{N!\Lambda^{3N}} \sum_{\text{int}} \exp\left\{-\frac{\mathcal{H}_{\text{int}}}{kT}\right\} \int \cdots \int \exp\left\{-\frac{\Phi_N}{kT}\right\} d\mathbf{r}_1 \cdots d\mathbf{r}_N$$

where $\Lambda = h/(2\pi mkT)^{1/2}$ is the thermal de Broglie wavelength and arises in the course of integration over the momentum projection. The summation over the internal states relates to both the classical and quantum mechanical contributions: the former may, of course, be replaced by their appropriate integrals—we shall, however, retain the summation notation for the time being.

The two-particle density distribution is then given as[2]

$$\rho_{(2)}(\mathbf{r}_1, \mathbf{r}_2; \mathbf{I}_1, \mathbf{I}_2) = \frac{N(N-1)}{N!Z_N\Lambda^{3N}} \sum_{\text{int}}'' \exp\left\{-\frac{\mathcal{H}_{\text{int}}}{kT}\right\} \int \cdots \int \exp\left\{-\frac{\Phi_N}{kT}\right\} d\mathbf{r}_3 \cdots d\mathbf{r}_N$$

(3.1.3)

where $\sum''$ indicates that the summation excludes the internal states of particles 1 and 2, as do the integrations. Throughout this chapter we shall assume that internal excitations involve only the continuum of orientational states, i.e. $\mathbf{I}_i \rightarrow \mathbf{e}_i(\theta_i, \varphi_i, \chi_i)$ at least for rigid molecular systems, and that these develop independently of the other internal components of the Hamiltonian. In this case the summation may be replaced by an integral over the orientational variables. Such an approximation is likely to be entirely satisfactory provided the rotational and vibrational levels $\gg kT$. Otherwise, these motions will contribute directly to the thermodynamic functions of the molecular liquid surface.

As in the case of classical fluids interacting through spherically symmetric pairwise potentials, rigorous generalizations of the van der Waals direct correlation (2.10.19) and Kirkwood–Buff pressure tensor (1.4.24) formulations may be developed describing the principal thermodynamic functions of polyatomic systems which correctly reduce to their spherical counterparts as the anisotropy of the interaction becomes negligible in comparison with the central component.

It is inappropriate here to present a detailed discussion of intermolecular forces. These are treated at length in a number of texts to which the reader requiring more detail is directed.[1]

3.2 Generalization of the Kirkwood–Buff equation to polyatomic systems[2,3]

Consider an inhomogeneous equilibrium system resolved into a stable liquid–vapour coexistence by a weak gravitational field normal to the planar interface. The surface tension is related to the Helmholtz free energy of the system A by (Table 1.4.1)

$$\gamma = \left(\frac{\partial A}{\partial \mathcal{A}}\right)_{NVT}$$
$$= -kT\left(\frac{\partial \ln Z_N}{\partial \mathcal{A}}\right)_{NVT} \qquad (3.2.1)$$

where $\mathcal{A}$ is the area of the interface. We suppose that the system is confined to a rectangular parallelopiped as shown in Figure 3.2.1, and investigate the change in free energy per unit area involved by increasing the interfacial area by an amount $d\mathcal{A}$. This is achieved by means of an isothermal reversible distortion of the volume $l_x \to l_x + dl_x$, $l_z \to l_z - dl_z$ such that the

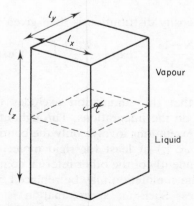

Figure 3.2.1 Geometry for determination of the surface tension

constant volume condition

$$dl_z = -(l_z/l_x)\, dl_x \qquad (3.2.2)$$

is satisfied. Thus,

$$\gamma = -\frac{kT}{l_y}\left(\frac{\partial \ln Z_N}{\partial l_x}\right)_{l_z, l_y, N, T} + \frac{kT l_z}{l_x l_y}\left(\frac{\partial \ln Z_N}{\partial l_z}\right)_{l_x, l_y, N, T} \qquad (3.2.3)$$

As in the case of simple liquids we may readily identify the two terms in (3.2.3) as the tangential and normal components of the pressure tensor (c.f. Section 1.4), whereupon

$$\gamma = -\frac{1}{2} \int_{e_1} \int_{e_2} \int_{-\infty}^{\infty} dz_1 \int \rho_{(2)}(\mathbf{r}_1, \mathbf{r}_2; \mathbf{e}_1, \mathbf{e}_2)$$
$$\times \left\{ z_{12} \frac{\partial \Phi(\mathbf{r}_{12}; \mathbf{e}_1, \mathbf{e}_2)}{\partial z_{12}} - x_{12} \frac{\partial \Phi(\mathbf{r}_{12}; \mathbf{e}_1, \mathbf{e}_2)}{\partial x_{12}} \right\} d\mathbf{r}_{12}\, d\mathbf{e}_1\, d\mathbf{e}_2 \qquad (3.2.4)$$

which reduces correctly to the Kirkwood–Buff stress-tensor result

$$\gamma = \frac{1}{2} \int_{-\infty}^{\infty} dz_1 \int \rho_{(2)}(\mathbf{r}_1, \mathbf{r}_2) \frac{d\Phi(r_{12})}{dr_{12}} \frac{(x_{12}^2 - z_{12}^2)}{r_{12}} d\mathbf{r}_{12} \qquad (3.2.5)$$

in the case of centrally symmetric interactions. The quantity $\mathbf{e}_i$ denotes the Eulerian angles $(\theta_i, \varphi_i, \chi_i)$ of molecule i (Figure 3.1.1), and $d\mathbf{e}_i = \sin\theta_i\, d\theta_i\, d\varphi_i\, d\chi_i$ with $0 < \theta_i < \pi$, $0 < \varphi_i < 2\pi$ and $0 < \chi_i < 2\pi$. If the molecule has axial symmetry, then for this particular choice of coordinate frame the χ variable is redundant and may be integrated to yield a factor of $2\pi = \Omega$.

An alternative but equivalent expression to (3.2.4) may be obtained by transforming to spherical coordinates, yielding the result

$$\gamma = \gamma_A + \gamma_B \qquad (3.2.6)$$

where

$$\gamma_A = -\frac{1}{2} \int_{-\infty}^{\infty} dz_1 \int P_2(\cos\theta) \int_{e_1} \int_{e_2} \rho_{(2)}(\mathbf{r}_1, \mathbf{r}_2; \mathbf{e}_1, \mathbf{e}_2)$$
$$\times \frac{\partial \Phi(\mathbf{r}_{12}; \mathbf{e}_1, \mathbf{e}_2)}{\partial r_{12}} d\mathbf{e}_1\, d\mathbf{e}_2 r_{12}\, d\mathbf{r}_{12} \qquad (3.2.7)$$

$$\gamma_B = \frac{3}{4} \int_{-\infty}^{\infty} dz_1 \int \sin\Theta \cos\Theta \int_{e_1} \int_{e_2} \rho_{(2)}(\mathbf{r}_1, \mathbf{r}_2; \mathbf{e}_1, \mathbf{e}_2)$$
$$\times \frac{\partial \Phi(\mathbf{r}_{12}; \mathbf{e}_1, \mathbf{e}_2)}{\partial \Theta} d\mathbf{e}_1\, d\mathbf{e}_2\, d\mathbf{r}_{12}. \qquad (3.2.8)$$

Again, (3.2.6) reduces to the spherical molecule result (3.2.5) when $\Phi(\mathbf{r}_{12}; \mathbf{e}_1, \mathbf{e}_2) \to \Phi(r_{12})$: the derivative in the γ_B term vanishes and γ_A reduces to the Kirkwood–Buff result.

These results apply to *rigid* molecules of arbitrary shape, the only assumption being pairwise additivity of the molecular interaction. Nonrigid, polymeric systems will be considered in Chapter 8.

3.3 The Fowler–Kirkwood–Buff equation for nonspherical molecules

The specific difficulty which arises in the evaluation of either version of the expression for the surface tension (3.2.4, 3.2.6) is the unknown form for $\rho_{(2)}(\mathbf{r}_1, \mathbf{r}_2; \mathbf{e}_1, \mathbf{e}_2)$. Precisely the same problem arose in the case of centrally symmetric systems: here, however, the problem is exacerbated by the complex angular dependences of the pair distribution. For a system whose inhomogeneity is confined to the z direction we may re-express the pair density function in terms of the single-particle distributions:

$$\rho_{(2)}(\mathbf{r}_1, \mathbf{r}_2; \mathbf{e}_1, \mathbf{e}_2) = \frac{1}{\Omega} \rho_{(1)}(z_1, \mathbf{e}_1) \rho_{(1)}(z_2, \mathbf{e}_2) g_{(2)}(z_1, \mathbf{r}_{12}; \mathbf{e}_1, \mathbf{e}_2) \quad (3.3.1)$$

exposing both the spatial and angular dependence of the density profile. The factor Ω represents the unit volume in angular space

$$\Omega = \int_0^{2\pi} \int_0^{\pi} \sin\theta \, d\theta \, d\varphi = 4\pi \quad \text{or} \quad \int_0^{2\pi} \int_0^{2\pi} \int_0^{\pi} \sin\theta \, d\theta \, d\varphi \, d\chi = 8\pi^2$$

for linear or nonlinear molecules, respectively. The simplest approximation is that of Fowler[4] which assumes that the density of the bulk phase retains its liquid value up to a planar surface of density discontinuity, coincident with the Gibbs surface, and that the vapour is of negligible density. These assumptions imply[4, 5]

$$\rho_{(2)}(\mathbf{r}_1, \mathbf{r}_2; \mathbf{e}_1, \mathbf{e}_2) \sim \frac{\rho_{(1)}(z_1) \rho_{(1)}(z_2)}{(8\pi^2)^2} g_{(2)}^L(\mathbf{r}_{12}; \mathbf{e}_1, \mathbf{e}_2) \quad (3.3.2)$$

and

$$\begin{aligned} \rho_{(1)}(z_i) &= \rho_L \qquad z_i < 0 \\ &= 0 \qquad z_i > 0 \end{aligned} \quad (3.3.3)$$

where we have adopted the value $\Omega = 8\pi^2$ appropriate to nonlinear molecules. This approximation, which has been discussed and criticized extensively elsewhere (Sections 2.2, 2.3) contains the additional assumption that the single-particle distribution remains constant up to the dividing surface, and that there is no surface modification of the orientational distribution, which is what equation (3.3.3) implies.

Insertion of (3.3.2, 3.3.3) in (3.2.4) yields[2]

$$\gamma = -\frac{\rho_L^2}{2(8\pi^2)^2} \int_{z_{12}>0} \int_{-\infty}^{0} \int_{\mathbf{e}_1} \int_{\mathbf{e}_2} g_{(2)}^L \left\{ z_{12} \frac{\partial \Phi(\mathbf{r}_{12}; \mathbf{e}_1, \mathbf{e}_2)}{\partial z_{12}} \right.$$

$$\left. -x_{12} \frac{\partial \Phi(\mathbf{r}_{12}; \mathbf{e}_1, \mathbf{e}_2)}{\partial x_{12}} \right\} \mathrm{d}\mathbf{e}_1 \, \mathrm{d}\mathbf{e}_2 \, \mathrm{d}z_1 \, \mathrm{d}\mathbf{r}_{12}$$

$$-\frac{\rho_L^2}{2(8\pi^2)^2} \int_{z_{12}>0} \int_{-\infty}^{z_{12}} \int_{\mathbf{e}_1} \int_{\mathbf{e}_2} g_{(2)}^L \left\{ z_{12} \frac{\partial \Phi(\mathbf{r}_{12}; \mathbf{e}_1, \mathbf{e}_2)}{\partial z_{12}} \right.$$

$$\left. -x_{12} \frac{\partial \Phi(\mathbf{r}_{12}; \mathbf{e}_1, \mathbf{e}_2)}{\partial x_{12}} \right\} \mathrm{d}\mathbf{e}_1 \, \mathrm{d}\mathbf{e}_2 \, \mathrm{d}z_1 \, \mathrm{d}\mathbf{r}_{12}. \qquad (3.3.4)$$

Since the stress tensor is isotropic in the bulk liquid the first term in (3.2.6) vanishes, whilst the second term reduces to

$$\gamma = \frac{\rho_L^2}{2(8\pi^2)^2} \int_{z_{12}>0} \int_{\mathbf{e}_1} \int_{\mathbf{e}_2} g_{(2)}^L \left\{ z_{12} \frac{\partial \Phi(\mathbf{r}_{12}; \mathbf{e}_1, \mathbf{e}_2)}{\partial z_{12}} \right.$$

$$\left. -x_{12} \frac{\partial \Phi(\mathbf{r}_{12}; \mathbf{e}_1, \mathbf{e}_2)}{\partial x_{12}} \right\} z_{12} \, \mathrm{d}\mathbf{e}_1 \, \mathrm{d}\mathbf{e}_2 \, \mathrm{d}\mathbf{r}_{12}. \qquad (3.3.5)$$

(The notational transformation $\int \mathrm{d}\mathbf{e}_1 \, \mathrm{d}\mathbf{e}_2(\,)/(8\pi^2)^2 \to \sum_{\text{int}}(\,)$ enables the more general case involving discrete, nonclassical internal degrees of freedom to be discussed.) Once again, for the special case of spherical molecules (3.3.5) reduces to the usual Kirkwood–Buff expression (2.3.2).

3.4 A Perturbation expansion about a Pople reference fluid[10]

Provided the molecular anisotropy is small, a perturbation expansion of the Helmholtz free energy about that of a reference system, *not necessarily spherical*, is possible. The perturbation theories for both simple and molecular fluids may be embedded in a general formalism in which the properties of interest are expanded about those of an arbitrary reference fluid in a functional Taylor series. In such an expansion the anisotropic functions, such as the pair potential, are averaged over the known reference fluid distributions. We consider the anisotropic pair potential to be a sum of a reference and a perturbing term:[9]

$$\Phi(\mathbf{r}_{12}; \mathbf{e}_1, \mathbf{e}_2) = \Phi^{(0)}(r_{12}) + \lambda \Phi^{(1)}(\mathbf{r}_{12}; \mathbf{e}_1, \mathbf{e}_2) \qquad (3.4.1)$$

where λ is a perturbation parameter such that when $\lambda = 0$ we recover the reference potential, whilst $\lambda = 1$ represents the full potential. Thus, expanding the Helmholtz free energy of the two phase system in powers of λ, and setting $\lambda = 1$, gives:

$$A = A^{(0)} + A^{(1)}$$

where

$$A^{(1)} = \frac{1}{2} \int\int\int\int \rho_{(2)}^{(0)}(z_1, \mathbf{r}_{12}; \mathbf{e}_1, \mathbf{e}_2) \left(\frac{\partial \Phi(\mathbf{r}_{12}; \mathbf{e}_1, \mathbf{e}_2 \mid \lambda)}{\partial \lambda} \right)_{\lambda=0} d\mathbf{e}_1\, d\mathbf{e}_2\, d\mathbf{r}_1\, d\mathbf{r}_2.$$

(3.4.2)

$\rho_{(2)}^{(0)}(z_1, \mathbf{r}_{12}; \mathbf{e}_1, \mathbf{e}_2)$ is the interfacial angular reference distribution. Since,

$$\gamma = \left(\frac{\partial A}{\partial \mathscr{A}} \right)_{NVT} = \gamma^{(0)} + \gamma^{(1)} + \cdots$$

(3.4.3)

then

$$\gamma^{(1)} = \frac{1}{2} \left\{ \frac{\partial}{\partial \mathscr{A}} \int\int\int\int \rho_{(2)}^{(0)}(z_1, \mathbf{r}_{12}; \mathbf{e}_1, \mathbf{e}_2) \left(\frac{\partial \Phi(\mathbf{r}_{12}; \mathbf{e}_1, \mathbf{e}_2 \mid \lambda)}{\partial \lambda} \right)_{\lambda=0} \right.$$
$$\left. \times d\mathbf{e}_1\, d\mathbf{e}_2\, d\mathbf{r}_1\, d\mathbf{r}_2 \right\}_{NVT}$$

(3.4.4)

Integrating (3.4.4) over x_1, y_1 and transforming $\mathbf{r}_2$ to $\mathbf{r}_{12}$ gives

$$\gamma^{(1)} = \frac{1}{2} \int\int_{-\infty}^{\infty} \mathscr{A} \left(\frac{\partial}{\partial \mathscr{A}} \left\langle \rho_{(2)}^{(0)}(z_1, \mathbf{r}_{12}; \mathbf{e}_1, \mathbf{e}_2) \left(\frac{\partial \Phi(\mathbf{r}_{12}; \mathbf{e}_1, \mathbf{e}_2 \mid \lambda)}{\partial \lambda} \right)_{\lambda=0} \right\rangle_{\mathbf{e}_1 \mathbf{e}_2} \right) d\mathbf{r}_{12}\, dz_1$$

(3.4.5)

where

$$\langle \cdots \rangle_{\mathbf{e}_1 \mathbf{e}_2} \equiv \int_{\mathbf{e}_1} \int_{\mathbf{e}_2} (\cdots)\, d\mathbf{e}_1\, d\mathbf{e}_2 = F(z_1, \mathbf{r}_{12})$$

and

$$\mathscr{A} \frac{\partial}{\partial \mathscr{A}} \langle \cdots \rangle = \mathscr{A} \frac{\partial \mathbf{r}_{12}}{\partial \mathscr{A}} \frac{\partial F(z_1, \mathbf{r}_{12})}{\partial \mathbf{r}_{12}} + \mathscr{A} \frac{\partial F(z_1, \mathbf{r}_{12})}{\partial z_1} \frac{\partial z_1}{\partial \mathscr{A}}$$
$$= \frac{\mathbf{r}_{12}}{2} \frac{\partial F(z_1, \mathbf{r}_{12})}{\partial \mathbf{r}_{12}} - \frac{3}{2} z_{12} \frac{\partial F(z_1, \mathbf{r}_{12})}{\partial z_{12}} - z_1 \frac{\partial F(z_1\, \mathbf{r}_{12})}{\partial z_1}.$$

(3.4.6)

Equation (3.4.5) then becomes

$$\gamma^{(1)} = -\frac{1}{2} \int\int_{-\infty}^{\infty} \left\{ z_1 \frac{\partial F(z_1, \mathbf{r}_{12})}{\partial z_1} + \frac{3}{2} z_{12} \frac{\partial F(z_1, \mathbf{r}_{12})}{\partial z_{12}} \right.$$
$$\left. - \frac{\mathbf{r}_{12}}{2} \frac{\partial F(z_1, \mathbf{r}_{12})}{\partial \mathbf{r}_{12}} \right\} d\mathbf{r}_{12}\, dz_1$$

(3.4.7)

the second two terms of which may be integrated by parts and shown to cancel, whereupon

$$\gamma^{(1)} = -\frac{1}{2} \int\int_{-\infty}^{\infty} z_1 \left\langle \left(\frac{\partial \Phi(\mathbf{r}_{12}; \mathbf{e}_1, \mathbf{e}_2 \mid \lambda)}{\partial \lambda}\right)_{\lambda=0} \frac{\partial \rho_{(2)}^{(0)}(z_1, \mathbf{r}_{12}; \mathbf{e}_1, \mathbf{e}_2)}{\partial z_1}\right\rangle_{\mathbf{e}_1\mathbf{e}_2} dz_1\, d\mathbf{r}_{12}.$$

(3.4.8)

This represents the rigorous first-order correction arising from the departure of the system from the reference fluid which need not necessarily be spherical. The result is exact to within the assumption of pairwise additivity of the molecular interaction.

In the absence of an accurate and explicit knowledge of the function $\rho_{(2)}^{(0)}(z_1, \mathbf{r}_{12}; \mathbf{e}_1, \mathbf{e}_2)$, the simplest approximation is again the Kirkwood–Buff–Fowler step model (3.3.2), (3.3.3). An explicit development of (3.4.8) in terms of the KBF model is not warranted here: we shall satisfy ourselves simply by quoting the final result:

$$\gamma_{KBF}^{(1)} = -\frac{\rho_L^2 \pi}{2(8\pi^2)^2} \int_0^{\infty} \left\langle \left(\frac{\partial \Phi(\mathbf{r}_{12}; \mathbf{e}_1, \mathbf{e}_2 \mid \lambda)}{\partial \lambda}\right)_{\lambda=0} g_{(2)}^{(0)}(\mathbf{r}_{12})\right\rangle_{\mathbf{e}_1\mathbf{e}_2} r_{12}^3\, dr_{12}.$$

(3.4.9)

A detailed development of this expression has been given by Haile.[3]

Higher order terms contain multibody distributions: $\gamma^{(2)}$ for example, involves the four-particle distribution $\rho_{(4)}$. Clearly, a rapidly convergent perturbation expansion is essential and the choice of reference fluid is crucial. A particularly convenient choice of isotropic reference potential is that proposed by Pople.[6]

$$\Phi^{(0)}(r_{12}) = \langle \Phi(\mathbf{r}_{12}; \mathbf{e}_1, \mathbf{e}_2)\rangle_{\mathbf{e}_1\mathbf{e}_2}$$

which amounts to a 'sphericalization' or unweighted orientational average over the two interacting particles, whereupon

$$\Phi(\mathbf{r}_{12}; \mathbf{e}_1, \mathbf{e}_2 \mid \lambda) = \Phi^{(0)}(r_{12}) + \lambda \Phi^{(1)}(\mathbf{r}_{12}; \mathbf{e}_1, \mathbf{e}_2)$$

(3.4.10)

(c.f. 3.4.1). Now, it follows,

$$\left\langle \left(\frac{\partial \Phi(\mathbf{r}_{12}; \mathbf{e}_1, \mathbf{e}_2 \mid \lambda)}{\partial \lambda}\right)_{\lambda=0}\right\rangle_{\mathbf{e}_1\mathbf{e}_2} = \langle \Phi^{(1)}(\mathbf{r}_{12}; \mathbf{e}_1, \mathbf{e}_2)\rangle_{\mathbf{e}_1\mathbf{e}_2} = 0$$

in which case we have from (3.4.2), (3.4.4)

$$A^{(1)} = \gamma^{(1)} = 0$$

(3.4.11)

i.e.

$$\left. \begin{aligned} A &= A^{(0)} + A^{(2)} + A^{(3)} + \cdots \\ \gamma &= \gamma^{(0)} + \gamma^{(2)} + \gamma^{(3)} + \cdots \end{aligned} \right\}.$$

(3.4.12)

Thus encouraged, formal expressions for the higher order components $A^{(i)}(i \geq 2)$ may be written down for a fluid whose inhomogeneity is confined to the z-axis:[7]

$$A = \sum_i A^{(i)} = \sum_i \frac{1}{2i} \int \int \rho_{(1)}^{(0)}(z_1) \rho_{(1)}^{(0)}(z_2) \langle \Phi^{(1)}(\mathbf{r}_{12}; \mathbf{e}_1, \mathbf{e}_2)$$
$$\times g^{(i-1)}(\mathbf{r}_{12}; \mathbf{e}_1, \mathbf{e}_2) \rangle_{\mathbf{e}_1 \mathbf{e}_2} d\mathbf{r}_1 d\mathbf{r}_2 \qquad (3.4.13)$$

where $g^{(i-1)}(\mathbf{r}_{12}; \mathbf{e}_1, \mathbf{e}_2)$ is the $(i-1)$th order term in a perturbation expansion of the angular pair correlation function. The corresponding terms in the expansion of the surface tension $\gamma^{(i)}(i \geq 2)$ may be found by forming

$$\gamma^{(i)} = \left(\frac{\partial A^{(i)}}{\partial \mathscr{A}} \right)_{NVT}. \qquad (3.4.14)$$

We continue to form our expansions about a Pople reference, in which case $g^{(1)}(\mathbf{r}_{12}; \mathbf{e}_1, \mathbf{e}_2)$ is given by[8]

$$g^{(1)}(\mathbf{r}_{12}; \mathbf{e}_1, \mathbf{e}_2) = -\beta \Phi^{(1)}(\mathbf{r}_{12}; \mathbf{e}_1, \mathbf{e}_2) g_{(2)}^{(0)}(r_{12})$$
$$-\beta \rho \int [\langle \Phi^{(1)}(\mathbf{r}_{13}; \mathbf{e}_1, \mathbf{e}_3) \rangle_{\mathbf{e}_3} + \langle \Phi^{(1)}(\mathbf{r}_{23}; \mathbf{e}_2, \mathbf{e}_3) \rangle_{\mathbf{e}_3}] g_{(3)}^{(0)}(r_{12}, r_{23}, r_{13}) d\mathbf{r}_3 \qquad (3.4.15)$$

where $\beta = (kT)^{-1}$, whereupon we may form the second-order component of the free energy of the anisotropic system. Thus we have

$$A^{(2)} = A^{(2A)} + A^{(2B)} \qquad (3.4.16)$$

where

$$A^{(2A)} = -\frac{\beta}{4} \int \int \rho_{(1)}^{(0)}(z_1) \rho_{(1)}^{(0)}(z_2) g_{(2)}^{(0)}(z_1, \mathbf{r}_{12}) \langle \Phi^{(1)^2}(\mathbf{r}_{12}; \mathbf{e}_1, \mathbf{e}_2) \rangle_{\mathbf{e}_1 \mathbf{e}_2} d\mathbf{r}_1 d\mathbf{r}_2 \qquad (3.4.17a)$$

$$A^{(2B)} = -\frac{\beta}{2} \int \int \int \rho_{(1)}^{(0)}(z_1) \rho_{(1)}^{(0)}(z_2) \rho_{(1)}^{(0)}(z_3) g_{(3)}^{(0)}(z_1, \mathbf{r}_{12}, \mathbf{r}_{13})$$
$$\times \langle \Phi^{(1)}(\mathbf{r}_{12}; \mathbf{e}_1, \mathbf{e}_2) \Phi^{(1)}(\mathbf{r}_{13}; \mathbf{e}_1, \mathbf{e}_3) \rangle_{\mathbf{e}_1 \mathbf{e}_2 \mathbf{e}_3} d\mathbf{r}_1 d\mathbf{r}_2 d\mathbf{r}_3. \qquad (3.4.17b)$$

The second-order contributions to the surface tension follow from (3.4.14), and may be evaluated in a similar manner to (3.4.4), giving

$$\gamma^{(2A)} = \frac{\beta}{4} \int \langle \Phi^{(1)}(\mathbf{r}_{12}; \mathbf{e}_1, \mathbf{e}_2) \rangle_{\mathbf{e}_1 \mathbf{e}_2} \int_{-\infty}^{\infty} z_1 \frac{\partial \rho_{(2)}^{(0)}(z_1, \mathbf{r}_{12})}{\partial z_1} dz_1 d\mathbf{r}_{12} \qquad (3.4.18a)$$

$$\gamma^{(2B)} = \frac{\beta}{2} \int \int \langle \Phi^{(1)}(\mathbf{r}_{12}; \mathbf{e}_1, \mathbf{e}_2) \Phi^{(1)}(\mathbf{r}_{13}; \mathbf{e}_1, \mathbf{e}_3) \rangle_{\mathbf{e}_1 \mathbf{e}_2 \mathbf{e}_3}$$
$$\times \int_{-\infty}^{\infty} z_1 \frac{\partial \rho_{(3)}^{(0)}(z_1, \mathbf{r}_{12}, \mathbf{r}_{13})}{\partial z_1} dz_1 d\mathbf{r}_{12} d\mathbf{r}_{13}. \qquad (3.4.18b)$$

It is appropriate at this stage to observe that in a spherical harmonic multipolar expansion of the anisotropic interaction $\Phi^{(1)}(\mathbf{r}_{12}; \mathbf{e}_1, \mathbf{e}_2)$, *taken with respect to a Pople reference*, all components of order $l \neq 0$ vanish on symmetry grounds, as may be readily shown from (3.4.15). Under these circumstances, of course, the terms $A^{(2B)}$ and $\gamma^{(2B)}$ vanish. However, anisotropic core overlap and dispersion tail interactions do contain the $l = 0$ component, in which case these latter terms must be included. These observations will not generally obtain for perturbation expansions not taken with respect to a Pople reference.

The third-order contribution to the Helmholtz free energy of an inhomogeneous fluid follows from (3.4.13):

$$A^{(3)} = \frac{1}{6} \int \int \rho_{(1)}^{(0)}(z_2) \langle \Phi^{(1)}(\mathbf{r}_{12}; \mathbf{e}_1, \mathbf{e}_2) g^{(2)}(\mathbf{r}_{12}; \mathbf{e}_1, \mathbf{e}_2) \rangle_{\mathbf{e}_1 \mathbf{e}_2} \, d\mathbf{r}_1 \, d\mathbf{r}_2. \quad (3.4.19)$$

General expressions for the second-order term in the perturbation expansion of the angular pair correlation $g^{(2)}(\mathbf{r}_{12}; \mathbf{e}_1, \mathbf{e}_2)$ are not available. If, however, we restrict our discussion to spherical harmonic expansions of the anisotropic interaction $\Phi^{(1)}(\mathbf{r}_{12}; \mathbf{e}_1, \mathbf{e}_2)$ for which $l \neq 0$ then some simplification in $g^{(2)}(\mathbf{r}_{12}; \mathbf{e}_1, \mathbf{e}_2)$ ensues. The third-order expression corresponding to (3.4.16) is

$$A^{(3)} = A^{(3A)} + A^{(3B)} \quad (3.4.20)$$

where[3]

$$A^{(3A)} = \frac{\beta^2}{12} \int \int \rho_{(1)}^{(0)}(z_1) \rho_{(1)}^{(0)}(z_2) g_{(2)}^{(0)}(z_1, \mathbf{r}_{12}) \langle \Phi^{(1)^3}(\mathbf{r}_{12}; \mathbf{e}_1, \mathbf{e}_2) \rangle_{\mathbf{e}_1 \mathbf{e}_2} \, d\mathbf{r}_1 \, d\mathbf{r}_2$$

$$(3.4.21a)$$

$$A^{(3B)} = \frac{\beta^2}{6} \int \int \rho_{(1)}^{(0)}(z_1) \rho_{(1)}^{(0)}(z_2) \rho_{(1)}^{(0)}(z_3) g_{(3)}^{(0)}(z_1, \mathbf{r}_{12}, \mathbf{r}_{13})$$

$$\times \langle \Phi^{(1)}(\mathbf{r}_{12}; \mathbf{e}_1, \mathbf{e}_2) \Phi^{(1)}(\mathbf{r}_{13}; \mathbf{e}_1, \mathbf{e}_3) \Phi^{(1)}(\mathbf{r}_{23}; \mathbf{e}_2, \mathbf{e}_3) \rangle_{\mathbf{e}_1 \mathbf{e}_2 \mathbf{e}_3} \, d\mathbf{r}_1 \, d\mathbf{r}_2 \, d\mathbf{r}_3. \quad (3.4.21b)$$

In a similar manner to that outlined for the second order component, we have for the third order contribution to the surface tension

$$\gamma^{(3A)} = -\frac{\beta^2}{12} \int \int_{-\infty}^{\infty} z_1 \frac{\partial \rho_{(2)}^{(0)}(z_1, \mathbf{r}_{12})}{\partial z_1} \langle \Phi^{(1)^3}(\mathbf{r}_{12}; \mathbf{e}_1, \mathbf{e}_2) \rangle_{\mathbf{e}_1 \mathbf{e}_2} \, dz_1 \, d\mathbf{r}_{12} \quad (3.4.22a)$$

$$\gamma^{(3B)} = -\frac{\beta^2}{6} \int \int \int_{-\infty}^{\infty} z_1 \frac{\partial \rho_{(3)}^{(0)}(z_1, \mathbf{r}_{12}, \mathbf{r}_{13})}{\partial z_1}$$

$$\times \langle \Phi^{(1)}(\mathbf{r}_{12}; \mathbf{e}_1, \mathbf{e}_2) \Phi^{(1)}(\mathbf{r}_{13}; \mathbf{e}_1, \mathbf{e}_3) \Phi^{(1)}(\mathbf{r}_{23}; \mathbf{e}_2, \mathbf{e}_3) \rangle_{\mathbf{e}_1 \mathbf{e}_2 \mathbf{e}_3} \, dz_1 \, d\mathbf{r}_{12} \, d\mathbf{r}_{13}.$$

$$(3.4.22b)$$

It should be emphasized that, as for the second-order terms, these third-order results assume pairwise additivity of the molecular interaction and a Pople reference. Additional simplification has been made in restricting the anisotropic interaction to contain only terms with spherical harmonics of order $l = 0$.

Finally we mention once again that throughout this perturbative development reference fluid distributions $\rho_{(1)}^{(0)}$, $\rho_{(2)}^{(0)}$, $\rho_{(3)}^{(0)}$ *only* arise, circumventing the explicit introduction of the exact, but unknown, angular distributions—though not, of course, without approximation. An exact decoupling of translational and orientational states for certain kinds of interaction has recently been given by Parsons.[19] We shall consider his work in more detail in Chapter 11.

3.5 The KBF approximation for the surface tension in a Pople reference expansion

The results for $A^{(2)}$, $A^{(3)}$ (3.4.16, 3.4.20) and $\gamma^{(2)}$, $\gamma^{(3)}$ (3.4.18, 3.4.22) in the perturbation expansion of the Helmholtz free energy and surface tension developed with respect to a spherically symmetric Pople reference are purely formal. They are exact to within the assumption of pairwise additivity, with the additional restriction in the case of the third-order term that only spherical harmonics $l \neq 0$ arise in the multi-polar expansion of $\Phi^{(1)}(\mathbf{r}_{12}; \mathbf{e}_1, \mathbf{e}_2)$. Inevitably, further progress depends upon the expression of the two- and three-particle distributions in the Pople fluid in terms of known functions. The simplest approximation is the Kirkwood–Buff–Fowler step model:

$$\rho_{(2)}^{(0)}(z_1, \mathbf{r}_{12}) = \rho_L^2 g_{(2)}^{L(0)}(r_{12})$$
$$\rho_{(3)}^{(0)}(z_1, \mathbf{r}_{12}, \mathbf{r}_{13}) = \rho_L^3 g_{(3)}^{L(0)}(r_{12}, r_{23}, r_{13}). \tag{3.5.1}$$

Insertion of these expressions in (3.4.18, 3.4.22) yields, after some manipulation,[3]

$$\gamma_{KBF}^{(2A)} = \frac{\pi \beta \rho_L^2}{4} \int_0^\infty g_{(2)}^{L(0)}(r_{12}) \langle \Phi^{(1)2}(\mathbf{r}_{12}; \mathbf{e}_1, \mathbf{e}_2) \rangle_{\mathbf{e}_1 \mathbf{e}_2} r_{12}^3 \, dr_{12} \tag{3.5.2a}$$

$$\gamma_{KBF}^{(2B)} = \frac{\pi^2 \beta \rho_L^3}{2} \int_0^\infty r_{12} \int_0^\infty r_{13} \int_{|r_{12}-r_{13}|}^{r_{12}+r_{13}} r_{23} g_{(3)}^{L(0)}(r_{12}, r_{13}, r_{23})$$
$$\times \langle \Phi^{(1)}(\mathbf{r}_{12}; \mathbf{e}_1, \mathbf{e}_2) \Phi^{(1)}(\mathbf{r}_{13}; \mathbf{e}_1, \mathbf{e}_3) \rangle_{\mathbf{e}_1 \mathbf{e}_2 \mathbf{e}_3} (r_{12} + r_{13} + r_{23}) \, dr_{23} \, dr_{13} \, dr_{12} \tag{3.5.2b}$$

and

$$\gamma_{KBF}^{(3A)} = -\frac{\pi \beta^2 \rho_L^2}{12} \int_0^\infty g_{(2)}^{L(0)}(r_{12}) \langle \Phi^{(1)3}(\mathbf{r}_{12}; \mathbf{e}_1, \mathbf{e}_2) \rangle_{\mathbf{e}_1 \mathbf{e}_2} r_{12}^3 \, dr_{12} \tag{3.5.3a}$$

$$\gamma_{KBF}^{(3B)} = -\frac{\pi^2 \beta^2 \rho_L^3}{6} \int_0^\infty r_{12} \int_0^\infty r_{13} \int_{|r_{12}-r_{13}|}^{r_{12}+r_{13}} r_{23} g_{(3)}^{L(0)}(r_{12}, r_{13}, r_{23})$$
$$\times \langle \Phi^{(1)}(\mathbf{r}_{12}; \mathbf{e}_1, \mathbf{e}_2) \Phi^{(1)}(\mathbf{r}_{13}; \mathbf{e}_1, \mathbf{e}_3) \Phi^{(1)}(\mathbf{r}_{23}; \mathbf{e}_2, \mathbf{e}_3) \rangle_{\mathbf{e}_1 \mathbf{e}_2 \mathbf{e}_3} (r_{12} + r_{13}$$
$$+ r_{23}) \, dr_{23} \, dr_{13} \, dr_{12}. \tag{3.5.3b}$$

Numerical evaluation of these expressions requires the specification of the molecular interaction, in particular the Pople reference together with its associated distributions $\rho_{(1)}^0$, $\rho_{(2)}^{(0)}$, and $\rho_{(3)}^{(0)}$. For the single-particle distribution the KBF model adopts the bulk liquid density ρ_L, whilst for the triplet function a superposition approximation may be invoked. For the Pople reference which, we remember, amounts to a central sphericalized interaction (3.4.10) it is convenient to adopt a parametrically appropriate Lennard–Jones function:

$$u^{(0)}(r_{12}) = 4\varepsilon\left\{\left(\frac{\sigma}{r_{12}}\right)^{12} - \left(\frac{\sigma}{r_{12}}\right)^6\right\}. \tag{3.5.4}$$

The associated reference distributions may be readily determined. The specification of the anisotropic term $u^{(1)}(\mathbf{r}_{12})$ remains, however; there are a variety of possibilities corresponding to dipole, quadrupole, higher multipole, anisotropic overlap, and anisotropic dispersion potentials. Haile[3] has calculated the effects of various forms of anisotropy on the surface tension on the basis of second-order theory $\gamma^{(0)} + \gamma^{(2)}$, third-order theory $\gamma^{(0)} + \gamma^{(2)} + \gamma^{(3)}$, and a resumming in terms of a Padé approximant:

$$\gamma = \gamma^{(0)} + \frac{\gamma^{(2)}}{1 - \gamma^{(3)}/\gamma^{(2)}}. \tag{3.5.5}$$

Such Padé approximants have yielded improved results over the linear perturbation theory in the case of bulk fluid properties:[11] in the present case the Padé approximant effectively interpolates between the second- and third-order results, in good agreement with the machine simulations.

Lennard–Jones + dipole

Figure 3.5.1 shows the reduced surface tension for a fluid of axially symmetric molecules interacting with a Lennard–Jones plus dipole model potential.

The reduced surface tension for the reduced dipole moment $\mu^* = \mu/(\varepsilon\sigma^3)^{1/2} = 0$ corresponds to a spherically symmetric reference Lennard–Jones fluid having the reduced parameters shown. The Padé results agree well with machine simulations over the range of dipole moment investigated, although there appears to be a worsening of agreement with increasing temperature which is undoubtedly associated with the KBF approximation.

Lennard–Jones + quadrupole

For a given value of multipole strength the quadrupole potential generally has a greater effect on the surface tension than the dipole term as in the case of bulk thermodynamic properties. In Figure 3.5.2 the reduced surface

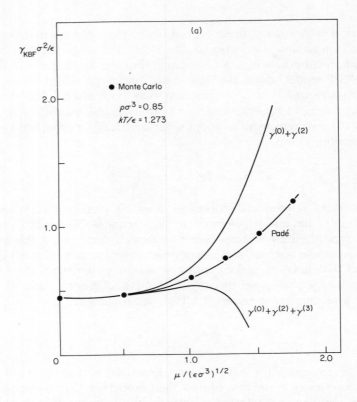

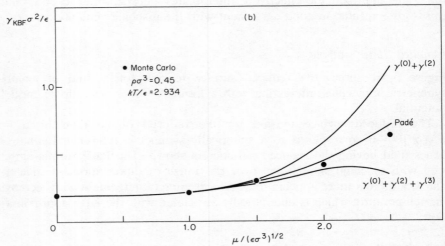

Figure 3.5.1 Reduced surface tension $\gamma^*_{KBF} = \gamma_{KBF}\sigma^2/\epsilon$ for an axially symmetric fluid of Lennard–Jones + dipole molecules as a function of reduced dipole moment $\mu^* = \mu/(\epsilon\sigma^3)^{1/2}$ at $T^* = kT/\epsilon = $ (a) 1.273, (b) 2.934. The Padé approximant to the zeroth, second, and third order contributions to the surface tension are seen to be in good agreement with Monte Carlo simulations of the fluid

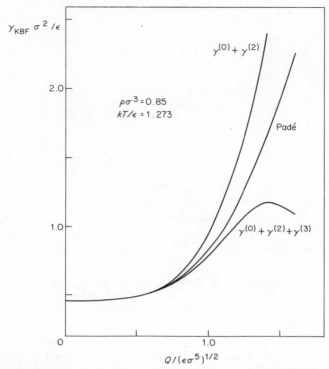

Figure 3.5.2 Reduced surface tension $\gamma^*_{\text{KBF}} = \gamma_{\text{KBF}} = \sigma^2/\varepsilon$ for a Lennard–Jones + quadrupole fluid as a function of reduced quadrupole moment $Q^* = Q/(\varepsilon\sigma^5)^{1/2}$

tension γ^*_{KBF} is shown as a function of reduced quadrupole moment $Q^* = Q/(\varepsilon\sigma^5)^{1/2}$ in the second-order, third-order, and Padé approximations. The curve bears direct comparison with Figure 3.5.1a from which the greater sensitivity of the surface tension to Q^* is clearly apparent.

Lennard–Jones + overlap and dispersion

Finally we consider the effect of anisotropic overlap and anisotropic dispersion on the reduced surface tension as a function of the appropriate strength constant (Figure 3.5.3). The curves are shown for the reduced Lennard–Jones parameters $\rho\sigma^3 = 0.85$, $kT/\varepsilon = 1.273$; the Padé dipole and quadrupole results from Figures 3.5.1a and 3.5.2 are included for comparison.

Before discussing the application of these results to specific molecular systems we should, perhaps, review the approximations involved in their development. Firstly the approach is essentially perturbative, and whilst the adoption of a Pople reference represents a felicitous choice in as far as the first-order term in the Zwanzig expansion vanishes, the convergence of the sum appears slow, though hastened by the formation of a Padé approximant

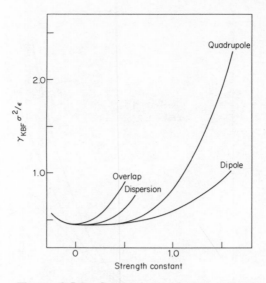

Figure 3.5.3 Comparison of the reduced surface tension $\gamma^*_{\text{KBF}} = \gamma_{\text{KBF}} \sigma^2/\varepsilon$ as a function of the various strength constants at $T^* = 1.273$

(Figures 3.5.1, 3.5.2). Thus the incorporation of multipolar terms beyond the quadrupole for high density systems is likely to be impracticable for complex molecular systems. Since the range of a multipolar interaction of order l falls off as $r^{-(l+1)}$, low density systems of moderate complexity may still be amenable to the technique.

The introduction of the Kirkwood–Buff–Fowler approximation for the anisotropic reference distributions $\rho^{(0)}_{(2)}(z_1, \mathbf{r}_{12})$, $\rho^{(0)}_{(3)}(z_1, \mathbf{r}_{12}, \mathbf{r}_{13})$ (equation 3.5.1) is a necessary expedient, given the overall complexity of the problem. Whilst it enables a calculation of the Helmholtz free energy and the surface tension to proceed, it must be said that such a closure device cannot characterize a stable liquid–vapour coexistence. This criticism has been discussed at length in Section 2.7. Moreover, the adoption of an infinitely sharp density discontinuity at the Gibbs surface leads to a worsening estimate of the surface tension with increasing temperature as the profile fails to relax. The immediate consequence is an overestimate of the Helmholtz free energy and surface tension at higher temperatures, as we see from the Padé curves in Figure 3.5.1 when compared with the Monte Carlo results.

Again, estimates of the surface excess internal energy per unit area u_s based on the classical Gibbs–Helmholtz equation

$$U_s = \gamma - T \left(\frac{\partial \gamma}{\partial T}\right)_{NV} \tag{3.5.6}$$

cannot yield accurate estimates for the surface excess energy, although a Padé approximant may be formed from the conjunction of (3.5.5) and (3.5.6):

$$U_s = U_s^{(0)} + U_s^{(1)} \qquad (3.5.7)$$

where $U_s^{(0)}$ is the isotropic reference contribution, and[3]

$$U_s^{(1)} = \frac{\gamma^{(2)} - \gamma^{(3)} - T(\partial \gamma^{(3)}/\partial T) - T(\partial \gamma^{(2)}/\partial T)(1 - 2\gamma^{(3)}/\gamma^{(2)})}{(1 - \gamma^{(3)}/\gamma^{(2)})^2}. \qquad (3.5.8)$$

Haile[3] has tested the Gibbs–Helmholtz equation in the perturbative KBF approximation for a *Lennard–Jones + quadrupole* fluid with the results shown in Table 3.5.1. The numerical coincidence of the calculated and simulated results does not represent an adequate criterion for assessment of the model, unfortunately, and as Croxton[12] has observed, much of the agreement must be attributed to cancellation of errors between the energy (γ) and entropy ($T \partial \gamma / \partial T$) terms in (3.5.6).

3.6 Surface tension calculations for real molecular fluids

The Padé approximant (3.5.5) to the perturbation expansion of the surface tension including multipolar terms up through the qudrupole–quadrupole interaction, plus anisotropic dispersion and overlap contributions, has been used to estimate the pure liquid surface tensions for CO_2, C_2H_2, and HBr.[3] The KBF approximation is made to reduce the expressions to a form amenable to numerical evaluation. The reference fluid was taken to interact through the Mie $(n, 6)$ potential:

$$\Phi_{(n,6)}(r) = \frac{n\varepsilon}{(n-6)} \left(\frac{n}{6}\right)^{6(n-6)^{-1}} \left\{ \left(\frac{\sigma}{r}\right)^n - \left(\frac{\sigma}{r}\right)^6 \right\} \qquad (3.6.1)$$

where σ, ε, n were determined by fitting perturbation theory calculations of liquid densities and pressures to experimental values along the orthobaric line for the fluids under consideration.

The relevant potential parameter values are given in Table 3.6.1.[3, 16]

The Padé results are compared with experiment for the three systems in

Table 3.5.1. Surface excess free energy per unit area for a *Lennard-Jones + quadrupole* fluid in the perturbative KBF approximation[3]

$Q/(\varepsilon\sigma^5)^{1/2}$	$\rho\sigma^3$	kT/ε	$U_s^{KBF}\sigma^2/\varepsilon$	
			Padé	MD
0.5	0.85	1.277	1.913	1.923 ± 0.016
0.707	0.931	0.765	2.670	2.656 ± 0.012
1.0	0.85	1.294	2.505	2.475 ± 0.019

Table 3.6.1[3, 16]

Fluid	ε/k(°K)	σ(Å)	n	dipole $\mu(10^{18})$ (esu cm)	quadrupole $Q(10^{24})$ (esu cm²)	anisotropic overlap δ	dispersion κ
CO_2	244.31	3.687	16		-4.30^{17}	-0.1	0.257
C_2H_2	253.66	3.901	16		5.01^{18}	0.3	0.270
HBr	248.47	3.790	12	0.788	4.0^{17}		

Figures 3.6.1 and 3.6.2. The overall agreement is quite good considering the nature of the approximations made. In the case of CO_2 the agreement actually *improves* with increasing temperature despite the use of the KBF step-model approximation which, for central symmetric interactions, invariably overestimates the surface tension as the discontinuous density variation becomes progressively inappropriate. Whilst it is difficult to be specific, Haile[3] observes that the formation of the Padé approximant tends to partially correct for errors introduced by the KBF approximation. Moreover, the decrease in density as the temperature increases results in a rapid decline in the higher order multipolar contributions whose range falls off as $r^{-(l+1)}$, so the convergence of the expansion would be expected to be correspondingly more rapid. As we have noted on several occasions, numerical coincidence of experimental and calculated estimates of the surface

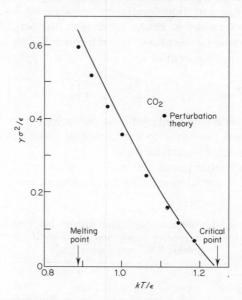

Figure 3.6.1 Comparison of Padé estimates of the reduced surface tension of CO_2 as a function of reduced temperature with experiment

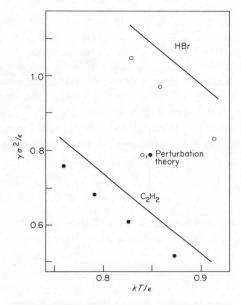

Figure 3.6.2 Comparison of Padé estimates of the reduced surface tensions of HBr and C_2H_2 as a function of reduced temperature with experiment

tension provides a necessary but not sufficient criterion for the adequacy of the theoretical treatment. Nevertheless, we should be encouraged by the broad measure of qualitative success which evidently underlies these results.

3.7 The single-particle density-orientation profile $\rho_{(1)}(z_1, \mathbf{e}_1)$

The formal expression for the surface tension of a rigid molecularly asymmetric system was developed in Section 3.2 in terms of the interfacial one- and two-particle correlation functions:

$$\rho_{(2)}(z_1, \mathbf{r}_{12}; \mathbf{e}_1, \mathbf{e}_2) = \rho_{(1)}(z_1, \mathbf{e}_1)\rho_{(1)}(z_2, \mathbf{e}_2)g_{(2)}(z_1, \mathbf{r}_{12}; \mathbf{e}_1, \mathbf{e}_2).$$

Analytic and numerical expediency suggested the replacement of $\rho_{(1)}(z_1; \mathbf{e}_1)$ by *an* orientational average, though not necessarily *the* orientational average $\rho_{(1)}(z_1) = \langle \rho_{(1)}(z_1; \mathbf{e}_1) \rangle_{\mathbf{e}_1}$. Moreover, the distribution of angular orientation is assumed identical to that of the bulk, and not to modify in the vicinity of the liquid surface. A further simplification, $g_{(2)}(z_1, \mathbf{r}_{12}; \mathbf{e}_1, \mathbf{e}_2) \rightarrow g_{(2)}^{L}(\mathbf{r}_{12}; \mathbf{e}_1, \mathbf{e}_2)$, corresponding to the Kirkwood–Buff–Fowler approximation is then made; however, as we have observed on a number of occasions, the adoption of the bulk liquid distribution throughout the interfacial zone is inconsistent with the development of a stable liquid–vapour coexistence.

Relatively little attention has been devoted to the density orientation profile $\rho_{(1)}(z_1; \mathbf{e}_1)$ for asymmetric molecular systems, although it is clear that in regions of structural anisotropy, such as that obtaining at boundaries, more specifically throughout the interfacial zone, angular averages will modify with respect to those in either homogeneous bulk phase with the consequential development of a surface excess or *interfacial torque field*. Homogeneity and isotropy of the bulk fluid phases ensures that no long range orientational order develops. At the surface the symmetry is broken, however, and we may expect the development of preferred molecular orientations most clearly reflected in the density-orientation profile: its formal statistical mechanical expression is straightforward:

$$\rho_{(1)}(z_1, \mathbf{e}_1) = \frac{N}{Z_Q} \int \cdots \int \exp \left\{ -\frac{\Phi_N(\mathbf{r}^N; \mathbf{e}^N)}{kT} \right\} d\mathbf{r}_2 \cdots d\mathbf{r}_N \, d\mathbf{e}_2 \cdots d\mathbf{e}_N$$

$$(3.7.1)$$

which represents the unnormalized probability of finding a molecule at z_1 with orientation $\mathbf{e}_1$. Z_Q is the configurational partition function

$$Z_Q = \int \cdots \int \exp \left\{ -\frac{\Phi_N(\mathbf{r}^N; \mathbf{e}^N)}{kT} \right\} d\mathbf{r}_1 \cdots d\mathbf{r}_N \, d\mathbf{e}_1 \cdots d\mathbf{e}_N$$

and $\Phi_N(\mathbf{r}^N : \mathbf{e}^N)$ is the total N-body configurational energy.

Croxton and Osborn[13] introduce the potential of mean force and torque $\Psi(\mathbf{r}_1, \mathbf{r}_2; \mathbf{e}_1, \mathbf{e}_2)$ which is defined by and related to the two-body distribution $g_{(2)}(\mathbf{r}_1, \mathbf{r}_2; \mathbf{e}_1, \mathbf{e}_2)$ as follows:

$$
\begin{aligned}
g_{(2)}(\mathbf{r}_1, \mathbf{r}_2; \mathbf{e}_1, \mathbf{e}_2) &= g_{(2)}(\mathbf{r}_1, \mathbf{r}_2)_{\mathbf{e}_1 \mathbf{e}_2} g_{(2)}(\mathbf{e}_1, \mathbf{e}_2)_{\mathbf{r}_1 \mathbf{r}_2} \\
&= \exp \left\{ -\frac{\Psi(\mathbf{r}_1, \mathbf{r}_2; \mathbf{e}_1, \mathbf{e}_2)}{kT} \right\}.
\end{aligned}
$$

$$(3.7.2)$$

The distribution $g_{(2)}(\mathbf{r}_1, \mathbf{r}_2; \mathbf{e}_1, \mathbf{e}_2)$ specifies the spatial $(\mathbf{r}_1, \mathbf{r}_2)$ and orientational $(\mathbf{e}_1, \mathbf{e}_2)$ configuration of the two molecules, whilst the conditional distributions $g_{(2)}(\mathbf{r}_1, \mathbf{r}_2)_{\mathbf{e}_1 \mathbf{e}_2}$, $g_{(2)}(\mathbf{e}_1, \mathbf{e}_2)_{\mathbf{r}_1 \mathbf{r}_2}$ represent the spatial distribution of the molecular centres of gravity *for the specified orientations* $\mathbf{e}_1, \mathbf{e}_2$ and vice versa, respectively.

The *mean force* experienced by particle 1 due to its neighbours will be, for specified $\mathbf{e}_1, \mathbf{e}_2$, taking the spatial gradient $-\nabla \Psi$

$$
\begin{aligned}
-\rho_L^2 g_{(2)}(\mathbf{r}_1, \mathbf{r}_2; \mathbf{e}_1, \mathbf{e}_2) &\nabla_1 \Psi(\mathbf{r}_1, \mathbf{r}_2; \mathbf{e}_1, \mathbf{e}_2) \\
&= -\rho_L^2 g_{(2)}(\mathbf{r}_1, \mathbf{r}_2; \mathbf{e}_1, \mathbf{e}_2) \nabla_1 \Phi(\mathbf{r}_1, \mathbf{r}_2; \mathbf{e}_1, \mathbf{e}_2) \\
&\quad - \rho_L^3 \int_{\mathbf{e}_3} \int_{\mathbf{r}_3} \nabla_1 \Phi(\mathbf{r}_1, \mathbf{r}_3; \mathbf{e}_1, \mathbf{e}_3) g_{(3)}(\mathbf{r}_1, \mathbf{r}_2, \mathbf{r}_3; \mathbf{e}_1, \mathbf{e}_2, \mathbf{e}_3) \, d\mathbf{r}_3 \, d\mathbf{e}_3 \quad (3.7.3)
\end{aligned}
$$

where ρ_L is the bulk number density of the molecular fluid. Using the superposition approximation in (3.7.3)

$$g_{(3)}(\mathbf{r}_1, \mathbf{r}_2, \mathbf{r}_3; \mathbf{e}_1, \mathbf{e}_2, \mathbf{e}_3) \sim g_{(2)}(\mathbf{r}_1, \mathbf{r}_2; \mathbf{e}_1, \mathbf{e}_2) g_{(2)}(\mathbf{r}_2, \mathbf{r}_3; \mathbf{e}_2, \mathbf{e}_3) g_{(2)}(\mathbf{r}_1, \mathbf{r}_3; \mathbf{e}_1, \mathbf{e}_3)$$

$$(3.7.4)$$

we obtain immediately, transforming to relative coordinates $(\mathbf{r}_{12}; \mathbf{e}_{12})$:

$$-kT\nabla_1 \ln g_{(2)}(\mathbf{r}_{12})_{\mathbf{e}_{12}} = g_{(2)}(\mathbf{r}_{12})_{\mathbf{e}_{12}}\nabla_1\Phi(\mathbf{r}_{12})_{\mathbf{e}_{12}}$$

$$+\rho_L \int\int \nabla_1\Phi(\mathbf{r}_{13})_{\mathbf{e}_{13}}g_{(2)}(\mathbf{r}_{23})_{\mathbf{e}_{23}}g_{(2)}(\mathbf{r}_{13})_{\mathbf{e}_{13}}\,\mathrm{d}\mathbf{r}_{13}\,\mathrm{d}\mathbf{e}_{13}. \quad (3.7.5)$$

An explicit expression is given in (A10.1.7). Similarly, the *mean torque* experienced by particle 1 due to its neighbours will be, taking the angular gradient $-\nabla_{\Omega_1}\Psi$,

$$-kT\nabla_{\Omega_1} \ln g_{(2)}(\mathbf{e}_{12})_{\mathbf{r}_{12}} = g_{(2)}(\mathbf{e}_{12})_{\mathbf{r}_{12}}\nabla_{\Omega_1}\Phi(\mathbf{e}_{12})_{\mathbf{r}_{12}}$$

$$+\rho_L \int\int \nabla_{\Omega_1}\Phi(\mathbf{e}_{13})_{\mathbf{r}_{13}}g_{(2)}(\mathbf{e}_{23})_{\mathbf{r}_{23}}g_{(2)}(\mathbf{e}_{13})_{\mathbf{r}_{13}}\,\mathrm{d}\mathbf{r}_{13}\,\mathrm{d}\mathbf{e}_{13}. \quad (3.7.6)$$

Again, an explicit expression is given in A10.1.8. The function $g_{(2)}(\mathbf{r}_{12}; \mathbf{e}_{12})$ may be formed from the iterative solution of (3.7.5) and (3.7.6) in conjunction with (3.7.2):

$$\ln g_{(2)}(\mathbf{r}_{12}; \mathbf{e}_{12}) = \ln g_{(2)}(\mathbf{r}_{12})_{\mathbf{e}_{12}} + \ln g_{(2)}(\mathbf{e}_{12})_{\mathbf{r}_{12}}. \quad (3.7.7)$$

The exact closure appropriate to a region of molecular inhomogeneity is

$$\rho_{(2)}(z_1, \mathbf{r}_{12}; \mathbf{e}_{12}) = \rho_{(1)}(z_1; \mathbf{e}_1)\rho_{(1)}(z_2; \mathbf{e}_2)g_{(2)}(z_1, \mathbf{r}_{12}; \mathbf{e}_{12}) \quad (3.7.8)$$

where $\mathbf{e}_{12} = \mathbf{e}_1 - \mathbf{e}_2$ and $\rho_{(1)}(z_1; \mathbf{e}_{12})$ is the single-particle density-orientation profile. Defining the single-particle potential of mean force and torque $\Psi(z_1; \mathbf{e}_1)$ as

$$\rho_{(1)}(z_1; \mathbf{e}_1) = \rho_L\left\{-\frac{\Psi(z_1; \mathbf{e}_1)}{kT}\right\} \quad (3.7.9)$$

and setting

$$\rho_{(1)}(z_1; \mathbf{e}_1) = \rho_{(1)}(z_1)_{\mathbf{e}_1}\rho_{(1)}(\mathbf{e}_1)_{z_1} \quad (3.7.10)$$

then we have at once

$$\rho_{(1)}(z_1; \mathbf{e}_1) = \rho_L \exp\left\{-\frac{1}{kT}\int_{\mathbf{r}_2}\int_{\mathbf{e}_2}\left[\int_{-\infty}^{z_1}\nabla_1\Phi(\mathbf{r}_1, \mathbf{r}_2; \mathbf{e}_1, \mathbf{e}_2)\right.\right.$$

$$\times\rho_{(1)}(z_2; \mathbf{e}_2)g_{(2)}(\mathbf{r}_1, \mathbf{r}_2; \mathbf{e}_1, \mathbf{e}_2)\,\mathrm{d}\mathbf{r}_1 + \int_0^{\mathbf{e}_1}\nabla_{\Omega_1}\Phi(\mathbf{r}_1, \mathbf{r}_2; \mathbf{e}_1, \mathbf{e}_2)$$

$$\left.\left.\times\rho_{(1)}(z_2; \mathbf{e}_2)g_{(2)}(\mathbf{r}_1, \mathbf{r}_2; \mathbf{e}_1, \mathbf{e}_2)\,\mathrm{d}\mathbf{e}_1\right]\mathrm{d}\mathbf{e}_2\,\mathrm{d}\mathbf{r}_2\right\}. \quad (3.7.11)$$

The first term in the square bracket represents the mean force acting on the centre of gravity of a linear molecule located at z_1 and expresses the development of a surface constraining field. The second term expresses the *torque field* which develops in a region of structural inhomogeneity which may be understood in terms of a competitive interplay between orientational energy and entropy contributions.

Evaluation of equation (3.7.11) requires a knowledge of the anisotropic pair distribution (3.7.8). However, if we adopt the closure

$$\rho_{(2)}(\mathbf{r}_1, \mathbf{r}_2; \mathbf{e}_1, \mathbf{e}_2) \sim \rho_{(1)}(z_1; \mathbf{e}_1)\rho_{(1)}(z_2; \mathbf{e}_2)g_{(2)}^{L}(\mathbf{r}_{12}; \mathbf{e}_{12}) \qquad (3.7.11a)$$

where $g_{(2)}^{L}(\mathbf{r}_{12}; \mathbf{e}_{12})$ is the homogeneous *bulk* distribution (3.7.7), then the profile $\rho_{(1)}(z_1; \mathbf{e}_1)$ may be determined.

Haile,[3] on the other hand, has developed a perturbation approach to the single-particle density-orientation function. The total potential may be expanded in terms of an isotropic reference plus an anisotropic perturbation:[9]

$$\Phi_N(\mathbf{r}^N; \mathbf{e}^N) = \Phi^{(0)}(\mathbf{r}^N) + \lambda \Phi^{(1)}(\mathbf{r}^N; \mathbf{e}^N) \qquad (3.7.12)$$

where λ is a perturbation parameter as in (3.4.1), whilst $\rho_{(1)}(z_1; \mathbf{e}_1)$ may be Taylor expanded with respect to the parameter λ about the reference (i.e. $\lambda = 0$):

$$\rho_{(1)}(z_1; \mathbf{e}_1) = \frac{\rho_{(1)}^{(0)}(z_1)}{\Omega} + \lambda \left[\frac{\partial \rho_{(1)}(z_1; \mathbf{e}_1)}{\partial \lambda}\right]_{\lambda=0} + \frac{\lambda^2}{2}\left[\frac{\partial^2 \rho_{(1)}(z_1, \mathbf{e}_1)}{\partial \lambda^2}\right]_{\lambda=0} + \cdots . \qquad (3.7.13)$$

Working to first order we have

$$\rho_{(1)}(z_1; \mathbf{e}_1) = \frac{\rho_{(1)}^{(0)}(z_1)}{\Omega} + \rho^{(1)}(z_1; \mathbf{e}_1) \qquad (3.7.14)$$

where

$$\rho^{(1)}(z_1; \mathbf{e}_1) \equiv \lambda \left[\frac{\partial \rho_{(1)}(z_1; \mathbf{e}_1)}{\partial \lambda}\right]_{\lambda=0} . \qquad (3.7.15)$$

Taking the derivative of (3.7.1) with respect to λ, and assuming pairwise additivity of the interaction, yields

$$\begin{aligned}
\left[\frac{\partial \rho_{(1)}(z_1; \mathbf{e}_1)}{\partial \lambda}\right]_{\lambda=0} = &-\frac{\beta}{\Omega}\int \rho_{(2)}^{(0)}(z_1, \mathbf{r}_{12})\langle\Phi^{(1)}(\mathbf{r}_{12}; \mathbf{e}_1, \mathbf{e}_2)\rangle_{\mathbf{e}_2}\, d\mathbf{r}_2 \\
&-\frac{\beta}{2\Omega}\iint \rho_{(2)}^{(0)}(z_1, \mathbf{r}_{12}, \mathbf{r}_{13})\langle\Phi^{(1)}(\mathbf{r}_{23}; \mathbf{e}_2, \mathbf{e}_3)\rangle_{\mathbf{e}_2\mathbf{e}_3}\, d\mathbf{r}_2\, d\mathbf{r}_3 \\
&+\frac{\beta}{2}\rho_{(1)}^{(0)}(z_1)\iint \rho_{(2)}^{(0)}(z_1, \mathbf{r}_{12})\langle\Phi^{(1)}(\mathbf{r}_{12}; \mathbf{e}_1, \mathbf{e}_2)\rangle_{\mathbf{e}_1\mathbf{e}_2}\, d\mathbf{r}_1\, d\mathbf{r}_2 .
\end{aligned} \qquad (3.7.16)$$

The choice of a Pople reference interaction, amounting to an unweighted angular average, or sphericalizing, of the interaction as defined in (3.4.10), causes the last two terms in (3.7.16) to vanish, leaving

$$\left[\frac{\partial \rho_{(1)}(z_1; \mathbf{e}_1)}{\partial \lambda}\right]_{\lambda=0} = -\frac{\beta}{\Omega}\int \rho_{(2)}^{(0)}(z_1, \mathbf{r}_{12})\langle\Phi^{(1)}(\mathbf{r}_{12}; \mathbf{e}_1, \mathbf{e}_2)\rangle_{\mathbf{e}_2}\, d\mathbf{r}_2 . \qquad (3.7.17)$$

Again, spherical harmonic expansions of the anisotropic interaction $\Phi^{(1)}(\mathbf{r}_{12}; \mathbf{e}_1, \mathbf{e}_2)$ having $l \neq 0$ components, such as multipolar terms, vanish upon forming the angular average:

$$\langle \Phi^{(1)}(\mathbf{r}_{12}; \mathbf{e}_1, \mathbf{e}_2)\rangle_{\mathbf{e}_2} = 0 \qquad (3.7.18)$$

in which case the density-orientation profile is, to first order, simply the reference profile $\rho_{(1)}^{(0)}(z_1)$. Anisotropic overlap and dispersion interactions do, however, contain the $l = 0$ harmonic, in which case the full expression (3.7.17) must be used. Combining (3.7.14) with (3.7.17) and setting $\lambda = 1$ to recover the total interaction we have

$$\rho_{(1)}(z_1; \mathbf{e}_1) = \frac{\rho_{(1)}^{(0)}(z_1)}{\Omega} - \frac{\beta}{\Omega} \int \rho_{(2)}^{(0)}(\mathbf{r}_{12})\langle \Phi^{(1)}(\mathbf{r}_{12}; \mathbf{e}_1, \mathbf{e}_2)\rangle_{\mathbf{e}_2} \, d\mathbf{r}_2. \qquad (3.7.19)$$

For the anisotropic overlap and dispersion interactions Haile[3] takes

$$\langle \Phi_{over}^{(1)}(\mathbf{r}_{12}; \mathbf{e}_1, \mathbf{e}_2)\rangle_{\mathbf{e}_2} = 8\delta\varepsilon \left(\frac{\sigma}{r_{12}}\right)^{12} P_2(\cos\theta_1) P_2(\cos\theta_{12}) \qquad (3.7.20a)$$

$$\langle \Phi_{dis}^{(1)}(\mathbf{r}_{12}; \mathbf{e}_1, \mathbf{e}_2)\rangle_{\mathbf{e}_2} = -4\kappa\varepsilon \left(\frac{\sigma}{r_{12}}\right)^{6} P_2(\cos\theta_1) P_2(\cos\theta_{12}) \qquad (3.7.20b)$$

where δ, κ are the appropriate interaction parameters. Insertion of these expressions in (3.7.19) yields

$$\rho_{(1)}(z_1; \mathbf{e}_1) = C P_2(\cos\theta_1)\rho_{(1)}^{(0)}(z_1) \int \left(\frac{r_{12}}{\sigma}\right)^{-n} P_2(\cos\theta_{12})\rho_{(1)}^{(0)}(z_2) g_{(2)}^{(0)}(z_1, \mathbf{r}_{12}) \, d\mathbf{r}_{12}$$
$$(3.7.21)$$

where $C = -8\beta\delta\varepsilon/\Omega$, $n = 12$ for overlap, and $C = 4\beta\kappa\varepsilon/\Omega$, $n = 6$ for dispersion.

In the case of rigid linear molecules $\mathbf{e} = \{\theta_1\}$ only, since from Figure 3.1.1 it is clear that rotation about the molecular axis χ_1 is irrelevant, whilst symmetry considerations eliminate any ϕ_1-dependence, at least far from any interfacial xy boundaries. The density profile $\rho_{(1)}(z_1)$ is simply related to the density-orientation profile $\rho_{(1)}(z_1; \mathbf{e}_1)$ through the orientational average:

$$\rho_{(1)}(z_1) = \langle \rho_{(1)}(z_1; \mathbf{e}_1)\rangle_{\mathbf{e}_1}. \qquad (3.7.22)$$

Thus, for weakly anisotropic systems, for which first-order perturbation theory might be expected to apply, integration of (3.7.21) over $\mathbf{e}_1$ vanishes, in which case the interfacial density profile $\rho_{(1)}(z_1)$ is just that of the reference fluid $\rho_{(1)}^{(0)}(z_1)$. However, in the case of strongly anisotropic molecular fluids, such as the liquid crystals, the density profile would be expected to differ substantially from the reference profile. Indeed, as we shall see, such systems cannot be adequately treated in terms of a perturbation approach and alternative methods are enforced.[19]

Haile[3] has determined $\rho_{(1)}(z_1; \theta_1)$ on the basis of equation (3.7.21) for linear molecules for two model interactions which we shall consider separately. In each case specification of the anisotropic two-particle distribution $g_{(2)}^{(0)}(z_1, \mathbf{r}_{12})$ was in terms of Toxvaerd's closure (2.7.2) which is itself open to some criticism, and has been reviewed at some length in Section 2.7. The density profile $\rho_{(1)}^{(0)}(z_1)$ of the reference Lennard–Jones fluid was determined by solving the single-particle Born–Green–Yvon equation: Verlet's molecular dynamics results[14] were used for $g_{(2)}^{(0)}(r_{12})$ which arises within the BGY integral.

Lennard–Jones + anisotropic dispersion

The $\rho_{(1)}(z_1; \theta_1)$ profile is shown in Figure 3.7.1 for a Lennard–Jones + anisotropic dispersion interaction ($\kappa = 0.25$). The density profile is seen to be monotonic and virtually identical to Toxvaerd's Lennard–Jones profile at T^*, as expected, since Toxvaerd's closure forms the basis of the central closure. Any discrepancy may be attributed to Toxvaerd's use of a Percus–Yevick pair distribution $g_{(2)}^{(0)}(r_{12})$ whilst Haile adopted Verlet's machine correlations.

It is clear from Figure 3.7.1 that these linear molecules show a preference for alignment *parallel* to the interfacial plane on the liquid side of the equimolar surface, whilst those on the vapour side prefer a *perpendicular* alignment. Deep in either bulk interior phase all alignments appear equally probable, as they should be.

This somewhat surprising result Haile[3] associates with Toxvaerd's finding that the surface excess pressure $[P_\perp - P_\parallel(z)]$ is *negative* corresponding to a compression in the vapour region. Thus, Haile suggests[3] 'the surface tension in the liquid tends to pull the molecules towards orientations in which the molecules lie in the interface. On the vapour side, the surface compression tends to push the molecules together, forcing them to stand in the interface.'

The surface tension integral $\gamma = \int_{-\infty}^{\infty} [P_\perp - P_\parallel(z)]\,dz$ is, of course, subject to the constraint that $\gamma > 0$, and whilst there is no *a priori* objection to the development of negative regions in the integrand, these are more likely artifacts of an imperfect specification of $P_\parallel(z)$,[15] sensitive as it is to the anisotropic distribution $g_{(2)}^{(0)}(z_1, \mathbf{r}_{12})$, as is, indeed, the orientation of the interfacial torque field itself.

Lennard–Jones + anisotropic overlap

Two specific cases arise according as the overlap parameter δ is positive (rod-like molecules) or negative (plate-like molecules). For $\delta > 0$ the behaviour of $\rho_{(1)}(z_1; \theta_1)$ is precisely the opposite of that for the dispersion interaction (Figure 3.7.2), that is, the molecules tend to align perpendicular to the interface on the liquid side of the dividing surface, whilst on the vapour side the molecules tend to align parallel. For $\delta < 0$ the converse is the case.

It is not altogether clear whether Haile's qualitative association of the molecular orientation with the sign of the excess pressure $[P_\perp - P_\parallel(z)]$ is sustained in the present case since, to first order, the excess pressure profile

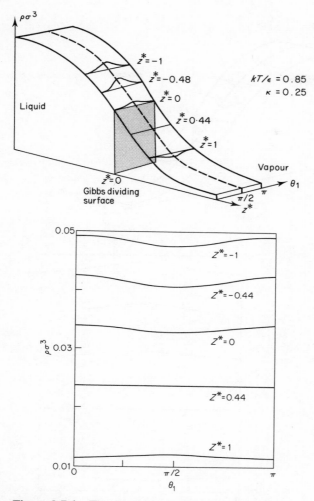

Figure 3.7.1 The density-orientation profile $\rho_{(1)}(z_1; \theta_1)$ for a Lennard–Jones + anisotropic dispersion interaction ($\kappa = 0.25$) at $T^* = 0.85$. These linear molecules show a preference for alignment *parallel* to the interfacial plane on the liquid side of the Gibbs dividing surface, and *perpendicular* on the vapour side. $z^* = z/\sigma$

remains unmodified depending only upon the reference density profile. The orientation has, however, switched through $\pi/2$ on either side of the dividing surface.

Lennard–Jones + anisotropic overlap and dispersion

Although such a system has not been investigated in the present approximation, it is interesting to speculate on the form of the density orientation

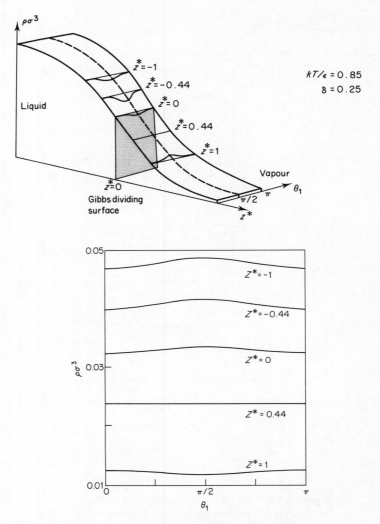

Figure 3.7.2 The density-orientation profile $\rho_{(1)}(z_1; \theta_1)$ for a Lennard–Jones + anisotropic dispersion interaction $(\delta = 0.25)$ at $T^* = 0.85$. These molecules tend to align *perpendicular* to the interface on the liquid side of the dividing surface, and *parallel* on the vapour side. $z^* = z/\sigma$

profile on the basis of the dispersion and overlap profiles determined above. Since each of the two forms of interaction, taken individually, adopt preferred orientations differing by $\pi/2$, it is presumably possible to contrive an interaction of the form (3.7.20) which, by a suitable choice of parameters $\kappa, \delta > 0$, shows *no* preferred orientation on either side of the dividing surface, despite the axially symmetric rodlike nature of the molecule. Indeed, each of the systems considered so far develops a zone of no specific orientation in the vicinity of $z \sim +0.44\sigma$, which is somewhat surprising in itself.

We shall not anticipate the discussion of Section 9.2; however, it is useful to compare Haile's conclusions with those of Parsons, and with machine simulations of Section 10.10.

Parsons considers a system of rigid axially symmetric molecules interacting through a permanent dipole potential, corresponding most closely with the anisotropic overlap interaction of Haile. On the basis of a step-discontinuity model of the surface Parsons concludes that a minimum interfacial free energy per unit area is achieved for an orientation *parallel* to the dividing surface. The introduction of an anisotropic dispersion interaction appears not to modify Parson's conclusion.

The simulations of Thompson *et al.* (Section 10.10) for heteronuclear diatomic systems whose interaction may be described as of a bicentric Lennard–Jones form having both anisotropic overlap and dispersion components appear to show a perpendicular alignment on the liquid side of the dividing zone switching to a parallel orientation in the vapour phase. Such a density-orientation profile is in qualitative agreement with Haile's *Lennard–Jones + anisotropic overlap* model. The machine results are not entirely unambiguous, however, the possible intervention of boundary condition effects cannot be neglected, and a discussion is given in Section 10.10.

References

1. J. O. Hirschfelder, C. F. Curtiss, and R. B. Bird, *Molecular Theory of Gases and Liquids*, Wiley, New York (1954).
   J. O. Hirschfelder, *Adv. Chem. Phys.*, **12,** (1967).
   D. D. Fitts, *A. Rev. Phys. Chem.*, **17,** 59 (1966).
   E. B. Smith, *Rep. Prog. Chem.*, **63,** 13 (1966).
   H. Margenau and N. R. Kestner, *Theory of Intermolecular Forces*, Pergamon, New York (1969).
2. H. T. Davis, *J. Chem. Phys.*, **62,** 3412 (1975).
3. J. M. Haile, *Ph.D. Thesis*, University of Florida, Gainesville (1976).
4. R. H. Fowler, *Proc. Roy. Soc.*, **A159,** 229 (1937).
5. J. G. Kirkwood and F. P. Buff, *J. Chem. Phys.*, **17,** 338 (1949).
6. J. A. Pople, *Proc. Roy. Soc.* **A221,** 498 (1954).
7. M. S. Ananth, K. E. Gubbins, and C. G. Gray, *Mol. Phys.*, **28,** 1005 (1974).
8. K. E. Gubbins and C. G. Gray, *Mol. Phys.*, **23,** 187 (1972).
9. R. W. Zwanzig, *J. Chem. Phys.*, **22,** 1420 (1954).
10. W. R. Smith, *Canad. J. Phys.*, **52,** 2022 (1974).
    W. R. Smith, W. G. Madden, and D. D. Fitts, *Chem. Phys. Lett.*, **36,** 195 (1975)
    W. R. Smith, *Chem. Phys. Lett.*, **40,** 313 (1976).
    W. G. Madden, D. D. Fitts, and W. R. Smith, *Mol. Phys.* **35,** 1017 (1978).
    W. R. Smith, I. Nezbeda, T. W. Melnyk, and D. D. Fitts, *Faraday Disc. Chem. Soc.*, 66/8 (1978).
11. G. Stell, J. C. Rasaiah, and H. Narang, *Mol. Phys.*, **27,** 1393 (1974).
12. C. A. Croxton, *Adv. Phys.*, **22,** 385 (1973).

114

C. A. Croxton, *Liquid State Physics—A Statistical Mechanical Introduction,* Cambridge University Press, London (1974).

13. C. A. Croxton and T. R. Osborn, *Phys. Lett.* **55A,** 415 (1976).
14. L. Verlet, *Phys. Rev.*, **165,** 201 (1968).
15. C. A. Croxton (Ed.), *Progress in Liquid Physics*, Wiley, Chichester (1978). Ch. 2.
16. C. H. Turn, *Ph.D. Thesis*, University of Florida, Gainesville (1976).
17. D. E. Stogryn and A. P. Stogryn, *Mol. Phys.*, **11,** 371 (1966).
18. T. H. Spurling and E. A. Mason, *J. Chem. Phys.*, **46,** 322 (1967).
19. J. D. Parsons, *Phys. Rev.*, **19,** 1225 (1979).

CHAPTER 4

Multicomponent Liquid Surfaces

4.1 Introduction

There is no difficulty in principle in extending the theory of molecular distribution functions of Chapter 2 to include multicomponent systems of molecules of differing size and interaction: indeed, this we shall do in Section 4.3. However, it should be realized from the outset that both for bulk phase and interfacial quantities the thermodynamic functions of mixtures do not generally consist of simple linear combinations of the molar properties of the pure components at the same temperature and pressure, weighted according to composition. Thus, the representation of the principal thermodynamic functions of a mixture (components $\alpha, \ldots, \omega$) in terms of molecular distribution functions will involve not only the *self* distributions $g_{(2)}^{\alpha\alpha}(z_\alpha, \mathbf{r}_{\alpha\alpha}), \ldots, g_{(2)}^{\omega\omega}(z_\omega, \mathbf{r}_{\omega\omega})$, but also the *cross* distributions $g_{(2)}^{\kappa\lambda}(z_\kappa, \mathbf{r}_{\kappa\lambda})$ which, through the appropriate pair interaction $\Phi_{\kappa\lambda}$, provide a distinct supplement to the molar sum of thermodynamic quantities arising from the self distributions. Of course, such terms can only arise in multicomponent systems which, on account of the rapid divergence in the number of distinct cross distributions with increasing number of species, generally restricts discussion to binary mixtures, and primarily in terms of the *ideal mixture*—a hypothetical system which provides a convenient standard of normal behaviour:

$$\mu_\alpha(p, T, x_\alpha) = \mu_\alpha^0(p, T) + RT \ln x_\alpha \qquad (\alpha, \ldots, \omega) \qquad (4.1.1)$$

where $\mu_\alpha^0(p, T)$ is the chemical potential of the pure component of mole fraction x_α at the same pressure and temperature as the mixture being studied.

The conditions under which a mixture becomes ideal in the sense (4.1.1) will be discussed in Section 4.2. All solutions tend towards ideal behaviour when they are sufficiently dilute. (Of course, it does not necessarily follow that for the given set of thermodynamic coordinates (p, T) each of the *pure* systems would necessarily exist, and this poses some difficulty at high pressures and low temperatures. However, for the range of classical liquid mixtures with which we shall be concerned in this chapter this difficulty is unlikely to arise.)

The free energy of such an ideal mixture follows as:

$$A(p, T; n_\alpha, \ldots, n_\omega) = \sum_{\lambda=\alpha}^{\omega} n_\lambda \mu_\lambda^0(p, T) + RT \sum_{\lambda=\alpha}^{\omega} n_\lambda \ln x_\lambda \qquad (4.1.2)$$

whilst for the entropy

$$S(p, T; n_\alpha, \ldots, n_\omega) = \sum_{\lambda=\alpha}^{\omega} n_\lambda S_\lambda^0(p, T) - R \sum_{\lambda=\alpha}^{\omega} n_\lambda \ln x_\lambda. \qquad (4.1.3)$$

The superscript 0 refers in each case to the molar property of the pure components at the same (p, T) as the mixture.

We see quite clearly in each of (4.1.2), (4.1.3) that A and S, as we anticipated earlier, are not simply linear molar sums of the pure component functions but are supplemented by the logarithmic ideal *free energy of mixing* and *entropy of mixing*, respectively, which amounts to the difference between the mixed and unmixed pure components. Obviously the ideal free energy of mixing will always be negative and the free entropy of mixing positive. The formation of an ideal mixture from its pure components is therefore spontaneous and irreversible. Whether, below a certain temperature, azeotropic segregation of the components becomes energetically advantageous does not concern us here. However, assuming the bulk liquid and vapour phases have adopted their lowest free energies, it follows that the contribution of the *interphasal surface region* will be minimized by surface migration of the species having the lowest surface tension, or free energy per unit area, at that temperature. Of course, such surface migration is subject to the overall constraints (4.1.2, 4.1.3) governing segregation. It is clear then that in the interfacial zone we are concerned with the surface adsorption and partitioning of the various component species, and we may anticipate that the surface tension of multicomponent systems will vary not only with temperature, but, of course, with molar composition as well. Two distinct surface adsorption processes need to be considered. The first, *physical adsorption,* arises in systems exhibiting no chemical interaction or bonding, such as mixtures of van der Waals fluids. It is with these that we shall be concerned in this chapter. *Chemisorption,* on the other hand, involves chemical activity between the molecular species adsorbed at the surface.[4] The difference in the heats of adsorption for the two processes is considerable, being ~0–20 kJ per mole of adsorbate for physical adsorption, whilst values ranging from 80–400 kJ per mole are typical for chemisorption, reflecting the development of chemical bonding.

We have already encountered expressions for the total differential of surface excess Helmholtz free energy[1–3] $A_s(T, V, \mathscr{A}; n_\alpha, n_\beta, \ldots, n_\nu)$ due to infinitesimal changes of the independent variables

$$dA_s = -S_s \, dT + \gamma \, d\mathscr{A} + \sum_\alpha^\nu \mu_\lambda \, dn_{\lambda_s} \qquad (4.1.4)$$

where n_{λ_s} is the surface excess number density of component λ and μ_λ is the (constant) chemical potential of the species. For an isothermal extension of the surface we have, integrating (4.1.4) holding constant the intensive properties T, μ_λ, and γ,

$$A_s = \gamma\mathcal{A} + \sum_\alpha^\nu \mu_\lambda n_{\lambda_s} \tag{4.1.5}$$

or, dividing through by $\mathcal{A}$

$$a_s = \gamma + \sum_\alpha^\nu \mu_\lambda \Gamma_\lambda. \tag{4.1.6}$$

Γ_λ are surface excess concentrations. For a single-component system, or if there were no physical adsorption, we recover equation (1.4.5): $A_s = \gamma\mathcal{A}$.

We may differentiate (4.1.5) to obtain

$$d\mathcal{A}_s = \gamma\,d\mathcal{A} + \mathcal{A}\,d\gamma + \sum_\alpha^\nu \mu_\lambda\,dn_{\lambda_s} + \sum_\alpha^\nu n_{\lambda_s}\,d\mu_\lambda. \tag{4.1.7}$$

But we see from (4.1.4) that at constant temperature (4.1.7) reduces to

$$-d\gamma = \sum_\alpha^\nu \Gamma_\lambda\,d\mu_\lambda \tag{4.1.8}$$

which is the general form of the Gibbs adsorption equation. The above expression can conveniently be written in terms of absolute activities Λ since $\mu_\lambda = \mu_\lambda^0 + NkT \ln \Lambda_\lambda$:

$$-\frac{1}{NkT}\,d\gamma = \sum_\alpha^\nu \Gamma_\lambda\,d(\ln \Lambda_\lambda). \tag{4.1.9}$$

These important equations relate surface concentration or excess of a species to the depression in the surface tension $d\gamma$ and the change in the bulk activity $d\mu_\lambda$ of the adsorbate. For a binary system, for example, we have

$$-d\gamma = \Gamma_1\,d\mu_1 + \Gamma_2\,d\mu_2 \tag{4.1.10}$$

and we may locate the dividing surface such that $\Gamma_1 = 0$ (Figure 4.1.1),

$$\Gamma_1 = 0 = \mathcal{A}^{-1}\int_{-\infty}^0 (\rho^L - \rho_{(1)}(z))_1\,dz - \int_0^\infty (\rho_{(1)}(z) - \rho_V)_1\,dz \tag{4.1.11}$$

whereupon (4.1.10) reduces to $-d\gamma = \Gamma_2^{(1)}\,d\mu_2$. $\Gamma_2^{(1)}$ is defined as the *relative adsorption*—the preferential adsorption of species 2 relative to species 1. With the same origin of coordinates as in (4.1.11):

$$\Gamma_2^{(1)} = \mathcal{A}^{-1}\int_{-\infty}^0 (\rho^L - \rho_{(1)}(z))_2\,dz - \int_0^\infty (\rho_{(1)}(z) - \rho^V)_2\,dz.$$

Rearranging (4.1.10) in terms of the relative adsorption we have

$$-\frac{d\gamma}{d\mu_2} = \Gamma_2^{(1)}. \tag{4.1.12a}$$

118

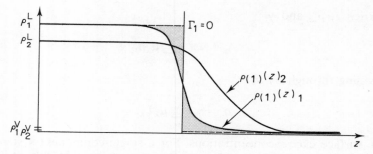

Figure 4.1.1 Partial component transition profiles for a binary mixture with the dividing surface located such that the superficial excess of component 1 vanishes

Now, for small concentrations, μ_2 is proportional to x_2 and (4.1.12a) relates the change of surface tension with bulk concentration to the relative adsorption of species 1 with respect to species 2, $\Gamma_2^{(1)}$. More generally, the surface tension of a multicomponent system may be said to have assumed an *extreme value*, including a maximum, minimum, or point of inflection, if an isothermal displacement from the state under consideration yields $(\delta\gamma)_T = 0$ for all possible variations in the set of component chemical potentials, $\delta\mu_i$:

$$(\delta\gamma)_T = -\sum_i \Gamma_i^{(1)}\delta\mu_i. \tag{4.1.12b}$$

For this to be the case it is necessary and sufficient that all $\Gamma_i^{(1)} \equiv 0$, that is, all relative adsorptions are zero, the surface concentrations of each species being identical to its bulk value.

It is readily verified that the quantity

$$\Gamma_2^{(1)} = \Gamma_2 - \frac{x_2}{1-x_2}\Gamma_1 \tag{4.1.13}$$

is invariant to the location of the dividing surface. x_2 denotes the mole fraction of species 2 in the liquid phase: $x_i = n_i/\sum_i n_i$. More generally we may locate the dividing surface such that the superficial number density of a given species, say α, vanishes. The superficial density of the remaining species defined in this way relative to species α is denoted $\Gamma_\lambda^{(\alpha)}$, and instead of (4.1.8) we have

$$-d\gamma = \sum_{\lambda=\beta}^{\omega} \Gamma_\lambda^{(\alpha)} \, d\mu_\lambda. \tag{4.1.14}$$

Provided the density of the vapour phases is sufficiently low that the partial pressures obey the ideal gas law, the chemical potential may be written

$$\mu_\lambda = \phi_\lambda + kT \ln p_\lambda \tag{4.1.15}$$

where ϕ_λ is a function of temperature only and p_λ is the partial vapour pressure of species λ. In that case, (4.1.14) becomes

$$-d\gamma = kT \sum_{\lambda=\beta}^{\omega} \Gamma_\lambda^{(\alpha)} \, d\ln p_\lambda. \qquad (4.1.16)$$

Evidently $\Gamma_\lambda^{(\alpha)}$ is positive or negative according as γ decreases or increases with bulk concentration of species λ, keeping the other species at fixed concentration. The above expression applies to imperfect systems provided p is then interpreted as a fugacity.

4.2   Ideal mixtures †

Our starting point is, as usual, the configurational integral for the mixture, which by means of a straightforward generalization from a pure substance becomes

$$Z_Q = \frac{1}{\prod\limits_{\lambda=\alpha}^{\omega} N_\lambda} \int \cdots \int \exp\left(-\Phi_N/kT\right) \, d\mathbf{r}_1 \cdots d\mathbf{r}_N$$

where

$$\sum_{\lambda=\alpha}^{\omega} N_\lambda = N.$$

Already an important distinction between single and multicomponent systems has arisen in that Φ_N will depend not only upon the configuration of the N particles, *but also upon the assignment of species to those locations*. This introduces an enormous complication into the statistical theory of mixtures, but fortunately one which is of little practical importance since we are primarily concerned with thermodynamic *excess* functions of the mixture relative to the component species. Nonetheless, even this limited objective may be obtained only for certain forms of interaction and in conjunction with further approximations, as we shall detail below.

We consider two molecular systems in the same configurational state whose N constituent particles of type A and of type B are sufficiently similar in size and shape that they are interchangeable between the two configurations without disturbing either configuration (Figure 4.2.1). Clearly this becomes progressively more difficult for strongly dissimilar species at high densities. The strength of the interactions, however, are assumed to be considerably different. If, for simplicity, we neglect interactions between non-nearest neighbours then the total interaction energy of a representative

---

† We should alert the reader to various uses of the words *ideal* and *perfect* which arise in the context of the description of mixtures in the literature. In this book we shall describe a mixture as being *ideal* when its chemical potential is defined by (4.1.1), and as being *perfect* when ideal at all concentrations.

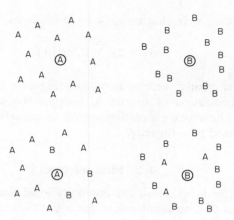

Figure 4.2.1

molecule for the two pure systems is

$$\Phi_N = -z(\Phi_{AA} + \Phi_{BB}) \qquad \Phi_{AA} \neq \Phi_{BB} \qquad (4.2.1)$$

where z is the coordination number in either fluid and Φ_{AA}, Φ_{BB} is the interaction energy between two AA and two BB particles, respectively. Suppose now we exchange $n(\leq z)$ nearest neighbour particles between the two systems. The total interaction energy then becomes

$$\begin{aligned}
\Phi'_N &= -(z-n)(\Phi_{AA} + \Phi_{BB}) - 2n\Phi_{AB} \\
&= -z(\Phi_{AA} + \Phi_{BB}) - n(2\Phi_{AB} - \Phi_{AA} - \Phi_{BB}) \qquad (4.2.2) \\
&= -z(\Phi_{AA} + \Phi_{BB}) - w
\end{aligned}$$

where w defines an *interchange energy*. An ideal solution is defined as one for which $w = 0$, which requires

$$\Phi_{AB} = \frac{\Phi_{AA} + \Phi_{BB}}{2}. \qquad (4.2.3)$$

That is, the interaction energy of an AB pair is the arithmetic mean of the AA and BB pair energies. Moreover, the total energy of the system remains unchanged. Although we have restricted the discussion to nearest neighbours and have assumed all interaction energies between particles of the same species to be equal, it is clear that provided we consider two identical molecular configurations for systems A and B then we shall always obtain condition (4.2.3) defining ideality. The arithmetic mean approximation for Φ_{AB} (4.2.3) has been investigated theoretically for a variety of simple liquid systems[5] and appears to be a reasonable approximation provided the two molecular types are sufficiently alike in size and shape to be able to pack in the same manner when packed as in the pure liquids.

It follows from (4.2.2) that if exchanging particles involves no energy of interchange, then the energy of a mixed system is simply

$$\frac{N_A}{2}\Phi_{AA} + \frac{N_B}{2}\Phi_{BB} \qquad (N_A + N_B = N) \qquad (4.2.4)$$

regardless of how the two kinds of molecules are distributed. The number of distinguishable ways of distributing the N_A and N_B particles within the configuration is

$$\frac{N!}{N_A! \, N_B!} \qquad (4.2.5)$$

whereupon we may write for the configurational partition function

$$Z_Q = \frac{N!}{N_A! \, N_B!} \exp\{(N_A\Phi_{AA} + N_B\Phi_{BB})/2kT\}. \qquad (4.2.6)$$

The free energy of the mixed system follows immediately as[3]

$$A = -kT \ln Z_Q = -\tfrac{1}{2}(N_A\Phi_{AA} + N_B\Phi_{BB}) - kT \ln (N!/N_A! \, N_B!)$$

$$= -\tfrac{1}{2}(N_A\Phi_{AA} + N_B\Phi_{BB}) + kT\{N_A \ln (N_A/N) + N_B \ln (N_B/N)\} \qquad (4.2.7)$$

$$= -\frac{N}{2}\{(1-x)\Phi_{AA} + x\Phi_{BB}\} + NkT\{(1-x) \ln (1-x) + x \ln x\} \qquad (4.2.8)$$

where we have used Stirling's approximation and set $N_A = N(1-x)$, $N_B = Nx$ as we may for a binary system. The first term represents the free energy of the two pure isolated systems A and B whilse the second term

$$\Delta A_m = NkT\{(1-x) \ln (1-x) + x \ln x\} \qquad (4.2.9)$$

represents the free energy of mixing. Clearly both x and $(1-x)$ are <1, in which case the free energy of mixing is negative which we understand as arising from the *increase* in the entropy of mixing, ΔS_m. Differentiation of (4.2.9) with respect to temperature yields

$$\Delta S_m = -\left(\frac{\partial(\Delta A_m)}{\partial T}\right) = -Nk\{(1-x) \ln (1-x) + x \ln x\}. \qquad (4.2.10)$$

It follows directly from (4.2.6) that the chemical potential per molecule of species A in the mixture is

$$\mu_A = \left(\frac{\partial A}{\partial[N(1-x)]}\right) = -\frac{\Phi_{AA}}{2} + kT \ln \frac{N_A}{N} = -\frac{\Phi_{AA}}{2} + kT \ln (1-x) \qquad (4.2.11)$$

where the excess associated with mixing is

$$\mu_A - \mu_A^0 = kT \ln (1-x) \qquad (4.2.12)$$

where μ_A^0 is the chemical potential of the pure system A ($x=0$). Equation

(4.2.11) is often taken as the working definition of an ideal mixture,[15] as we have seen (4.1.1).

If the vapour mixture may be regarded as a perfect gas then the partial pressures will be the molar fractions of the 'pure' pressures:

$$\left.\begin{aligned} p_A &= p_A^0(1-x) \\ p_B &= p_B^0 x \\ p &= p_A + p_B \end{aligned}\right\} \qquad (4.2.13)$$

which are statements of Raoult's Law, and have been shown to be approximately correct for mixtures of very similar systems.[6] Raoult's law remains formally correct for imperfect systems if p is interpreted instead as a fugacity.

The above results are readily adapted to multicomponent systems, whereupon

$$\left.\begin{aligned} \Delta A_m &= NkT \sum_\lambda x_\lambda \ln x_\lambda \\ \Delta S_m &= -Nk \sum_\lambda x_\lambda \ln x_\lambda \\ \mu_\lambda - \mu_\lambda^0 &= kT \ln x_\lambda \\ p_\lambda &= p_\lambda^0 x \end{aligned}\right\}. \qquad (4.2.14)$$

We are now in a position to determine the surface tension of such an ideal solution solely on the basis of a knowledge of the surface tensions of the pure components and the respective molar fractions.

The procedure, first developed by Defay, Prigogine, Bellemans, and Everett,[1] consists in equating the bulk and surface chemical potentials which, of course, characterizes both the total and the partial equilibrium of the system.

The bulk chemical potential of component λ in an ideal solution is, of course, equal to the work done in 'charging up' a particle of species λ in the bulk:

$$\mu_\lambda = \mu_\lambda^0 + NkT \ln x_\lambda \qquad (4.2.15)$$

where μ_λ^0 is the chemical potential of the pure component. If we now regard the surface interfacial zone as a monolayer on either side of which bulk properties prevail, we may write for the surface chemical potential

$$\mu_{\lambda_s} = \mu_{\lambda_s}^0 + NkT \ln x_{\lambda_s} - \gamma a_\lambda. \qquad (4.2.16)$$

The extra term $-\gamma a_\lambda$ arises from the mechanical work done by surface tension in charging up a particle of species λ in the monolayer. a_λ is the partial molar area occupied within the monolayer by the fully charged up λ component. Since, for equilibrium, we have $\mu_\lambda = \mu_{\lambda_s}$ then from (4.2.15) and

(4.2.16) we have

$$\mu_\lambda^0 - \mu_{\lambda_s}^0 = NkT \ln \frac{x_\lambda}{x_{\lambda_s}} - \gamma a_\lambda. \tag{4.2.17}$$

And, since

$$\mu_\lambda^0 - \mu_{\lambda_s}^0 = -\gamma_\lambda^0 a_\lambda \tag{4.2.18}$$

where γ_λ^0 is the surface tension of the pure component, we have at once

$$\gamma = \gamma_\lambda^0 + \frac{NkT}{a_\lambda} \ln \frac{x_{\lambda_s}}{x_\lambda}. \tag{4.2.19}$$

For a binary system

$$\gamma = \gamma_A^0 + \frac{NkT}{a_A} \ln \frac{x_{A_s}}{x_A}$$

$$= \gamma_B^0 + \frac{NkT}{a_B} \ln \frac{x_{B_s}}{x_B}. \tag{4.2.20}$$

If $a_A = a_B = a$ which is approximately the case for perfect solutions, then we have for the adsorbed surface molar fractions

$$x_{A_s} = x_A \exp\left[a(\gamma - \gamma_A^0)/NkT\right]$$

$$x_{B_s} = x_B \exp\left[a(\gamma - \gamma_B^0)/NkT\right] \tag{4.2.21}$$

These equations are due to Schuchowitsky.[17] If a, γ_A^0 and γ_B^0 are known, the surface composition may be determined as a function of the bulk molar constitution. x_A and x_B are not independent of course, but are related as

$$x_A = (1-x)$$

$$x_B = x. \tag{4.2.22}$$

Clearly, preferential adsorption of the species of lower surface tension will occur, as we anticipated from the outset. Equation (4.2.21) may be rearranged into a more convenient form

$$\exp\left(-\frac{\gamma a}{kT}\right) = (1-x)\exp\left(-\frac{\gamma_A^0 a}{kT}\right) + x\exp\left(-\frac{\gamma_B^0 a}{kT}\right) \tag{4.2.23}$$

first proposed by Guggenheim.[3] In particular, for an equimolecular mixture of such a binary system ($x = \frac{1}{2}$), (4.2.23) may be written

$$\gamma = \frac{\gamma_A^0 + \gamma_B^0}{2} - \frac{kT}{a} \ln \cosh\left(\frac{\gamma_B^0 - \gamma_A^0}{2kT}\right). \tag{4.2.24}$$

Guggenheim approximates this expression, making the observation that for pairs of liquids forming ideal mixtures $(\gamma_B^0 - \gamma_A^0)/\frac{1}{2}(\gamma_A^0 + \gamma_B^0)$ is unlikely to exceed 0.1, whereupon ln cosh is replaced by its leading term in a power series:

$$\gamma = \frac{\gamma_A^0 + \gamma_B^0}{2} - \frac{(\gamma_B^0 - \gamma_A^0)^2 a}{8kT}. \tag{4.2.25}$$

Belton and Evans[7] have compared the predictions of (4.2.24) with experiment for a variety of binary systems and the results are seen from Table (4.2.1) to be in good agreement.

4.3 Surface tension of mixtures of monomeric molecules of different sizes

As usual, the specification of the surface excess free energy is central to any analysis of the interfacial region between two bulk phases, except that in the present case we are confronted with two additional problems. Firstly, the calculation of the entropy of the mixture, that is, finding the number of ways of arranging the molecules in space, taking due account of their shape and volume, and secondly the calculation of the configurational energy, which involves the number and nature of intermolecular contacts within the transition region. Of course, the two problems are closely interrelated, but provided the interaction energies of the component species are not too dissimilar, it has been shown that they may be treated independently. A very closely related problem concerns the surface adsorption of polymer chains in a solvent: we shall return to its discussion in Chapter 8.

Flory[18] has considered a highly simplified model in which the sites of a regular crystalline half-lattice are populated by single molecules of types A and B, representing the liquid mixture, whilst the vapour phase is entirely empty. The density discontinuity is infinitely sharp.

Table 4.2.1. Experimental and calculated surface tensions for binary liquid mixtures having $x = \frac{1}{2}$. (From Defay, et al.[1] after Belton and Evans[7]).

Mixture	$T\,°K$	$a\,(\text{Å}^2)$	γ_A	γ_B	γ expt.	γ calc.
			(dyn/cm)		(dyn/cm)	(dyn/cm)
benzene + m-xylene	291.2	37.7	28.40	28.40	28.40	28.40
methanol + ethanol	273	22.35	23.64	23.09	23.40	23.37
,,	303	22.84	21.06	20.76	20.91	20.91
,,	323	23.07	19.45	19.20	19.24	19.34
$H_2O + D_2O$	298	11.2	72.06	70.91	71.48	71.48
m-xylene + o-xylene	293	41.3	28.90	30.17	29.20	29.52
m-xylene + p-xylene	293	41.7	28.90	28.37	28.50	28.63
chlorobenzene + bromobenzene	293	37.0	33.11	36.60	34.65	34.86

If N_A and N_B represent the numbers of molecules of the two species in the solution, then the total number of configurations is

$$g(N_A, N_B) = \frac{(N_A + N_B)!}{N_A! \, N_B! \, (N_A + N_B)} \tag{4.3.1}$$

whilst the configurational energy is

$$\Phi_N = \Phi_{N_A}^0 + \Phi_{N_B}^0 + \frac{\alpha}{\mathcal{N}} \frac{N_A N_B}{N_A + N_B} \tag{4.3.2}$$

$\mathcal{N}$ is Avogadro's number. $\Phi_{N_A}^0$ represents the configurational energy of N_A atoms of pure A; similarly $\Phi_{N_B}^0$ for pure B, whilst

$$\alpha = \mathcal{N} z [\varepsilon_{AB} - \tfrac{1}{2}(\varepsilon_{AA} + \varepsilon_{BB})] \tag{4.3.3}$$

where z is the lattice coordination number and ε_{AB}, ε_{AA}, and ε_{BB} are the interaction parameters developed between the various species. From (4.3.1), (4.3.2) we may calculate the free energy, and hence by differentiation

$$\begin{aligned}
\mu_A(p, T) &= \mu_A^0(p, T) + RT \ln x_A + \alpha (x_B)^2 \\
\mu_B(p, T) &= \mu_B^0(p, T) + RT \ln x_B + \alpha (x_A)^2
\end{aligned} \tag{4.3.4}$$

which represent the chemical potential in the solution. The corresponding expressions at the surface are, for an athermal solution ($\alpha = 0$),

$$\begin{aligned}
\mu_{A_s}(p, T) &= \mu_{A_s}^0(p, T) + RT \ln x_{A_s} - \gamma a \\
\mu_{B_s}(p, T) &= \mu_{B_s}^0(p, T) + RT \ln x_{B_s} - \gamma a
\end{aligned} \tag{4.3.5}$$

where γ is the surface tension of the mixture and a is the surface area per mole. x_{A_s}, x_{B_s} represent the surface molar fraction of species A and B, respectively. Equating the bulk and surface chemical potentials (4.3.4, 4.3.5), we have at once

$$\begin{aligned}
\gamma &= \gamma_A^0 + \frac{RT}{a} \ln (x_{A_s}/x_A) \\
\gamma &= \gamma_B^0 + \frac{RT}{a} \ln (x_{B_s}/x_B)
\end{aligned} \tag{4.3.6}$$

where γ^0 represents the surface tension of the pure component. The composition of the surface and the bulk may be related by subtracting one equation from the other in (4.3.6):

$$\frac{x_{A_s}}{x_A} = \left(\frac{x_{B_s}}{x_B} \right) \exp [(\gamma_B - \gamma_A)a/RT]. \tag{4.3.7}$$

Alternatively, multiplying (4.3.6) by x_A and x_B respectively, and adding, we obtain the adsorption dependence

$$\gamma = (x_A \gamma_A^0 + x_B \gamma_B^0) + \frac{RT}{a} \left[x_A \ln \frac{x_{A_s}}{x_A} + x_B \ln \frac{x_{B_s}}{x_B} \right] \tag{4.3.8}$$

which reduces to the linear relationship $\gamma = (x_A \gamma_A^0 + x_B \gamma_B^0)$ for small surface adsorptions—x_{A_s}/x_A, $x_{B_s}/x_B \sim 1$. Clearly, if these ratios are to be close to unity, it follows that the exponent in (4.3.7) must be $\ll 1$, whereupon it is straightforward to show, to first order,

$$\gamma = (x_A \gamma_A^0 + x_B \gamma_B^0) - \frac{1}{2}\left(\frac{x_A x_B}{x_A + x_B}\right)\frac{(\gamma_B - \gamma_A)^2 a}{RT} \qquad (4.3.9)$$

from which we see that the surface adsorption contributes *negatively* to the surface tension roughly as $(\gamma_B - \gamma_A)^2$, having its maximum value when

$$x_A = \tfrac{1}{2}.$$

If now we extend the analysis to non-athermal solutions, we have for the interfacial chemical potentials of the two components

$$\mu_{A_s} = \mu_{A_s}^0(p, T) + RT \ln x_{A_s} + \alpha l(x_{B_s})^2 + \alpha m (x_B)^2 - \gamma a$$
$$\mu_{B_s} = \mu_{B_s}^0(p, T) + RT \ln x_{B_s} + \alpha l(x_{A_s})^2 + am (x_A)^2 - \gamma a \qquad (4.3.10)$$

where, in this lattice model, each site has z nearest neighbours of which lz are in the same plane and mz are in the two adjacent planes, and of course $2m + l = 1$. Again, equating chemical potentials of each species in the bulk and surface phases, we have

$$\gamma = \gamma_A^0 + \frac{RT}{a}\ln\left(\frac{x_{A_s}}{x_A}\right) + \frac{\alpha}{a}\{l(x_{B_s})^2 - (l+m)(x_B)^2\}$$
$$\gamma = \gamma_B^0 + \frac{RT}{a}\ln\left(\frac{x_{B_s}}{x_B}\right) + \frac{\alpha}{a}\{l(x_{A_s})^2 - (l+m)(x_A)^2\}. \qquad (4.3.11)$$

Again, if $(\gamma_B - \gamma_A)a/RT$ and α/RT are both $\ll 1$, we may obtain the following development of (4.3.9):

$$\gamma = x_A \gamma_A^0 + x_B \gamma_B^0 - \frac{\alpha m}{a} x_A x_B - \frac{1}{2}\frac{x_A x_B}{x_A + x_B}$$

$$\times \left[(\gamma_B - \gamma_A) + \frac{\alpha m}{a}(x_B - x_A)\right]^2 \frac{a}{RT}. \qquad (4.3.12)$$

Since we have shown that an extremum in the surface tension occurs when the relative adsorptions are zero (4.1.12b), we require $x_{A_s} = x_A$, $x_{B_s} = x_B$, whereupon from (4.3.11)

$$\gamma_{extr} = \gamma_A^0 - \frac{m\alpha}{a}(x_B)^2$$

$$= \gamma_B^0 - \frac{m\alpha}{a}(x_A)^2.$$

Eliminating γ_{extr} yields

$$x_A - x_B = \frac{(\gamma_B - \gamma_A)a}{m\alpha}$$

whereupon

$$-1 \leqslant \frac{(\gamma_B - \gamma_A)a}{m\alpha} \leqslant 1$$

since $x_A - x_B$ must, of course, lie between these extremes.

4.4 Multicomponent transition profiles: the BGY equation

There has been very little attention devoted to the statistical mechanical description of the transition profile in classical, simple, nonmetallic multicomponent systems. The profiles would, of course, be expected to exhibit features of surface migration in accordance with the Gibbs adsorption theorem (4.1.8), and for an ideal binary solution should reproduce the qualitative features of Guggenheim's equation (4.2.23). Of course, in the derivation of the latter equation surface modification of the bulk molar compositions was confined to an interfacial monolayer, whilst presumably a statistical mechanical analysis would yield a family of partial profiles making immediate comparison rather more difficult. Nevertheless, we shall establish a set of linked partial integro-differential equations of Born–Green–Yvon form for each of the components, and identify the thermodynamic 'forces' operating on the various component species in the vicinity of the transition zone.

Suppose the bulk fluid is composed of species $\alpha, \beta, \ldots, \omega$ present in number densities $\rho_{L\alpha}, \rho_{L\beta}, \ldots, \rho_{L\omega}$ such that $\rho_L = \sum_{\lambda=\alpha}^{\omega} \rho_{L\lambda}$. It is then straightforward to write down the single-particle BGY equation for the λth partial profile[8,9]

$$kT \nabla_\lambda \rho_{(1)}(z_\lambda) = - \sum_{\mu=\alpha}^{\omega} \int \nabla_\lambda \Phi_{\lambda\mu}(r_{\lambda\mu}) \rho_{(2)}^\dagger(z_\lambda, \mathbf{r}_{\lambda\mu}) \, d\mathbf{r}_\mu. \qquad (4.4.1)$$

Here, $\rho_{(1)}(z_\lambda)$ represents the single-particle distribution of species λ, whilst $\Phi_{\lambda\mu}(r_{\lambda\mu})$ and $\rho_{(2)}^\dagger(z_\lambda, \mathbf{r}_{\lambda\mu})$ represent the pair potential and the *anisotropic* two-particle density distribution developed between particles of species λ and μ respectively, in the presence of the other species $\neq \lambda, \mu$. For $\Phi_{\lambda\mu}$ we may adopt the arithmetic mean

$$\Phi_{\lambda\mu} = \frac{\Phi_{\lambda\lambda} + \Phi_{\mu\mu}}{2} \qquad (4.4.2)$$

which, of course, is the ideal mixture condition (Section 4.2). For $\rho_{(2)}^\dagger(z_\lambda, \mathbf{r}_{\lambda\mu})$ we may write the exact closure

$$\rho_{(2)}^\dagger(z_\lambda, \mathbf{r}_{\lambda\mu}) = \rho_{(1)}(z_\lambda) \rho_{(1)}(z_\mu) g_{(2)}^\dagger(z_\lambda, \mathbf{r}_{\lambda\mu}). \qquad (4.4.3)$$

In the absence of any explicit knowledge of the two-component radial distribution $g_{(2)}^\dagger(z_\lambda, \mathbf{r}_{\lambda\mu})$ we may, following Green,[16] assume

$$\rho_{(2)}^\dagger(z_\lambda, \mathbf{r}_{\lambda\mu}) = \rho_{(1)}(z_\lambda) \rho_{(1)}(z_\mu) g_{(2)}^{\dagger L}(r_{\lambda\mu}) \qquad (4.4.4)$$

that is, adopt the *isotropic* bulk liquid radial distribution $g_{(2)}^{\dagger L}(r_{\lambda\mu})$ of the species μ about λ in the presence of the remaining species. We should emphasize that the adoption of $g_{(2)}^{\dagger L}(r_{\lambda\mu})$ in (4.4.4) neglects the functional dependence of the pair distribution upon the local number densities $\rho_{(1)}(z_\lambda)$ and $\rho_{(1)}(z_\mu)$, and to that extent is an approximation additional to that of Green. There are, however, more fundamental objections to this closure (Section 2.2). Alternatively, we may adopt the more realistic midpoint closure (2.4.24):

$$g_{(2)}^{\dagger}(z_\lambda, \mathbf{r}_{\lambda\mu}) = 0.5 g_{(2)}^{\dagger}[r_{\lambda\mu} \,|\, \rho_{(1)}^{*}(z_\lambda)] + 0.5 g_{(2)}^{\dagger}[r_{\lambda\mu} \,|\, \rho_{(1)}^{*}(z_\mu)] \qquad (4.4.5)$$

where $\rho_{(1)}^{*}(z_\lambda)$ and $\rho_{(1)}^{*}(z_\mu)$ relate to the $g_{(2)}(r_{\lambda\mu})$ radial distributions appropriate to the local density and composition at z_λ and z_μ. Clearly this is only one of many possible model closures, but, in addition to angularity, it does possess the essential feature of symmetry with respect to particle exchange which is of central importance in multicomponent systems. Indeed, violation of this condition has been at the basis of our criticism of certain of the earlier closure devices (Section 2.8).

For simplicity we shall continue with the closure of (4.4.4), although there is no formal difficulty associated with the use of more complex closures such as (4.4.5). Insertion of (4.4.4) in (4.4.1) yields, in a simplified but obvious notation,

$$kT\frac{\nabla_\lambda \rho_{(1)}(z_\lambda)}{\rho_{(1)}(z_\lambda)} = -\sum_{\mu=\alpha}^{\omega} \int \nabla_\lambda \Phi(r_{\lambda\mu})\rho_{(1)}(z_\mu) g_{(2)}^{\dagger L}(r_{\lambda\mu})\,d\mathbf{r}_\mu. \qquad (4.4.6)$$

Integrating subject to the partial boundary condition $\rho_{(1)}(-\infty_\lambda) = \rho_{L\lambda}$ we have for the λ profile

$$\rho_{(1)}(z_\lambda) = \rho_{L\lambda} \exp\left\{-\frac{1}{kT}\int_{-\infty}^{z_\lambda} \sum_{\mu=\alpha}^{\omega} \int \nabla_\lambda \Phi(r_{\lambda\mu})\rho_{(1)}(z_\mu) g_{(2)}^{\dagger L}(r_{\lambda\mu})\,d\mathbf{r}_\mu\,dz_\mu\right\}. \tag{4.4.7}$$

Theoretical distributions $g_{(2)}^{\dagger L}(r_{\lambda\mu})$ for binary Lennard–Jones systems based on the 'ideal' approximation (4.4.2) for the pair interaction are generally available,[5] and will be discussed in the following section.

The specific difficulty which arises in the solution of equations (4.4.6) or (4.4.7) is the cross-dependence of the profile $\rho_{(1)}(z_\lambda)$ upon all the other unknown partial profiles $\rho_{(1)}(z_\mu)$. A possible, if tedious, approach might be to solve the $\alpha, \ldots, \omega$ coupled equations by initially adopting partial transition profiles of step function form according to

$$\begin{aligned}\rho_{(1)}(z_\mu) &= \rho_{L\mu} & z \leq 0 \\ &= 0 & z > 0\end{aligned} \tag{4.4.8}$$

and then determine the selfconsistent distributions $\rho_{(1)}(z_\alpha), \ldots, \rho_{(1)}(z_\omega)$ in a similar manner to that proposed for the ionic and electronic partial distributions at the surface of a liquid metal (Section 5.3).

More formally, we may express (4.4.7) in matrix form. Setting

$$\hat{\boldsymbol{\rho}} = \begin{bmatrix} -kT \ln \rho_{(1)}(z_1)_\alpha \\ \cdot \\ \cdot \\ \cdot \\ -kT \ln \rho_{(1)}(z_1)_\omega \end{bmatrix}$$

the square symmetric matrix

$$\boldsymbol{\kappa} = \begin{bmatrix} \nabla_\alpha \Phi(r_{\alpha\alpha}) g_{(2)}^{\dagger L}(r_{\alpha\alpha}) & \cdots & \nabla_\alpha \Phi(r_{\alpha\omega}) g_{(2)}^{\dagger L}(r_{\alpha\omega}) \\ \cdot & & \cdot \\ \cdot & & \cdot \\ \cdot & & \cdot \\ \nabla_\omega \Phi(r_{\omega\alpha}) g_{(2)}^{\dagger L}(r_{\omega\alpha}) & \cdots & \nabla_\omega \Phi(r_{\omega\omega}) g_{(2)}^{\dagger L}(r_{\omega\omega}) \end{bmatrix}$$

and

$$\boldsymbol{\rho} = \begin{bmatrix} \rho_{(1)}(z_2)_\alpha \\ \cdot \\ \cdot \\ \cdot \\ \rho_{(1)}(z_2)_\omega \end{bmatrix}$$

then (4.4.7) may be written

$$\hat{\boldsymbol{\rho}} = \int_{-\infty}^{z_1} \int_2 \boldsymbol{\kappa}\boldsymbol{\rho} \, dz \, dz_1. \tag{4.4.9}$$

A useful discussion of the solution of simultaneous, nonlinear integral equations of the above form in the present context has been given by Plesner, Platz, and Christiansen.[10]

The structural features of the partial profiles may be discussed in terms of partial surface constraining fields, just as for single-component systems. Indeed, the partial constraining field appropriate to the component λ is, from (4.4.7),

$$\sum_{\mu=\alpha}^{\omega} \int \nabla_\lambda \Phi(r_{\lambda\mu}) \rho_{(1)}(z_2)_\mu g_{(2)}^{\dagger L}(r_{\lambda\mu}) \, d\mathbf{r}_\mu \tag{4.4.10}$$

in the present closure approximation. Whether this surface field drives the component λ towards or away from the interface underlies the process of positive or negative surface adsorption. If the surface were discontinuously sharp, then a surface migration of those components tending to reduce the surface excess free energy would be apparent at *all* molar fractions, and conversely for components tending to increase the surface free energy. In fact, of course, softer partial constraining fields prevail and the development or otherwise of surface partial densities greater than those in the bulk depends sensitively upon the detailed features of the functions appearing in (4.4.10). Whilst necessarily $\sum_{\mu=\alpha}^{\omega} \rho_{(1)}(-\infty)_\mu = \rho_{L\mu}$ for reasons of stability, the partial profiles may show preferential surface adsorption of one species with respect to another, and in consequence we anticipate 'nonclassical' thermal dependences of the surface excess functions, in particular the surface tension–temperature characteristic. As we have already observed from the Gibbs adsorption equation (4.1.8), dilute solutions almost always have surface tensions which differ from that of the pure solvent. Moreover, if surface adsorption or surface migration of the solute occurs, the surface tension *decreases*, whilst if it *increases* we conclude that the desorption of the solute is occurring with a migration away from the surface into the bulk of the liquid mixture. This has an important bearing upon the interpretation of experimental surface tension–temperature data for 'pure' liquids containing traces of impurity, as they inevitably do. We shall return to this point in the discussion of the surface tension of liquid metals (Section 5.4) where it will have considerable significance in relation to the 'inverted' form of the $\gamma(T)$ curve reported for a number of pure metallic systems.

Attempts have been made to verify the Gibbs equation, with somewhat inconclusive results. One of the earliest successful measurements was that of McBain *et al.*[11] who removed a very thin monolayer (~ 0.1 mm) from the surface of the solution and determined the difference in concentration of the solute between this and the bulk solution, thereby permitting a calculation of the surface excess. More recently, isotopically labelled molecules have been used in a direct measurement of surface adsorption. Again, although not very accurate, the results appear to confirm the Gibbs equation.

The extension of the single-component mechanical definition (1.4.17) of surface tension to multicomponent systems is readily made. For example, (1.4.24) becomes

$$\gamma = \frac{1}{2} \int_{-\infty}^{\infty} \sum_{\kappa=\alpha}^{\omega} \sum_{\lambda=\alpha}^{\omega} \int \nabla_\kappa \Phi(r_{\kappa\lambda}) \rho_{(2)}(z_\kappa, \mathbf{r}_{\kappa\lambda}) \left(\frac{x_{\kappa\lambda}^2 - z_{\kappa\lambda}^2}{r_{\kappa\lambda}} \right) d\mathbf{r}_{\kappa\lambda} \, dz_\kappa$$

$$(4.4.11)$$

where we draw attention to the fact that $\rho_{(2)}(z_\kappa, \mathbf{r}_{\kappa\lambda}) = \rho_{(1)}(z_k)\rho_{(1)}(z_\lambda)g_{(2)}(z_k, \mathbf{r}_{\kappa\lambda})$ is a function of temperature, density, *and composition*. For a binary system we require a knowledge of $g_{(2)}^{(\kappa,\kappa)}$, $g_{(2)}^{(\lambda,\lambda)}$, and $g_{(2)}^{(\kappa,\lambda)} (= g_{(2)}^{(\lambda,\kappa)}$ as is easily shown from a number balance of each component— the distributions must be symmetrical with respect to particle exchange). These functions will be discussed in more detail in Section 4.6.

4.5 Multicomponent transition profiles: constancy of the chemical potential

It has been demonstrated that the mechanical or hydrostatic stability expressed in the single-particle Born–Green–Yvon or the Kirkwood equation is formally consistent with constancy of the chemical potential throughout the transition zone and into each of the bulk phases (Section 2.6). Constancy of the *partial* chemical potentials of the component species of a multicomponent system is, of course, a further necessary condition for the development of stable partial profiles, and Plesner, Platz, and Christiansen[10] have extended the method of Section 2.4 in order to discuss a binary argon–nitrogen mixture at 84 °K for a range of mole fractions.

The model used is identical to that of Section 2.4, complicated only by the presence of two species. For the pair interaction Plesner *et al.* take the Sutherland potential (i, j refer to species throughout)

$$\Phi(r_{ij}) = -(\varepsilon_i\varepsilon_j)^{1/2}\left(\frac{\sigma_i\sigma_j}{r_{ij}}\right)^6 \qquad r \geq (\sigma_i\sigma_j)^{1/2}$$

$$= +\infty \qquad r < (\sigma_i\sigma_j)^{1/2} \qquad (4.5.1)$$

and the (inconsistent) radial distributions

$$g_{(2)}(r_{ij}) = 1 \qquad r \geq (\sigma_i\sigma_j)^{1/2}$$

$$= 0 \qquad r < (\sigma_i\sigma_j)^{1/2}. \qquad (4.5.2)$$

The geometric mean representation of the interaction energy $\varepsilon_{ij} = \varepsilon_i^{1/2}\varepsilon_j^{1/2}$ was first used by Berthelot in 1898. Theory offers little justification for its use, although the mean arises naturally in London's theory of dispersion forces in mixtures: its real justification is its empirical success, and has been closely investigated for Percus–Yevick mixtures by Throop and Bearman.[5] As in the single-component case, the above simplification ensures from the outset that only the gross structural features of the transition profiles will develop. Nevertheless, we proceed as in Section 2.4 to express the local chemical potential as the sum of a local 'entropy' term based upon the point density $\rho_{(1)}(z)$ and a 'potential energy' term which is coupled through the long range attractive branch to the inhomogeneous regions of the density profile. We anticipate that this term will be consistently underestimated on account of the simplification of the radial distribution function (4.5.2), particularly on the liquid side of the transition zone where the more highly developed correlation in the fluid is expected to govern the features of the density profile.

Adopting the equation of state of Lebowitz *et al.* for a hard sphere mixture, and supplementing the result with a potential term to account for the van der Waals tail (equation 4.5.1), the local chemical potential of species i in a binary mixture becomes

$$\ln\frac{\eta_i(z)}{(1-\alpha_3(z))} + \frac{\alpha_0(z)+3\alpha_1(z)+3\alpha_2(z)}{(1-\alpha_3(z))} + \frac{\frac{9}{2}\alpha_2^2(z)+3\alpha_1(z)\alpha_2(z)}{(1-\alpha_3(z))^2}$$

$$+ \frac{3\alpha_2^3(z)}{(1-\alpha_3(z))^3} - \Psi_i[\eta_i(z), \eta_j(z)] = \mu_i(z) = \text{constant} \qquad (4.5.3)$$

where

$$\left.\begin{array}{l}\alpha_0(z) = \eta_i(z)\theta^{-3} + \eta_j(z)\theta^{-3} \\ \alpha_1(z) = \eta_i(z)\theta^{-3} + \eta_j(z)\theta^{-1} \\ \alpha_2(z) = \eta_i(z)\theta^{-3} + \eta_j(z)\theta \\ \alpha_3(z) = \eta_i(z)\theta^{-3} + \eta_j(z)\theta^3 \\ \theta = (\sigma_j/\sigma_i)^{1/2} \\ \eta_i(z) = \tfrac{1}{6}\pi(\sigma_i\sigma_j)^{1/2}\rho_i(z) \\ \rho_i(z) = n_i(z)/V \end{array}\right\} \tag{4.5.4}$$

and the interaction energy of a molecule of type i at z with the rest of the fluid is (c.f. equation 2.4.6)

$$kT\Psi_i(z) = \int \rho_{(1)i}(z+z_{12})g_{(2)}(r)_{ii}\Phi(r)_{ii}\,d\mathbf{r} + \int \rho_{(1)j}(z+z_{12})g_{(2)}(r)_{ij}\Phi(r)_{ij}\,d\mathbf{r} \tag{4.5.5}$$

with similar expressions for $\mu_i(z)$ and Ψ_j obtained by interchanging η_i with η_j and substituting θ for θ^{-1}. The two equations for $\mu_i(z)$ and $\mu_j(z)$ form two simultaneous nonlinear integral equations to be solved for the densities of the two components. For details of the numerical solution of these equations the reader is referred to reference 10.

Having obtained the partial density profiles it is then a straightforward matter to calculate the surface tension, surface of tension, location of the surfaces at which the superficial densities Γ_i, Γ_j vanish, etc. For example, using the mechanical definition

$$\gamma = \int_{-\infty}^{\infty} [p - p_\parallel(z)]\,dz \tag{4.5.6}$$

and with the assumed equation of state

$$\frac{p_\parallel(z)}{kT} = \frac{6\alpha_0(z)}{1-\alpha_3(z)} + \frac{18\alpha_1(z)\alpha_2(z)}{(1-\alpha_3(z))^2} + \frac{18\alpha_2^3(z)}{(1-\alpha_3(z))^3} + 3\theta^{-3}[\Psi_i(z)\eta_i(z) + \Psi_j(z)\eta_j(z)]. \tag{4.5.7}$$

Plesner et al.[10] are able to calculate the surface tension of the binary mixture at all molar fractions.

The partial density profiles for an argon–nitrogen mixture at 84 °K are shown in Figure 4.5.1 over the entire composition range. Notice surface adsorption of N_2 for molar fractions of argon $x_{Ar} \geqslant 0.6$ arising as a consequence of the stronger interaction parameters of argon, despite the similar atomic diameters of the two species. At higher molar fractions of argon this species would be expected to remain preferentially in the bulk phase where it may take advantage of the larger number of neighbours with which it can interact. Consequently, the surface tension will be depressed below a simple

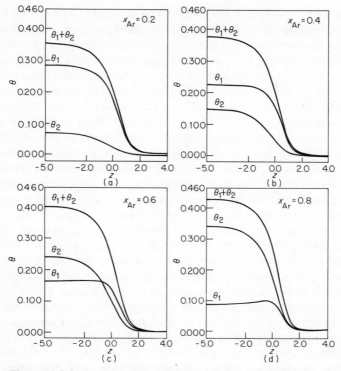

Figure 4.5.1 Density transition curves for AR–N_2 mixtures.
(a) $X_{Ar} = 0.2$; (b) $X_{Ar} = 0.4$; (c) $X_{Ar} = 0.6$; (d) $X_{Ar} = 0.8$

linear variation with bulk molar fraction between γ_{Ar} and γ_{N_2} on account of the modification of the surface composition. A plot of the excess surface tension defined by

$$\gamma_{ex} = \gamma - \gamma_{Ar}x_{Ar} - \gamma_{N_2}x_{N_2} \qquad (4.5.8)$$

where $x_{N_2} = 1 - x_{Ar}$ is shown in Figure 4.5.2, together with the experimental values.[13]

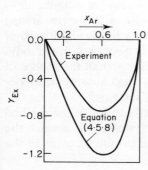

Figure 4.5.2 Excess surface tension as a function of mole fraction of argon. $T = 84\,°K$

Of course, these calculations are open to precisely the same objections as those discussed in Section 2.4: the inconsistent and inadequate radial distribution functions which, moreover, show no dependence upon composition, nor do they reflect the essentially anisotropic nature of the two-particle function in the anisotropic surface region. Although the 'potential' terms $\Psi_i(z)$, $\Psi_j(z)$ in the expression for the chemical potential are coupled through the attractive branch of the interaction to the structurally inhomogeneous regions of the fluid, the 'entropy' term is not, being a point function in the most restricted sense. The subdivision of the transition zone into elemental strata each of which constitutes a stable thermodynamic entity is clearly mechanically unstable since the (isotropic) pressure in each element ensures that $p_\perp(z) = p_\|(z)$ rather than the bulk phase pressure p.

Unfortunately, the drastic approximations embodied in the theory restrict conclusions to purely qualitative observations. Although the calculated surface tensions as a function of mole fraction are in reasonable agreement with experiment, their numerical coincidence is an insensitive criterion for the assessment of the calculated partial profiles. We do observe that the profiles bear comparison with the computer simulations of Chapela et al.[12] (Figure 4.5.3) to be discussed in more detail in Chapter 10.

The two molecular species in the equimolar binary mixture were chosen such that the ratio of the Lennard–Jones interaction parameters $(\varepsilon_{aa}/\varepsilon_{bb}) = 0.763$, with a geometric mean cross energy $\varepsilon_{ab} = (\varepsilon_{aa}\varepsilon_{bb})^{1/2}$. It is quite clear from Figure 4.5.3 that the more volatile component is preferentially adsorbed, and extends about one atomic diameter into the vapour phase, agreeing well with the profiles of Plesner, Platz, and Christiansen.

4.6 A pseudopotential approach to multicomponent transition profiles

In the specification of the partial profile of the λth component, say, we may regard the system as an *effectively* single-component fluid in which the λ particles interact through a pseudopotential arising from the screening of the λ particle by the other species.[14] If we designate the screened particle by λ^*, then it is evident from the outset that the total configurational energy of this system of λ^* particles will comprise a 'self-energy' term arising from the interaction between the λ particle and its non-λ screening neighbours—anisotropically screened if necessary—and a pairwise $\lambda^*\lambda^*$ term arising from the interaction of two screened λ particles. Clearly this is nothing other than a reformulation of the more familiar approach of Section 4.4, embodying a simultaneous knowledge of all the partial profiles and all the modes of species–species pair interaction. (The approach is analogous to that described in Section 5.3 for ion–electron liquid metal systems.)

For simplicity we shall consider a binary system AB, although the method bears straightforward extension to multicomponent systems. We assume the profile $\rho_{(1;B)}(z)$ is known: we seek to determine the profile $\rho_{(1;A)}(z)$ in the 'external' field of B particles.

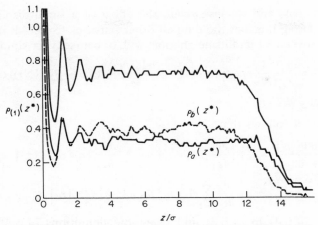

Figure 4.5.3 Density profile $\rho_{(1)}(z)$ for an equimolar binary total mixture at equivalent temperatures $T_a^* = 0.918$, $T_b^* = 0.701$. The component profiles are shown by the broken lines

The energy of the system in the configuration $\{N\}$ is

$$\Phi(\{N\}) = \sum_i \phi[\rho_{(1;B)}(z_i)] + \frac{1}{2}\sum_{i>j} \Phi^*[r_{ij} \mid \rho_{(1;B)}(z_i)] + \Phi^*[r_{ij} \mid \rho_{(1;B)}(z_j)]$$

(4.6.1)

where the first term represents the self energy of an A particle at z_i where the local B density is $\rho_{(1;B)}(z_i)$, and the second term represents the pseudo-potential developed between two differentially screened A particles at separation r_{ij}. Attention is drawn to the fact that (4.6.1) is an essentially 'local' expression of the configurational energy being expressed as a functional only of the point densities at z_i, z_j when in fact the A particles are coupled through the long-range part of the pair potential to the inhomogeneity of the transition zone, although (4.6.1) reduces correctly to its bulk form as $z_{i,j} \to \pm\infty$. Since, in general, the A and B profiles will not coincide, there should be an additional surface term arising between the interaction of a surface layer of (say) type B particles with the inhomogeneous mixture beneath. We shall neglect this contribution.

The formal expressions for the one- and two-particle configurational projections of species A are

$$\left.\begin{aligned} \rho_{(1;A)}(\mathbf{r}_1) &= \frac{1}{Z_Q(N-1)!} \int \cdots \int \exp\left(-\beta\Phi(\{N\})\right) d2 \cdots dN \\ \rho_{(2;A)}(\mathbf{r}_{12}) &= \frac{1}{Z_Q(N-2)!} \int \cdots \int \exp\left(-\beta\Phi(\{N\})\right) d3 \cdots dN \end{aligned}\right\}.$$

(4.6.2)

We assume, as usual, that the single-particle density inhomogeneity arises along the z-axis, and we investigate the effect of a small displacement of particle 1_A along this axis, holding all other particles fixed. All interparticle distances between 1_A and its neighbours will, of course, vary simultaneously:

$$\frac{\partial \rho_{(1;A)}(z_1)}{\partial z_1} = \frac{1}{Z_Q(N-1)!} \int \cdots \int \left(-\beta \frac{\partial \Phi(\{N\})}{\partial z_1}\right) \exp\left(-\beta \Phi(\{N\})\right) \mathrm{d}\mathbf{2} \cdots \mathrm{d}\mathbf{N}$$

$$\beta = (kT)^{-1}. \tag{4.6.3}$$

From (4.6.1) it follows that

$$-\beta \frac{\partial \Phi(\{N\})}{\partial z_1} = -\beta \left(\frac{\partial \phi}{\partial z_1} \rho_{(1;B)}(z_1) + \frac{1}{2} \sum_{j=2}^{N} \frac{\partial \Phi^*}{\partial z_1} [r_{ij} \mid \rho_{(1;B)}(z)]\right.$$

$$\left. + \frac{1}{2} \frac{\partial}{\partial \mathbf{r}_1} \sum_{j=2}^{N} \Phi^*[r_{ij} \mid \rho_{(1;B)}(z_1)] + \Phi^*[r_{ij} \mid \rho_{(1;B)}(z_j)]\right) \tag{4.6.4}$$

Inserting (4.6.4) in (4.6.3) and using the definitions (4.6.2) we finally obtain after some manipulation and transformation into polar coordinates:

$$-kT \frac{\partial \rho_{(1;A)}(z_1)}{\partial z_1} = \rho_{(1;A)}(z_1) \frac{\mathrm{d}}{\mathrm{d}z_1} \phi[\rho_{(1;B)}(z_1)]$$

$$+ \frac{1}{2} \int \frac{\partial \Phi^*}{\partial z_1} [r_{12} \mid \rho_{(1;B)}(z_1)] \rho_{(2;A)}(z_1, \mathbf{r}_{12}) \, \mathrm{d}\mathbf{r}_{12}$$

$$- \frac{1}{2} \int \frac{z_{12}}{\mathbf{r}_{12}} \left\{ \frac{\partial \Phi^*}{\partial r_{12}} [r_{12} \mid \rho_{(1;B)}(z_1)] \right.$$

$$\left. + \frac{\partial \Phi^*}{\partial r_{12}} [r_{12} \mid \rho_{(1;B)}(z_2)] \right\} \rho_{(2;A)}(z_1, \mathbf{r}_{12}) \, \mathrm{d}\mathbf{r}_{12} \tag{4.6.5}$$

which is, of course, the two-component analogue of (4.4.6), both equations being generically of BGY form. We should perhaps mention once again that equation (4.6.5) represents the response of the partial profile $\rho_{(1;A)}(z)$ to a *specified* 'external' field of B particles: the dependence is implicitly contained in the functions ϕ and Φ^*. A precisely analogous expression for $\rho_{(1;B)}(z)$ may be written down simply by exchanging the subscripts A and B in (4.6.5). For a sufficiently dilute system such that the B profile is virtually identical to that of the pure component, the A profile may be obtained directly from (4.6.5).

The force on a particle A located at z_1 is, quite simply,

$$\frac{kT}{\rho_{(1;A)}(z_1)} \frac{\partial \rho_{(1;A)}(z_1)}{\partial z_1} = -\frac{\mathrm{d}\phi}{\mathrm{d}z_1} [\rho_{(1;B)}(z_1)] - \frac{1}{2} \int \frac{\partial \Phi^*}{\partial z_1} [r_{12} \mid \rho_{(1;B)}(z_1)]$$

$$\times \rho_{(1;A)}(z_2) g_{(2;A)}(z_1, \mathbf{r}_{12}) \, \mathrm{d}\mathbf{r}_{12}$$

$$+ \frac{1}{2} \int \frac{z_{12}}{\mathbf{r}_{12}} \left\{ \frac{\partial \Phi^*}{\partial r_{12}} [r_{12} \mid \rho_{(1;B)}(z_1)] \right.$$

$$\left. + \frac{\partial \Phi^*}{\partial r_{12}} [r_{12} \mid \rho_{(1;B)}(z_2)] \right\} \rho_{(1;A)}(z_2) g_{(2;A)}(z_1, \mathbf{r}_{12}) \, \mathrm{d}\mathbf{r}_{12}. \tag{4.6.6}$$

The usual force term arising from the structural anisotropy at the liquid surface is supplemented by two others arising from the z-dependence of the ϕ and Φ^* functions. In the case of a pure component, of course, the self-energy term does not arise, and the pseudopotential Φ^* reverts to the conventional z-independent pair potential Φ. Under these circumstances equations (4.6.5) and (4.6.6) reduce to single-component form.

From the mechanical stress tensor formulation of surface tension, it follows immediately from (4.6.6) that

$$
\gamma = -\int_{-\infty}^{\infty} \frac{\mathrm{d}\phi}{\mathrm{d}z_1}[\rho_{(1;B)}(z_1)]\rho_{(1;A)}(z_1)z_1\,\mathrm{d}z_1 - \frac{1}{2}\int_{-\infty}^{\infty}\int \frac{\partial\Phi^*}{\partial z_1}[r_{12}\,|\,\rho_{(1;B)}(z_1)]
$$
$$
\times \rho_{(2;A)}(z_1,\mathbf{r}_{12})z_1\,\mathrm{d}\mathbf{r}_{12}\,\mathrm{d}z_1 + \frac{1}{2}\int_{-\infty}^{\infty}\int \frac{(x_{12}^2-z_{12}^2)}{r_{12}}
$$
$$
\times \frac{\partial\Phi^*}{\partial r_{12}}[r_{12}\,|\,\rho_{(1;B)}(z_1)]\rho_{(2;A)}(z_1,\mathbf{r}_{12})\,\mathrm{d}\mathbf{r}_{12}\,\mathrm{d}z_1 \tag{4.6.7}
$$

(where $\rho_{(1;A)}$ and $\rho_{(2;A)}$ are the one- and two-particle A distributions, respectively) whilst for the surface energy of the mixture

$$
U_s = \int_{-\infty}^{0} (\phi[\rho_{(1;B)}(z_1)]\rho_{(1;A)}(z_1) - \phi[\rho_{B_L}]\rho_{A_L})\,\mathrm{d}z_1
$$
$$
+ \int_{0}^{\infty} (\phi[\rho_{(1;B)}(z_1)]\rho_{(1;A)}(z_1) - \phi[\rho_{B_V}]\rho_{A_L})\,\mathrm{d}z_1
$$
$$
+ \frac{1}{2}\int_{-\infty}^{0}\int \{\Phi^*[r_{12}\,|\,\rho_{(1;B)}(z_1)]\rho_{(2;A)}(z_1,\mathbf{r}_{12})
$$
$$
- \Phi^*[r_{12}\,|\,\rho_{B_L}]\rho_{L_A}^2 g_{(2;A)}(r_{12})\}\,\mathrm{d}\mathbf{r}_{12}\,\mathrm{d}z_1
$$
$$
+ \frac{1}{2}\int_{0}^{\infty}\int \{\Phi^*[r_{12}\,|\,\rho_{(1;B)}(z_1)]\rho_{(2;A)}(z_1,\mathbf{r}_{12})
$$
$$
- \Phi^*[r_{12}\,|\,\rho_{B_v}]\rho_{V_A}^2 g_{(2;A)}(r_{12})\}\,\mathrm{d}\mathbf{r}_{12}\,\mathrm{d}z_1 \tag{4.6.8}
$$

where ρ_{A_L}, ρ_{A_V} represent the bulk liquid and vapour densities for species A, respectively. Similarly for species B. Equation (4.6.8) again reduces to classical form for a single-component system.

It does remain, of course, to specify the self energy $\phi[\rho_{(1;B)}(z)]$ and the pseudopotential $\Phi^*[r_{12}\,|\,\rho_{(1;B)}(z)]$. We consider a binary mixture AB of density $\rho = \rho_A + \rho_B$, and molar composition x_A, x_B where $x_A = \rho_A/\rho$, $x_B = (1-x_A) = \rho_B/\rho$. In such a system the self-energy of an A particle located at z where the local density is $\rho_{(1;B)}(z)$, and the appropriate pair density is $\rho_{(2;AB)}(r_{AB})$

$$
\phi[\rho_{(1;B)}(z)] = \int \Phi(r_{AB})\rho_{(2;AB)}(r_{AB})\,\mathrm{d}\mathbf{r}_{AB} \tag{4.6.9}
$$

whilst if, in such a system, the *bulk* AA radial distribution function is

$g_{(2;A)}(r_{AA})$, then the pseudopotential $\Phi^*(r_{AA})$ is defined as

$$\Phi^*(r_{AA}) = -kT \ln g_{(2;A)}(r_{AA}). \qquad (4.6.10)$$

Notice that we have adopted *isotropic* local distributions in these specifications: anisotropic functions would be more appropriate, but also more cumbersome.

Of particular interest is the temperature-dependence of such a binary mixture. Substitution of (4.6.7) and (4.6.8) into the Gibts–Helmholtz equation (1.4.9)

$$U_s = \gamma - \frac{T \partial \gamma}{\partial T}$$

and integration by parts yields

$$T\frac{\partial \gamma}{\partial T} = \int_{-\infty}^{\infty} \phi[\rho_{(1;B)}(z_1)] \frac{\partial \rho_{(1;A)}(z_1)}{\partial z_1} z_1 \, dz_1 + \text{pairwise contributions.}$$

$$(4.6.11)$$

Such an expression is extremely interesting since it shows that the surface excess entropy $-\partial \gamma/\partial T$ is sensitively dependent upon the gradient of the gradient $\partial \rho_{(1;A)}(z_1)/\partial z_1$. For conventional monotonically decreasing profiles, of course, the surface excess entropy is positive, and the first term in (4.6.11) contributes accordingly. But for systems which, under certain thermodynamic conditions of temperature and molar composition show preferential surface migration (see Figure 4.5.1), the $\gamma(T)$ characteristic may well show *positive* slopes with subsequent inversion to the familiar monotonic decreasing form. Indeed, for strongly adsorbed species only small molar fractions may be necessary to modify and even invert the slope of the surface tension–temperature curve. Such results have been reported for multicomponent systems, for alloys, and even for certain pure metals in which physical adsorption of impurities may be responsible for the positive slopes on the basis of such a mechanism described above.

References

For a general discussion of the surface thermodynamics of liquid mixtures see
1. R. Defay, I. Prigogine, A. Bellemans, and D. H. Everett, *Surface Tension and Adsorption*, Longmans, London (1966).
2. R. Aveyard and D. A. Haydon, *An Introduction to the Principles of Surface Chemistry*, Cambridge University Press (1973), Section 3.6.
3. E. A. Guggenheim, *Mixtures*, Oxford University Press (1952), Ch. 9.
4. D. O. Hayward and G. M. W. Trapnell, *Chemisorption*, 2nd Edition, Butterworths, London (1964).
5. G. J. Throop and R. J. Bearman, *J. Chem. Phys.*, **44,** 1423 (1966).
6. von Zawidgki, *Z. physikal. Chem.* **35,** 129 (1900).
7. J. W. Belton and M. G. Evans, *Trans. Faraday Soc.*, **41,** 1 (1945).
8. C. A. Croxton, *Phys. Lett.*, **60A,** 215 (1977).
9. C. A. Croxton (Ed.), In *Progress in Liquid Physics* Wiley, London (1978), Ch. 2.

10. I. W. Plesner, O. Platz, and S. E. Christiansen, *J. Chem. Phys.*, **48,** 5364 (1968).
11. J. W. McBain and C. W. Humphreys, *J. Phys. Chem.*, **36,** 300 (1932).
    J. W. McBain and R. C. Swain, *Proc. Roy. Soc.*, **A154,** 608 (1936).
12. G. A. Chapela, G. Saville, and J. S. Rowlinson, *Chem. Soc. Faraday Disc. No.* 59 (1975).
13. F. B. Sprow and J. M. Prauznitz, *Trans. Faraday Soc.*, **62,** 1105 (1966).
14. C. A. Croxton, *Mol. Phys.* (in press) (1980).
15. J. S. Rowlinson, *Liquids and Liquid Mixtures,* Butterworths, London (1969), p. 111.
16. H. S. Green, *Hand. Phys.*, **10,** 79 (1960).
17. A. Schuchowitsky, *Acta Physiochem. URSS*, **19,** 176 (1944).
18. P. J. Flory, *J. Chem. Phys.*, **9,** 660 (1941).

CHAPTER 5

The Liquid Metal Surface

5.1 Introduction

Despite extensive theoretical and technological interest in the surface tension and surface energy of liquid metals and their alloys, a rigorous statistical mechanical description remains largely undeveloped in anything other than purely formal terms, although some progress has been made for solid metal surfaces, and whilst we cannot depend too heavily upon these results they nevertheless provide a basis for comparison and assessment of our exploratory investigations of the mobile metallic liquid–vapour interface. And, of course, we do have a reasonably sound understanding of the simple dielectric liquid surface. The specific difficulties which arise at the metallic liquid surface may be best illustrated perhaps by making a brief comparison between the liquid metals and their simple liquid counterparts.

We have already seen in Section 2.11 that the amplitudes of thermally excited waves are generally comparable to or larger than the interatomic spacing, and would be expected to wash out any structural features of the interface. Certainly suppression of thermal oscillation in the vicinity of a smooth wall, or by application of boundary conditions (see Chapter 10), appears to result in the development of a structured transition zone. Macroscopic theory of thermally excited capillary waves applied to the description of the liquid argon surface at the triple point suggests that the r.m.s. amplitude normal to the surface is greater than an atomic diameter (Chapter 2, References 75, 76); the interfacial thickness may be attributed, in substantial part at least, to the thermal motion of the molecules.

Consider, now, a liquid metal. We recognize from the outset that the liquid metal system is essentially a two-component assembly: mobile positive ions and conduction electrons which are to a greater or lesser extent 'free' and which belong to the metal as a whole. Such a description of bulk liquid metals was introduced long ago by Kirkwood and his school, and has been subsequently developed by a number of workers.[1] Clearly, at the liquid metal surface we require the specification of both the ionic and the electronic distributions. The presence of a near discontinuity in the distribution of positive charge at the surface will induce oscillations in the essentially non-classical electron density near the surface. These two points will be elaborated considerably in the following sections. Since the ionic charges are mobile the positive charge distribution will respond to the excess electron density, and as Croxton[30] has anticipated, the mutually selfconsistent ion and electron distributions may both have oscillations in the transition region near the surface. Ionic screening by the electrons will, of course, make all interactions density dependent, and the ionic and electron distributions will generally be expected to be of different amplitudes and to differ from metal to metal. The precise nature of the ionic profile arises as a 'zero-point' contribution basically attributable to the nonzero wavelength of the electrons, their influence on the effective ion–ion potential, and the ionic response to the electronic distribution convolved with a thermal contribution.

The surface energies and surface tensions of liquid metal systems are invariably very much greater than for the dielectric fluids—the surface tensions ranging from $\sim 70 \, dyn \, cm^{-1}$ for Cs to $\sim 2700 \, dyn \, cm^{-1}$ for Re at their melting points.[22] Consequently it is concluded that the amplitudes of the thermally excited surface waves on a liquid metal are relatively small, less than an ionic diameter (e.g. the r.m.s. amplitude of motion normal to the Hg surface at $298 \, °K$ is $1.4 \, Å$,[58]), and that the oscillatory surface structure survives the capillary wave disruption (see, however, Section 2.11).

It is on this basis, together with some careful experiments of White on the surface tensions of liquid metals, that the possibility of low-entropy surface states associated with more highly ordered or structured interfacial regions may develop, with the consequence of surface tension–temperature characteristics having negative slopes in certain cases. Such observations have been made, and their existence must be accounted for in any theoretical description of the liquid metal surface.[2]

We should, perhaps, emphasize that the surface energetics of a liquid metal will differ substantially from those of its classical counterpart, and we may expect the consequences to be reflected in the thermodynamic functions of the surface. For example, the electronic and ionic distributions $\rho_{(1)}^{-}(z)$ and $\rho_{(1)}^{+}(z)$ cannot be expected to coincide (although overall charge neutrality must, of course, be preserved) and we therefore anticipate electrostatic double layer contributions to the surface energy. (Indeed, Frenkel[37] once attributed the entire surface energy of liquid mercury to this electrostatic contribution. His result, $472 \, dyn \, cm^{-1}$, is in excellent agreement with modern determinations!) The 'self-energy' term arising from the Coulombic or Debye–Hückel interaction of an ion with its neutralizing cloud of conduction electrons will evidently depend upon the location of the ion within the system—that is, will depend upon the local electron density, and so arises as one of the components of the surface excess energy which has no classical counterpart. Again, the oscillatory ion–ion interaction term is itself a density-dependent function, and its variation from bulk to surface will be reflected in contributions to the surface excess energy of the system. Moreover, the electronic excitations arising in the form of plasmon modes may be expected to differ from bulk to surface, constituting a further contribution to the surface excess energy which is specific to liquid metal systems.

It should be pointed out, however, that a distinction has to be drawn between nearly free electron metals, that is those with an s–p conduction band well-separated from the d states, and the transition liquid metal systems for which the filling of the d shell is of primary importance. Considerable progress has been made for the former systems, although some qualitative discussion may be given of the liquid transition metals based on tight binding arguments, as we shall see below.

5.2   The electron density profile $\rho_{(1)}^-(z)$ for nearly-free electrons

The purely electronic analyses[26] have concentrated on specifying how the uniform bulk electronic density modifies in the vicinity of the surface, and how they leak out of a solid ionic jellium at $T = 0$ when, of course, surface excess entropy contributions do not arise and $\gamma = u_s$. Bardeen[3] first suggested that as a useful initial approximation we may ask how the bulk electron density ρ_L^- varies as an infinite barrier model of the surface is approached. The z-component of the electronic wavefunction may be written in the form

$$\Psi_n(z) = (2/L)^{1/2} \sin(n\pi z/L) \qquad (5.2.1)$$

where L is the width of the box with infinite walls. The electron density $\rho_{(1)}^-(z)$ then follows as

$$\rho_{(1)}^-(z) = \sum_n |\Psi_n|^2 = \rho_L^- \left\{ 1 + \frac{3\cos(2k_F z)}{4k_F^2 z^2} - \frac{3\sin(2k_F z)}{8k_F^3 z^3} \right\} \qquad (5.2.2)$$

where $\sum$ is over all occupied states. k_F is the Fermi wave number related to the bulk electronic density in the usual way: $k_F = (3\pi^2 \rho_L^-)^{1/3}$. It is clear from (5.2.2) that Friedel oscillations of characteristic wavelength π/k_F are induced in the homogeneous electron gas in the vicinity of the 'surface'. For low electronic densities such as caesium for which the mean electronic spacing $r_s \sim 5a_0$ ($a_0 = $ Bohr radius) the infinite barrier model seems quite reasonable, but proves quite inadequate for Al, for example, in which $r_s \sim 2a_0$. Such high density systems require the more realistic semi-infinite jellium model in which the ions are replaced by a uniform semi-infinite positive charge density. Such a model has been developed by a number of workers[4] culminating in the fully selfconsistent profile of Lang and Kohn[5] shown in Figure 5.2.1 which again show the Friedel oscillations already present in the Bardeen result (5.2.1), but with the obvious development of an electrostatic double layer. The profile was determined on the basis of a density functional formalism (see Section 2.10) which is a modern version of the Thomas–Fermi method[6] based on the Hohenberg–Kohn–Sham theory of the inhomogeneous electron gas.

The surface energy for the semi-infinite jellium model has four components arising from the surface modification of the single particle kinetic energy, u_{ke} reflecting the fact that the surface electron density is more diffuse, plus contributions from exchange (u_{ex}), correlation (u_c), which

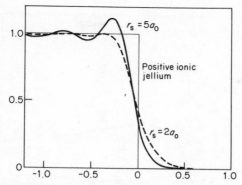

Figure 5.2.1 The electronic density profile $\rho_{\overline{(1)}}(z)$ at the surface of a positive ionic jellium

includes the surface modification of the bulk ion–ion, ion–electron, and electron–electron correlations, and an electrostatic term u_{es}:

$$U_s = u_{\text{ke}} + u_{\text{ex}} + u_c + u_{\text{es}}. \tag{5.2.3}$$

These four contributions are compared as a function of electron density in Figure 5.2.2 for the infinite barrier model and the Lang–Kohn semi-infinite jellium. The two models give quite different results for each of the components, particularly the kinetic, which in fact dominates the overall behaviour of U_s.

The net result of Lang and Kohn's surface energy calculations on the basis of a semi-infinite jellium are shown in Figure 5.2.3. Comparison is made

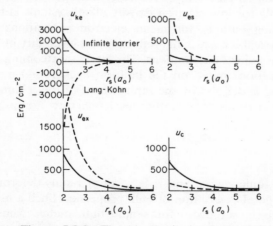

Figure 5.2.2 The kinetic (u_{ke}), electrostatic (u_{es}), exchange (u_{ex}), and correlation (u_c) energies for the infinite barrier and Lang–Kohn semi-infinite jellium models, as a function of mean electronic spacing, $r_s(a_0)$ ($a_0 =$ Bohr radius)

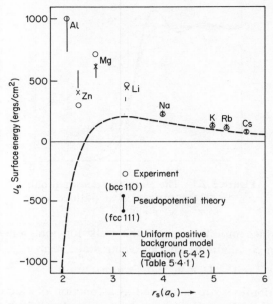

Figure 5.2.3 Net result of Lang and Kohn's surface energy calculation for a semi-infinite jellium as a function of mean electronic spacing, $r_s(a_0)$. Comparison is made with experimental surface tensions extrapolated to 0 °K

with experimental surface tensions linearly extrapolated to absolute zero temperature: only for low electron density alkali systems is the agreement reasonable, failing entirely for mean electron separations $r_s \lesssim 4a_0$, and becoming unphysical for $r_s \lesssim 2.4a_0$ reflecting the instability of the electronic and semi-infinite jellium distributions. This basic shortcoming which characterizes all calculations based on the uniform background model has been rectified by Lang and Kohn by the replacement of the jellium by a regular half lattice of ionic centres,[11] or rather local Ashcroft[10] ion pseudopotentials having the form

$$\left. \begin{array}{ll} \Phi_{ps}(\mathbf{r}) = 0 & r \leqslant r_c \\ \phantom{\Phi_{ps}(\mathbf{r})} = Z/r & r > r_c \end{array} \right\} \tag{5.2.4}$$

where Z is the ionic charge and r_c is a cutoff radius determined for each metal to give a good description of bulk properties. Such a model is known to be quantitatively successful for simple bulk metals whose conduction band has an s–p character·widely separated from d-like states.

Lang and Kohn recalculate the surface energy for regular half bcc and fcc lattices of pseudopotentials (5.2.4) on the basis of the electronic profile $\rho^-_{(1)}(z)$ obtained from the uniform positive background jellium calculation. The difference in surface energy of the two models is therefore entirely due to the differences in electrostatic interaction energies of the jellium and ionic representations of the background charge. There will, of course, be higher-order contributions to the surface energy arising from the response of the (jellium) electronic distribution to the ionic lattice, which Lang and Kohn do not take into account. We shall return to this point later. The results are shown in Figure 5.2.3 and are indicated by a vertical bar, the upper extremity relating to a bcc lattice, the lower to fcc. There is seen to be good semi-quantitative agreement, with particularly good agreement for the lower electron density alkalis. In particular the description at high electron densities, $r_s \lesssim 4a_0$, the inclusion of *discrete* ions is seen to be essential[11] for the accurate calculation of surface energies.

As we mentioned above, Lang and Kohn's calculation is only to first order, and does not involve the inhomogeneous electronic response function χ of $\rho_{(1)}^-(z)_{\text{jellium}}$ to the ionic lattice which is immersed in it. The response function for a *homogeneous* electron gas is better understood, and has been applied to Na, leaving the Lang–Kohn result essentially unmodified. For polyvalent Al however, the correction is large and negative, and Finnis[17] obtains a surface energy of 300 erg cm^{-2} which is to be compared with the experimental value of $\sim$1000 erg cm^{-2} and it would appear that the Lang–Kohn treatment is inadequate for polyvalent systems. More recently a number of other selfconsistent approaches to the calculation of $\rho_{(1)}^-(z)$ at the surface of a metallic crystal have been developed. Their discussion here would take us too far afield; we refer the reader to the literature for further details.[16]

Throughout the calculations Lang and Kohn assumed that the ionic lattice remains undistorted as the surface is approached: allowing the surface plane of ions to relax to a position of lower energy has a negligible effect on the magnitude of the surface energies, although there is some evidence to suggest that contraction occurs at the (110) face of Al[18] although for rare gas solids a detailed calculation appears to indicate that there is a slight expansion of the outermost layer.[19]

It has been suggested by Craig[7] and Schmit and Lucas[8] that the discrepancy between the semi-infinite jellium representation and the extrapolated experimental surface energies could be attributed almost entirely to the change in zero point energy of the plasma oscillations at the free surface with respect to the bulk. The results of Schmit and Lucas are shown in Figure 5.2.3, and appear convincing. We shall not enter into a detailed discussion here however, except to say that their result leads to a surface energy $U_s \propto r_s^{-5/2}$, which when multiplied by their coefficient of proportionality yields the result shown in the figure. The two approaches, that of Lang and Kohn and of Schmit and Lucas, are quite different, and initiated considerable controversy which focused initially on the very different weights accorded to the contribution of the correlation energy in the two procedures: Lang and Kohn found it to be quite large whilst Schmit and Lucas found it to be quite small.

A more recent result by Wikborg and Inglesfield[53] shows that whilst surface modification of plasmon modes does contribute to the surface energy, it is substantially less than that given by the Schmit–Lucas result, whilst others[9] regard the success of earlier analyses[7, 8] as spurious—being attributable to arbitrary cut-off procedures.

More recently, Mahan[12] has performed a variational calculation for the surface energy of the jellium model by introducing a variational parameter λ into the wavefunction (5.2.1): $\Psi_n(z, \lambda)$. The wavefunctions are chosen such that for $\lambda = 1$ we have a semi-infinite jellium, whilst for $\lambda = \infty$ we regain the infinite barrier model. Variations in λ only affected the surface energy and so a minimization of the total energy of the N-particle assembly having two free surfaces of area

$$U_{\text{total}}(\lambda) = 2AU_{\text{s}}(\lambda) + Nu_{\text{b}} \qquad (5.2.5)$$

minimizes the surface energy. u_{b} is the single-particle bulk energy. It turns out that it is the strong λ-dependence of the kinetic term which forces $u_{\text{s}}(\lambda)$ to minimize with $\lambda_0 \sim 1$ (Figure 5.2.4)—very close to the Lang–Kohn semi-infinite jellium result. Mahan omits the contribution of correlation energies in his calculation, but observes that the contribution of $u_{\text{c}}(\lambda)$ is likely to be relatively unimportant in comparison with $u_{\text{ke}}(\lambda)$, and the conclusion that the Lang–Kohn electronic profile is essentially correct remains (Figure 5.2.5). Replacement of the jellium by a discrete ionic lattice would yield an essentially similar conclusion, since the result depends only upon the electronic charge density which is virtually identical in the two cases.

A recent generalization of the method of Lang and Kohn has been reported by Allen and Rice[55] in which the electronic and ionic distributions are selfconsistently adjusted so as to yield a minimum surface energy. The procedure is repeated for each member of a given class of trial ion distributions until a pair of mutually selfconsistent profiles are obtained which represents the 'best' of its class in the sense that it minimizes the surface

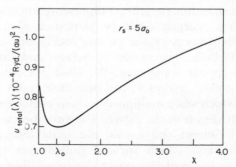

Figure 5.2.4 Variational minimization of the surface energy $U_{\text{total}}(\lambda)$ for the jellium model[12]

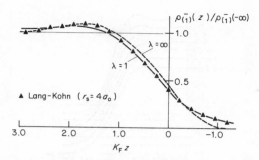

Figure 5.2.5 Comparison of the Lang–Kohn and variationally determined electronic profiles

energy. Various classes of trial function may then be considered. As with the variational principle, for sufficiently flexible trial functions one may come arbitrarily close to the exact result.

The trial ionic distribution is modelled by matching a tanh function† and a first order spherical Bessel function (limited to one full oscillation) in such a way that the sum is smoothly varying as a function of z. The components of the profile are weighted by variational parameters so that an entire family of functions, ranging from the monotonic tanh-like to the oscillatory Bessel-like, may be investigated in one variational operation.

For a given ionic distribution $\rho^+_{(1)}(z)$ the selfconsistent electron density profile $\rho^-_{(1)}(z)$ may then be determined.[55] There does remain one problem, however, which may be attributed directly to the nondiscrete nature of the ionic jellium. On the basis of the calculations of Allen and Rice there is a minimum in the *bulk* ground state energy per particle for $r_s = 4.3a_0$. Suppose we wish to study the surface transition zone for jellium at a density for which $r_s \neq 4.3a_0$. Clearly, if the form of the ion density function permits, there will be a tendency for the density in the surface zone to increase if $r_s > 4.3a_0$ or decrease if $r_s < 4.3a_0$, in an attempt to minimize the bulk component of the energy per particle. Examples are given for $r_s = 3.41a_0$, when the surface energy actually becomes *negative* and the ionic transition zone becomes anomalously diffuse, whilst at $r_s = 5a_0$ an anomalously high density ionic surface layer develops. In a real metal the electron–ion and ion–ion interactions would modify of course, but this is a degree of flexibility which is unavailable in the present calculations. It does, however, indicate a pressing need for a pseudopotential kind of calculation in which the discrete nature of the ions can simultaneously ensure bulk stability whilst providing an adequate description of the interphasal zone.

Allen and Rice[55] circumvent the problem by the *ad hoc* introduction of a stabilizing energy whose magnitude ensures that the total ground state energy per particle has a minimum at the desired bulk value of r_s.

---

† But *not* the same tanh function adopted in the description of classical interfaces.

148

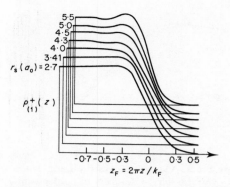

$r_s(a_0) = 2.7$

$\rho_{(1)}^+(z)$

$z_F = 2\pi z / k_F$

Figure 5.2.6 Variationally determined ionic density profiles $\rho_{(1)}^+(z)$ as a function of mean electronic spacing

The surface energy of the system is then determined for a given value of r_s, by variational adjustment of the three-parameter trial function. Variational minimization was not carried out beyond the first cycle since this seemed sufficient for the qualitative characterization of $\rho_{(1)}^+(z)$ in the transition zone. The profiles are shown in Figure 5.2.6 for a variety of densities r_s.

The general findings are as follows:

(i) Mutually selfconsistent electronic distributions are generally non-monotonic closely following the ionic profiles, in conformity with the Hellman–Feynmann theorem,[56] ensuring, virtually, local charge neutrality everywhere. (It follows that electrostatic contributions to the surface energy are likely to be minimal.) The positive charge distributions are, however, monotonic for high electron densities ($r_s < 4.3a_0$, e.g. Hg) and nonmonotonic for low electron densities ($r_s > 4.3a_0$, e.g. K). At $r_s = 5.5a_0$ the magnitude of the charge oscillation is such as to give nearly a 5% density variation with respect to the bulk.

(ii) The widths of the positive profiles, tabulated in Table 5.2.1, show a maximum of $\sim 6.7\,\text{Å}(\sim 2\sigma)$ at $4.5a_0 < r_s < 5a_0$ roughly in agreement with the width of liquid argon at the triple point.

(iii) The monotonic charge distributions differ substantially from the symmetrical hyperbolic tangent form characteristic of the transition zone of argon.

A more detailed analysis of the results is given in Table 5.2.2 where the difference in surface energy between the Lang–Kohn step ionic profile and the relaxed jellium are listed. Of the energy saved in relaxation of the ionic profile, 60–80% of the saving is due to the loss of electrostatic energy which results when the ionic and electronic profiles are nearly matched. The remainder of the saving arises as a result of the competition between the kinetic energy component and the exchange-correlation component.

Table 5.2.1. Width of ionic profiles for various mean electronic spacings.

$r_s(a_0)$	Width (Å)
2.7	3.75
3.41	4.7
4.0	5.9
4.3	6.3
4.5	6.63
5.0	5.36
5.5	5.0

A subsequent refinement by Allen and Rice[57] to incorporate the perturbative introduction of discrete ionic pseudopotentials into the positive jellium, albeit in the form of crystal planes parallel to the surface, appears to be in qualitative agreement with the oscillations found for the selfconsistently relaxed ionic jellium discussed above. In this analysis the surface of a liquid is modelled by close packed planes, formed by cleavages perpendicular to either the fcc (111) or the bcc (110) direction. When $r_s \lesssim 3a_0$ the surface ionic density decreases, reflected by the first plane moving outwards by 0.5% to 1% of the bulk spacing. For $r_s \gtrsim 3a_0$ the first two planes move together by 0.5% to 1.5%, whilst the second and third planes move apart by ~0.3%. This behaviour appears to hold regardless as to whether electronic distributions appropriate to a step or a relaxed ionic profile are used. We should point out that both the relaxed ionic jellium and its discrete ion derivative fail to account for the drastic modification of the ion–ion interaction as the surface is approached, ranging from the familiar bulk oscillatory function to its essentially van der Waals form in the vapour. These conclusions are related to a recent experimental investigation of the surface structure of Hg by Lu and Rice,[58] discussed in detail in Section 5.6.

Finally, an extension of the surface tension/isothermal compressibility relationship yields a qualitative measure of the thickness of the electronic

Table 5.2.2. Differences between Lang–Kohn and relaxed jellium estimates of the surface energy at various mean electronic spacings.

$r_s(a_0)$	$U_s(\text{LK}) - U_s(\text{relaxed jellium})$ (erg cm^{-2})
2.7	360
3.41	167
4.0	85
4.3	53
4.5	37
5.0	24
5.5	8

interface. As we have seen, the product of surface tension and bulk isothermal compressibility $\gamma\chi_T$ provides a measure of surface thickness for a wide variety of liquid systems (equation 1.4.37, Table 1.4.1). We now investigate the analogous relationship for a gas of conduction electrons.[44]

Early attempts to calculate the surface energy per unit area gave[48, 49]

$$\gamma \sim \frac{k_F^2 E_F}{80\pi} \qquad (5.2.6)$$

where E_F and k_F are the Fermi energy and wavenumber, respectively. This expression is in reasonable agreement with experiment for the lightest alkali metals. The simplest expression available for the isothermal compressibility of a liquid metal is that of Bohm and Staver:[50]

$$\chi_T \sim \frac{3}{2\rho_L^- E_F} \qquad (5.2.7)$$

where ρ_L^- is the bulk electron density $k_F^3/3\pi^2$. Their product follows immediately as

$$\gamma\chi_T = \frac{9\pi}{160 k_F} \qquad (5.2.8)$$

which is proportional to the de Broglie wavelength of an electron at the Fermi surface, which for simple metals, at least, is a measure of the thickness of the electronic transition profile at the metal surface. More sophisticated arguments appear to yield essentially the same result.[51]

5.3 The ionic density profile $\rho_{(1)}^+(z)$

Of course, the uniform positive jellium and even its replacement by a discrete half-lattice of pseudo-ions is little more than a caricature of the true ionic distribution prevailing at the surface of a liquid metal. It does appear that it is the *electronic* contributions to the surface energy and surface tension which dominate, primarily through the kinetic term. Indeed, as we observed in the previous section, surface modification of the lattice structure in terms of relocation to its equilibrium position had a negligible effect upon the numerical results, although spatial delocalization of the crystal surface, whilst in itself having little numerical effect, may modify the electronic distribution somewhat, perhaps with substantial consequences for u_{ke}. Clearly further detailed analysis is required and further comment at this stage would be purely speculative. Certainly, once the essential replacement of a jellium by a discrete ionic assembly has been made, the ionic and electronic distributions will selfconsistently adjust, of which some account will be taken below. If, however, we again retain the Lang–Kohn distribution $\rho_{(1)}^-(z)$, presumably the difference between an ordered and a disordered ionic structure will, for the purposes of calculation, be negligible, depending more sensitively upon coordination number than the brand of molecular organization. No attempt will be made here to extend the Lang–Kohn arguments for the electronic profile to disordered discrete ionic systems: the mathematical arguments are too specious, the reliability of the results too uncertain, and their relevance to problems of experimental interest too remote—although preliminary results for three-dimensional bulk disordered metallic systems have recently appeared.

Essentially ionic computations of the thermodynamic parameters of the liquid metal surface have been given by Johnson et al.[13] and more recently by Waseda and Suzuki.[14] These authors work in what is the KB–Fowler step model of the liquid surface adopting pair potentials determined from experimental bulk liquid structure factors. This approach takes no account of the electronic processes mentioned above, and although the surface tensions are in reasonable agreement with experimental determinations, the results must, unfortunately, be considered largely fortuitous. It is known[27] that the immersion of a homogeneous array $\{N\}$ of pseudo-ions into an initially homogeneous conduction electron gas of initially uniform density ρ_L^- yields a system of total energy $\Phi(\{N\})$, where

$$\Phi(\{N\}) = N\varphi[\rho_L^-] + \sum_{i>j} \Phi[r_{ij} \mid \rho_L^-]. \tag{5.3.1}$$

$\varphi[\rho_L^-]$ is the coulombic 'self-energy'[28] of a pseudo-ion interacting with its neutralizing cloud of electrons of local density ρ_L^-, and includes the kinetic, exchange, and correlation energies of the electron gas. $\Phi[r_{ij} \mid \rho_L^-]$ is the interaction energy of two such pseudo-ions at separation r_{ij}. In the vicinity of a free surface the retention of an initially homogeneous electronic distribution would mean that the bulk and surface self-energy terms would be identical and only surface modification of the ionic structure would contribute to the surface energy of the metal: the description of the liquid metal surface would reduce to the classical Kirkwood–Buff form. Indeed, this is precisely the basis of the calculations of Johnson et al.[13] and Waseda and Suzuki[14] outlined at the beginning of this section. There is reason to believe that the self-energy term, sensitively dependent upon the local electron density as it is, is of crucial importance for the calculation of surface energies of liquid metals[17] and is, moreover, one of the principal distinguishing features between the surface energetics of metallic and inert gas systems.

Clearly, equation (5.3.1) needs to be modified in regions of surface structural inhomogeneity in a way which involves both the self-energy and configurational contributions to the total energy. A device which reduces the two-component system of ions and conduction elect s to an effectively single-component system of metallic pseudo-ions follows directly from (5.3.1):[15]

$$\Phi(\{N\}) = \sum_i \phi[\rho_{(1)}^-(z_i)] + \sum_{i>j} [\Phi(r_{ij} \mid \rho_{(1)}^-(z_i)) + \Phi(r_{ij} \mid \rho_{(1)}^-(z_j))]$$

$$+ \text{electrostatic terms} \tag{5.3.2}$$

in which it is implicitly assumed that at the surface both ϕ and Φ exhibit the same functional dependence upon the local conduction electron density as in the bulk metal. In this local density approximation the self-energy ϕ is now a functional of the inhomogeneous electron density $\rho_{(1)}^-(z)$ whilst $\frac{1}{2}\{\Phi[r_{ij} \,|\, \rho_{(1)}^-(z_i)] - \Phi[r_{ij} \,|\, \rho_{(1)}^-(z_j)]\}$ represents the effective pairwise potential developed between two pseudo-ions at a separation r_{ij}. Quite clearly the configurational energy (5.3.2) of the inhomogeneous assembly reduces correctly to the homogeneous bulk liquid expression (5.3.1). However, in the low density vapour phase a nearly free electron representation is quite inappropriate and the total interaction would be better represented as a pairwise sum of short range, density-independent inert gas-like potentials: a metal to insulator transition. This shortcoming will be overlooked in much of what follows since the principal contributions to the surface structure and thermodynamics will be assumed to arise on the liquid side of the dividing surface. Nevertheless, this would appear to be a reasonable first-order representation of the pairwise density-dependent interaction between two pseudo-ions, and is in the same spirit as certain of the anisotropic two-particle closure devices for $g_{(2)}(z_1, \mathbf{r}_{12})$ described in Section 2.4 (2.4.24, for example). It is, however, open to the same objection, namely that the conduction electron density $\rho_{(1)}^-(z)$ should be linear and slowly varying over the range of the pseudopotential, which in this case falls off much more slowly, as r^{-1} in fact. Moreover, an immediate improvement in the estimate of ϕ would result from accounting for the *anisotropy* of the electron screening. The electrostatic term arises from the general noncoincidence of the ionic and electronic 'partial' profiles.

The one- and two-particle ionic surface distributions $\rho_{(1)}^+(z_1)$, $\rho_{(2)}^+(z_1, \mathbf{r}_{12})$ may be directly related by the metallic analogue of the first of the BGY hierarchy. Recalling the one- and two-particle configurational projections[41]

$$\rho_{(1)}^+(\mathbf{r}_1) = \frac{1}{Z_Q(N-1)!} \int \cdots \int \exp\left(-\beta \Phi(\{N\})\right) d\mathbf{2} \cdots d\mathbf{N}$$

$$\rho_{(2)}^+(\mathbf{r}_{12}) = \frac{1}{Z_Q(N-2)!} \int \cdots \int \exp\left(-\beta \Phi(\{N\})\right) d\mathbf{3} \cdots d\mathbf{N}$$

$$(5.3.3)$$

and assuming the single particle density inhomogeneity is along the z axis, we investigate the effect of a small displacement of particle 1 holding all other particles fixed. We note that such a displacement will also modify the inter-ionic separations r_{1j}, all others remaining constant:

$$\frac{\partial \rho_{(1)}^+(z_1)}{\partial z_1} = \frac{1}{Z_Q(N-1)!} \int \cdots \int \left(-\beta \frac{\partial \Phi(\{N\})}{\partial z_1}\right) \exp\left(-\beta \Phi(\{N\})\right) d\mathbf{2} \cdots d\mathbf{N}$$

$$(5.3.4)$$

$$\beta = (kT)^{-1}.$$

Now from (5.3.2), neglecting electrostatic contributions, we have

$$\Phi(\{N\}) = \phi[\rho^-_{(1)}(z_1)] + \phi[\rho^-_{(1)}(z_2)] + \cdots + \phi[\rho^-_{(1)}(z_N)]$$
$$+ \tfrac{1}{2}\{\Phi[r_{12} \mid \rho^-_{(1)}(z_1)] + \Phi[r_{12} \mid \rho^-_{(1)}(z_2)]$$
$$+ \Phi[r_{13} \mid \rho^-_{(1)}(z_1)] + \Phi[r_{13} \mid \rho^-_{(1)}(z_3)] + \cdots$$
$$+ \Phi[r_{23} \mid \rho^-_{(1)}(z_2)] + \Phi[r_{23} \mid \rho^-_{(1)}(z_3)] + \cdots\}$$

whereupon

$$-\beta \frac{\partial \Phi(\{N\})}{\partial z_1} = -\beta \left(\frac{\partial \phi}{\partial z_1} (\rho^-_{(1)}(z_1)) + \frac{1}{2} \sum_{j=2}^{N} \frac{\partial \Phi}{\partial z_1} [r_{1j} \mid \rho^-_{(1)}(z_1)] \right.$$
$$\left. + \frac{1}{2} \frac{\partial}{\partial \mathbf{r}_1} \sum_{j=2}^{N} (\Phi[r_{1j} \mid \rho^-_{(1)}(z_1)] + \Phi[r_{1j} \mid \rho^-_{(1)}(z_j)]) \right).$$

Inserting this expression in (5.3.4) and using (5.3.3) we finally obtain, after some manipulation and transformation to cylindrical polar coordinates,

$$-kT \frac{\partial \rho^+_{(1)}(z_1)}{\partial z_1} = \rho^+_{(1)}(z_1) \frac{d\phi}{dz_1} [\rho^-_{(1)}(z_1)] + \frac{1}{2} \int \frac{\partial \Phi}{\partial z_1} [r_{12} \mid \rho^-_{(1)}(z_1)] \rho^+_{(2)}(z_1, \mathbf{r}_{12}) \, d\mathbf{r}_{12}$$
$$- \frac{1}{2} \int \frac{z_{12}}{r_{12}} \left\{ \frac{\partial \Phi}{\partial r_{12}} [r_{12} \mid \rho^-_{(1)}(z_1)] + \frac{\partial \Phi}{\partial r_{12}} [r_{12} \mid \rho^-_{(1)}(z_2)] \right\} \rho^+_{(2)}(z_1, \mathbf{r}_{12}) \, d\mathbf{r}_{12}. \qquad (5.3.5)$$

There is no formal difficulty in determining the ionic distribution $\rho^+_{(1)}(z)$ given the electronic distribution: equation (5.3.5) expresses the mechanical stability of the ionic component immersed in the electronic distribution $\rho^-_{(1)}(z)$ at the liquid metal surface. The first term on the right expresses the force on the ionic component arising from the asymmetric screening by the inhomogeneous conduction electron gas. In addition to the usual structural inhomogeneity at the liquid surface, asymmetry in the pair potential itself accounts for the second force term acting on the ionic distribution. The final term is the conventional contribution arising from the configurational anisotropy of the ionic distribution prevailing at the surface. Equation (5.3.5) does, of course, reduce to its classical rare gas form in the absence of the specifically metallic terms $d\phi/dz$ and $d\Phi/dz$ (c.f. equation 2.5.4).

Since we have, as usual, for the anisotropic two-particle pseudo-ion distribution

$$\rho^+_{(2)}(z_1, \mathbf{r}_{12}) = \rho^+_{(1)}(z_1)\rho^+_{(1)}(z_2)g^+_{(2)}(z_1, \mathbf{r}_{12})$$

integration of (5.3.5) subject to the usual boundary condition $\rho^+_{(1)}(-\infty) = \rho^+_L$ yields the following equation for the ionic profile:

$$\rho^+_{(1)}(z_1) = \rho^+_L \exp \left\{ -\frac{1}{kT} \int_{-\infty}^{z_1} \left(\frac{d\phi}{dz_1} [\rho^-_{(1)}(z_1)] + \frac{1}{2} \int \rho^+_{(1)}(z_2) g^+_{(2)}(z_1, \mathbf{r}_{12}) \right. \right.$$
$$\times \frac{d\Phi}{dz_1} [r_{12} \mid \rho^-_{(1)}(z_1)] \, d\mathbf{r}_{12} - \frac{1}{2} \int \rho^+_{(1)}(z_2) g^+_{(2)}(z_1, \mathbf{r}_{12})$$
$$\left. \left. \times \frac{\partial}{\partial r_{12}} (\Phi[r_{12} \mid \rho^-_{(1)}(z_1)] + \Phi[r_{12} \mid \rho^-_{(1)}(z_2)]) \frac{z_{12}}{r_{12}} \, d\mathbf{r}_{12} \right) dz_1 \right\}. \qquad (5.3.6)$$

In an earlier paper, Evans[20] adopts the Green closure for the ionic density distribution

$$\rho_{(2)}^+(z_1, \mathbf{r}_{12}) \sim \rho_{(1)}^+(z_1)\rho_{(1)}^+(z_2)g_{(2)}^L(r_{12})$$

where $g_{(2)}^L(r_{12})$ is the experimental bulk liquid ionic radial distribution. This approximation assumes that contributions to the profile structure arising from two-body correlations arise on the liquid side of the transition zone. (However there are grounds for believing that this approximation cannot, in principle, yield stable profiles—Section 2.2.) For the initial ionic profile Evans adopted the form of the Lang–Kohn *electronic* distribution (Figure 5.2.1)

$$\rho_{(1)}^+(z_1) = \frac{\rho^-(z_1)}{Z} \qquad (5.3.7)$$

where Z is the valence. Assuming a similar electron density dependence for ϕ and Φ as in the bulk, good agreement with experiment was obtained for the surface energy of liquid alkali metals, although the calculations tend to overestimate u_s for polyvalent systems.[21] This is not altogether surprising, since the use of a *local* isotropic electron density functional in the specification of $\phi[\rho_{(1)}^-(z)]$ is appropriate only to slowly varying electronic profiles. We have seen that the characteristic thickness of the transition region is of the order of the de Broglie wavelength at the Fermi surface (5.2.8), which for polyvalent systems implies a relatively sharp spatial variation in the electron density at the liquid metal surface. A somewhat more realistic expression of ϕ which takes account of the local anisotropy in the electronic distribution may well lead to some numerical improvement for polyvalent metals. We shall return to these calculations in Section 5.4.

This pseudo-atom approach can, in principle, be extended to the liquid transition metals, but the failure of the perturbative description of the strong or tightly bound ion–electron interaction means that, unlike the simple metals, we have no way of establishing the pairwise or self-energy terms. The difficulty, of course, arises from the d states. They are not sufficiently tightly bound (nor sufficiently localized) that we may validly treat them as the same in the metal as in the free atom. At the same time they retain enough of their atomic character that if they are treated as conduction band states the pseudowavefunction is not smooth, the pseudopotential is very large, and perturbation theory is not applicable. (Harrison[52] has reformulated the pseudopotential model with a direct generalization to transition metals and obtains a modified but weak pseudopotential supplemented by an additional term which plays a role akin to s–d hybridization. The treatment seems appropriate to the noble metals, the alkaline earths, and possibly some transition metals in between.) At this stage the reliability of the results for transition metal systems is too uncertain to justify detailed consideration here: the reader must be satisfied with some brief references to the original literature.[54] Smith,[51] however, in a description of the work function and related surface properties in liquid transition metal systems, regarded the d electrons as being 'free' with some success. The resulting very high electron density offers an explanation for the very large surface tensions and energies which characterize the transition metal series. Such qualitative remarks must still be regarded as purely speculative at this stage, however.

In a subsequent paper, Evans and Kumaradivel[15] use a Weeks–Chandler–Anderson perturbative approach, generalized to the liquid metal surface, to minimize variationally the surface excess Helmholtz free energy (and incidentally determine the surface tension) with respect to a family of exponential profiles for both the electronic and ionic distributions.

If, following the standard WCA approach (Section 2.4), we split the effective pairwise ionic potential into its repulsive and attractive components, then instead of (5.3.2) we have

$$V(\{N\}) = \lambda \sum_i \phi[\rho_{(1)}^-(z_i)] + \frac{1}{2} \sum_{i<j} \{\Phi^{(0)}[r_{ij} \mid \rho_{(1)}^-(z_i)] + \Phi^{(0)}[r_{ij} \mid \rho_{(1)}^-(z_j)]$$
$$+ \lambda(\Phi^{(1)}[r_{ij} \mid \rho_{(1)}^-(z_i)] + \Phi^{(1)}[r_{ij} \mid \rho_{(1)}^-(z_j)]\} \qquad (5.3.8)$$

where

$$\Phi[r \mid \rho_{(1)}^-] = \Phi^{(0)}[r \mid \rho_{(1)}^-] + \Phi^{(1)}[r \mid \rho_{(1)}^-] \qquad (5.3.9)$$

λ is a constant: $0 \le \lambda \le 1$. $\lambda = 0$ corresponds, as usual, to the *reference* system (designated$^{(0)}$) which is assumed to determine the essential structural features of the system, whilst as $\lambda \to 1$ the perturbation (designated$^{(1)}$) arising from self-energy contributions is gradually increased until the full interaction (5.3.9) is achieved. By straightforward manipulation of the canonical partition function it is easy to show that the Helmholtz free energy of the true system (bulk and surface) is

$$A = A_0 + \frac{\int_0^1 d\lambda \int \cdots \int \exp[-\beta V(\{N\})](\frac{1}{2}\sum_{i<j}(\Phi^{(1)}[r_{ij}|\rho_{(1)}^-(z_i)] + \Phi^{(1)}[r_{ij} \mid \rho_{(1)}^-(z_j)])}{\int \cdots \int \exp[-\beta V(\{N\})] d\{N\}}$$

$$+ \frac{\sum_i \phi[\rho_{(1)}^-(z_i)]) d\{N\}}{\int \cdots \int \exp[-\beta V(\{N\})] d\{N\}}$$

which reduces to

$$A = A_0 + \frac{1}{2} \int_0^1 d\lambda \int\int \Phi^{(1)}[r_{12} \mid \rho_{(1)}^-(z_1)]\rho_{(2)}^+[\mathbf{r}_1, \mathbf{r}_2 \mid \lambda] d\mathbf{r}_1 d\mathbf{r}_2$$

$$+ \int_0^1 d\lambda \int \phi[\rho_{(1)}^-(z_1)]\rho_{(1)}^+[\mathbf{r}_1 \mid \lambda] d\mathbf{r}_1 \qquad (5.3.10)$$

where A_0 is the free energy of the reference system and $\rho_{(1)}^+[\mathbf{r}_1 \mid \lambda]$, $\rho_{(2)}^+[\mathbf{r}_1, \mathbf{r}_2 \mid \lambda]$ are the ionic distribution functions in the partially perturbed system. Adopting a local density approximation for the reference free energy, we have, as for classical inert gas fluids:

$$A_0 \sim \int \rho_{(1)}^+(z_1) a^0 [\rho_{(1)}^+(z_1)] \, d\mathbf{r}_1 \qquad (5.3.11)$$

where a^0 is the free energy per particle in a *uniform* reference fluid of density $\rho_{(1)}^+(z_1)$, which bears the implication that $\Phi^{(0)}$ does not vary strongly with $\rho_{(1)}^+(z_1)$. Neglecting the self-energy contributions to the pair correlation allows us to replace $\rho_{(2)}^+[\mathbf{r}_1, \mathbf{r}_2 \,|\, \lambda]$ by $\rho_{(2)}^{+(0)}(z_1, \mathbf{r}_{12})$, which is then further approximated by the midpoint density estimate:

$$\rho_{(2)}^{+(0)}(z_1, \mathbf{r}_{12}) = \rho_{(1)}^+(z_1) \rho_{(1)}^+(z_2) g_{(2)}^{+(0)}(r_{12} \,|\, \rho_{(1)}^+[\tfrac{1}{2}(z_1 + z_2)]). \qquad (5.3.12)$$

Such an approximation has been discussed in detail in Section 2.4.

With the aid of these approximations we finally obtain the following expression for the free energy of the system:

$$A = \int \rho_{(1)}^+(z_1) \left\{ a^0[\rho_{(1)}^+(z_1)] + \frac{1}{2} \int \rho_{(1)}^+(z_2) \Phi^{(1)}[r_{12} \,|\, \rho_{(1)}(z_1)] \right.$$
$$\times g_{(2)}^{+(0)}[r_{12} \,|\, \rho_{(1)}^+[\tfrac{1}{2}(z_1 + z_2)]] \, d\mathbf{r}_{12}$$
$$\left. + \phi[\rho_{(1)}^-(z_1)] \right\} d\mathbf{r}_1. \qquad (5.3.13)$$

The single-particle ionic and electronic density distributions are represented by the exponential trial functions

$$\begin{aligned}
\rho_{(1)}^+(z_1) &= \rho_L^+[1 - \tfrac{1}{2} \exp(z_1/L_+)] & z_1 &< 0 \\
&= \tfrac{1}{2} \rho_L^+ \exp(-z_1/L_+) & z_1 &\geqslant 0 \\
\rho_{(1)}^-(z_1) &= \rho_L^-[1 - \tfrac{1}{2} \exp(z_1/L_-)] & z_1 &< 0 \\
&= \tfrac{1}{2} \rho_L^- \exp(-z_1/L_-) & z_1 &\geqslant 0
\end{aligned} \right\} \qquad (5.3.14)$$

where L_+, L_- are measures of the extent of the ionic and electronic transition zones, respectively. For the purpose of numerical calculation two further approximations are introduced: the local radial distribution $g_{(2)}^{+(0)}[r_{12} \,|\, \rho_{(1)}^+[\tfrac{1}{2}(z_1 + z_2)]]$ is replaced by the bulk reference distribution $g_{(2)}^{+(0)}[r_{12} \,|\, \rho_L^+]$ which, for liquid metals with a relatively sharp density transition between the two bulk phases, is likely to be a reasonable simplification. The more complex dependence of the Coulombic screening and self-energy terms $\Phi^{(1)}$ and ϕ upon the local electron density is reduced by means of a Taylor expansion about the bulk electron density, retaining only the lowest order terms:

$$\Phi^{(1)}[r_{12} \,|\, \rho_{(1)}^-(z_1)] \sim \Phi^{(1)}(r_{12} \,|\, \rho_L^-) + (\rho_{(1)}^-(z_1) - \rho_L^-) \frac{\partial \Phi^{(1)}}{\partial \rho_L^-}(r_{12} \,|\, \rho_L^-)$$

$$(5.3.15)$$

and

$$\phi[\rho_{(1)}^-(z_1)] \sim \phi(\rho_L^-) + (\rho_{(1)}^-(z_1) - \rho_L^-) \frac{\partial \phi}{\partial \rho_L^-}(\rho_L^-).$$

Given the long-range nature of these essentially Coulombic interactions, substantially in excess of the characteristic transition width of the profile, *a priori* considerations would seem to suggest that Taylor linearization to low order is inadequate. Evans, however, maintains that for calculations of the surface energy u_s based on such expansions 'led to results which did not differ considerably from those obtained from more rigorous calculations.'[15] Nevertheless, as we shall see, gross overestimates of the surface energy in high electron density polyvalent metals appear to be directly attributable to misrepresentations of the self-energy component of the total energy u_s.

For the calculation of the characteristic thickness of the electronic profile, Evans and Kumaradivel fitted the exponential function (5.3.14) to Lang and Kohn's results, neglecting any Friedel oscillations which may occur in $\rho_{(1)}^-(z)$ and then minimized variationally the free energy with respect to the ion density parameter L_+. The optimum values for Na, K, and Al are shown in Table 5.3.1 together with the incidentally determined surface tension. Two bulk ionic packing fractions η_L^+ are listed according to the use of either the Vashista–Singwi[23] (VS) or the Kleinmann[24] (KL) ionic pair sphere diameters. We note that the ionic profile is considerably sharper than the electronic distribution, which may be directly attributed to the important self-energy term; the excess free energy is lowered for configurations of the pseudo-ions located in regions of maximum electron density. This model, despite its numerous assumptions, approximations, and largely fortuitous agreement with experiment, clearly illustrates the role of self-energy contributions of the pseudo-ions arising specifically in metallic systems.

The role of surface plasmon contributions is entirely neglected in this treatment. Certainly it is true that plasmon contributions do arise; collective

Table 5.3.1.

Metal	η_L^+	$L(\text{Å})^*$	$M(\text{Å})$	$\gamma(\text{dyn cm}^{-1})$	$\gamma_{\text{expt}}^\dagger(\text{dyn cm}^{-1})$
Na	VS 0.454	1.23	0.575	146	191
	KL 0.477		0.568	157	
K	VS 0.432	1.38	0.465	99	115
	KL 0.459		0.49	104	
Al	VS 0.522	0.86	0.154	900	915
	KL 0.536		0.162	1000	

* Determined by fitting (5.3.14) to the Lang–Kohn profiles[21]
† Values determined at the melting point[22]
VS Vashista and Singwi[23]
KL Kleinmann[24]

effects due to the change in zero point energy of surface with respect to bulk plasmons remain difficult to assess, and at the moment their contribution is the subject of some controversy.[7-9, 29]

5.4 Surface tension, surface energy, and their temperature dependences

On the basis of the metallic pseudo-ion representation of the effective interaction in an inhomogeneous conduction electron gas (5.3.2) the following formally exact expressions for the surface tension γ and the surface energy u_s of a liquid metal in contact with its vapour may be derived:[21]

$$
\gamma = \frac{1}{2} \int\limits_{-\infty}^{\infty} \int \frac{(x_{12}^2 - z_{12}^2)}{r_{12}} \frac{\partial \Phi}{\partial r_{12}} [r_{12} \mid \rho_{(1)}^-(z_1)] \rho_{(2)}^+(z_1, \mathbf{r}_{12}) \, \mathrm{d}\mathbf{r}_{12} \, \mathrm{d}z_1
$$

$$
- \int_{-\infty}^{\infty} \frac{\mathrm{d}\phi}{\mathrm{d}z_1} [\rho_{(1)}^-(z_1)] \rho_{(1)}^+(z_1) z_1 \, \mathrm{d}z_1
$$

$$
- \frac{1}{2} \int\limits_{-\infty}^{\infty} \int \frac{\partial \Phi}{\partial z_1} [r_{12} \mid \rho_{(1)}^-(z_1)] \rho_{(2)}^+(z_1, \mathbf{r}_{12}) z_1 \, \mathrm{d}\mathbf{r}_{12} \, \mathrm{d}z_1 \qquad (5.4.1)
$$

and

$$
U_s = \int_{-\infty}^{0} (\phi[\rho_{(1)}^-(z_1)] \rho_{(1)}^+(z_1) - \phi[\rho_L^-] \rho_L^+) \, \mathrm{d}z_1
$$

$$
+ \int_{0}^{\infty} (\phi[\rho_{(1)}^-(z_1)] \rho_{(1)}^+(z_1) - \phi[\rho_V^-] \rho_V^+) \, \mathrm{d}z_1
$$

$$
+ \frac{1}{2} \int\limits_{-\infty}^{0} \int \{\Phi[r_{12} \mid \rho_{(1)}^-(z_1)] \rho_{(2)}^+(z_1, \mathbf{r}_{12}) - \Phi(r_{12} \mid \rho_L^-) \rho_L^{+2} g_{(2)L}^+(r_{12})\} \, \mathrm{d}\mathbf{r}_{12} \, \mathrm{d}z_1
$$

$$
+ \frac{1}{2} \int\limits_{0}^{\infty} \int \{\Phi(r_{12} \mid \rho_{(1)}^-(z_1)] \rho_{(2)}^+(z_1, \mathbf{r}_{12}) - \Phi(r_{12} \mid \rho_V^-) \rho_V^{+2} g_{(2)V}^+(r_{12})\} \, \mathrm{d}\mathbf{r}_{12} \, \mathrm{d}z_1.
$$

$$
(5.4.2)
$$

The first term in the expression for the surface tension of a liquid metal is seen to be of conventional 'inert gas form', supplemented by two further terms which, in the language of the stress tensor formulation, express the modification of the normal pressure component in the metallic case arising from the desire of the pseudo-ions to reside in regions of maximum electron density, and in regions of isotropic screening. Alternatively, we may regard the electron density profile $\rho_{(1)}^-(z)$ as an 'external' field, modifying the hydrostatics of the ions. For an insulating fluid the self-energy term $\mathrm{d}\phi/\mathrm{d}z$ and the partial variation of the pair potential $\partial \Phi/\partial z$ both vanish and (5.4.1) reduces to its inert gas form. Similarly, the surface energy is resolved quite clearly into surface excess self-energy and surface excess pair energy components, the numerical value depending sensitively upon location of the Gibbs dividing surface. The appearance of terms depending upon the self-energy $\phi(\rho^-)$ in (5.4.2) is a result of the explicit dependence of the interaction potential on $\rho^-(z)$. If $\phi(\rho^-)$ were constant, independent of location with respect to the surface, the first two terms in (5.4.2) would sum to zero, provided the origin of coordinates is located on the Gibbs dividing surface. Equation (5.4.2) then reduces to its insulating fluid form since then Φ is independent of $\rho_{(1)}^-(z)$.

Provided we resign ourselves to exponential ionic and electronic profiles of the form (5.3.14), and abandon our determination of the liquid metal surface structure, numerical estimates of the surface energy on the basis of (5.4.2) are relatively easily obtained. For two special cases,[20, 21] corresponding to $L_+ = 0$ (sharp ionic profile) and $L_+ = L_-$ (neutral profile), surface energy estimates have been obtained on the basis of a number of assumptions and approximations, principally, (a) the replacement of $\rho_{(2)}^+(z_1, \mathbf{r}_{12})$ by $\rho_{(1)}^+(z_1)\rho_{(1)}^+(z_2)g_{(2)L}^+(r_{12})$, (b) the adoption of the Taylor expansion (5.3.15) for the dependence of ϕ and Φ upon $\rho_{(1)}^-(z)$, and (c) an exponential fit of the electronic profile (5.3.14) to the Lang–Kohn functions, suppressing any Friedel structure. Calculations of the surface energy of a number of liquid metal systems show surprisingly good, and possibly fortuitous, agreement with experiment (Table 5.4.1). Nevertheless, the results warrant direct comparison with the results of Lang and Kohn[5] for solid metals (Figure 5.2.3). Again, good agreement is obtained for the low electron density alkali metals, although it must be realized that the results are quite sensitive to the choice of the pseudo-ion core radius r_c (5.2.4) and several values are listed in the literature: Ashcroft and Langreth's[25] values were used throughout in the above computations, however. Moreover, Allen[22] claims that the surface energies are known only to about $\pm 60\%$ in which case agreement perhaps

Table 5.4.1. U_s calculated on the basis of an approximate version of equation (5.4.2) with the neutral profile $L_+ = L_-$

Metal	(°K)	U_s(calc)(dyn cm^{-1})	U_s(exp)(dyn cm^{-1})[22]
Cs	(303)	88	88
Rb	(313)	117	104
K	(338)	145	142
Na	(373)	229	228
Li	(453)	406	461
Mg	(948)	685	882
Zn	(723)	400	900
In	(433)	1144	595
Al	(1000)	1460	1264
Pb	(613)	1352	546
Sn	(523)	1842	579

becomes somewhat illusory. Nevertheless, Kumaradivel and Evans feel that their results provide a semi-quantitative description of the behaviour of U_s for liquid alkali metals, and observe that for these systems the pairwise contributions to u_s, whilst subordinate to the self-energy contributions, nevertheless make an important contribution to the total surface energy—a result previously observed by Johnson et al.[13] and Waseda and Suzuki.[14]

The rather poor agreement for the high electron density metals cannot, apparently, be attributed to inadequacies in the empty core pseudopotential. A change of $\pm 10\%$ in r_c for In, Sb, and Pb changed U_s by $\sim \pm 10\%$, modifying the pairwise contributions to the surface energy only. The gross overestimates must be attributed directly to the self-energy contributions, and although there can be no termwise comparison between the present treatment and that of Lang and Kohn, both theories substantially overestimate the surface energy of Pb, the latter authors reporting a value of $1400\ \mathrm{dyn\ cm^{-1}}$. The inescapable conclusion appears to be that the simple pseudopotential formalism developed here is inadequate for high electron density polyvalent metals. It is known experimentally that U_s is roughly proportional to the melting temperature, which is easy to understand in terms of conventional pairwise forces, but there appears to be no straightforward 'electronic' explanation as yet.

The temperature variation $\partial \gamma / \partial T$ of the surface tension is more sensitive to the structure of the transition zone than γ itself. For a single-component system we have seen (1.4.7)

$$\frac{\partial \gamma}{\partial T} = -S_s$$

where S_s is the surface excess entropy per unit area. Substitution of equations (5.4.1, 5.4.2) in the Gibbs–Helmholtz equation (1.4.9)

$$U_s = \gamma - T \frac{\partial \gamma}{\partial T}$$

and integration by parts yields

$$T \frac{\partial \gamma}{\partial T} = \int_{-\infty}^{\infty} \phi[\rho_{(1)}^-(z_1)] \frac{\partial \rho_{(1)}^+(z_1)}{\partial z_1} z_1\, \mathrm{d}z_1 + \text{pairwise contributions} \quad (5.4.3)$$

from which we see the great sensitivity of the temperature coefficient $\partial \gamma / \partial T$ to the development of density oscillations in the ionic profile $\rho_{(1)}^+(z)$: alternatively we may say that there is a substantial contribution to $\partial \gamma / \partial T$ from the configurational entropy. We shall return to this point again below. In addition to the purely classical initiation of oscillations by the surface constraining field, quantum interference effects on the electronic profile and Friedel oscillations in the ionic screening may combine to produce a strongly oscillatory density transition at temperatures just above the triple point,[30] particularly for those systems having characteristic Friedel wavelengths $\pi / k_F \sim r_c$, the ionic diameter.

Table 5.4.2. Surface entropies of liquid metals (after Faber[31]). (These data are at temperatures close to the melting point, except in the case of Hg).

Metal	$\partial\gamma/\partial T(\text{dyn cm}^{-1}\,^{\circ}\text{C}^{-1})$	surface entropy per atom (°K)
Li	−0.14	0.80
Na	−0.1	0.85
K	−0.06	0.78
Rb	−0.06	0.94
Cs	−0.05	0.78
Ag	−0.13	0.65
Al	−0.135	0.68
In	−0.096	0.61
Sn	−0.083	0.55
Cu	+0.75	−2.9
Zn	+0.5	−2.2
Hg (25 °C)	−0.20	+1.2

These low entropy states correspond to a positive temperature coefficient $\partial\gamma/\partial T$, as we see from equation (5.4.3). Similar low entropy surface states for strongly anisotropic molecular systems with associated positive $\partial\gamma/\partial T$ coefficients have already been discussed in Chapter 3. There it was suggested that a surface excess *orientational* order developed. In Table 5.4.2[31] figures are given for a number of liquid metals for which the slopes $\partial\gamma/\partial T$ should be fairly reliable.[32–34] Certain of the liquid metals—Cu, Zn, and Cd in particular—show positive slopes with subsequent inversion over a temperature range ~100° above the melting point (Figure 5.4.1). The crudest explanation is that the liquid surface of these metals is virtually crystalline,[31] and since the entropy of melting of Cu, Zn, and Cd is about $1.2k$ per atom, we should need to regard the top three layers as crystalline to account for the data in Table 5.4.2. At higher temperatures, of course, the liquid surface

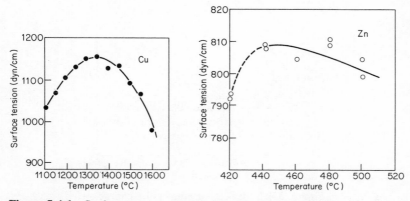

Figure 5.4.1 Surface tension of liquid Cu and Zn as a function of temperature. T_i denotes the inversion temperature

162

will thermally delocalize, and systems exhibiting positive $\gamma(T)$ slopes will ultimately invert to yield the familiar monotonic decreasing function of temperature, going to zero as $(T_c - T)^\mu$—the exponent is as yet unknown for liquid metals. Such a $\gamma(T)$ characteristic exhibiting an inversion is certainly contrary to the majority of experimental evidence, and probably all theoretical models.

The important series of *equilibrium* $\gamma(T)$ measurements by White[33, 34] has been reviewed adequately elsewhere,[30] and here we shall simply emphasize that, contrary to the majority of experimental determinations of surface tension, in particular those for liquid metals, the experiments of White were performed with the liquid in equilibrium with its saturated vapour so that there was no net flux of particles across the surface. The majority of $\gamma(T)$ determinations have neglected this outstandingly important point, and in consequence a non-equilibrium measurement of an equilibrium parameter is obtained.[35] Evaporation rates of only ~100 atomic layers per second appear to modify substantially the $\gamma(T)$ and $\partial\gamma/\partial T$ characteristics. This is at first sight a somewhat surprising result since the surface would reach equilibrium in a time ~10^{-11} s. However, a net flux of particles across the surface implies that constancy of the chemical potential across the interface does not prevail, and the single-particle transition profile cannot be of equilibrium form, even though it may be time-independent. In Figure 5.4.2 we show the effect on the $\gamma(T)$ characteristic of Zn under conditions of progressive

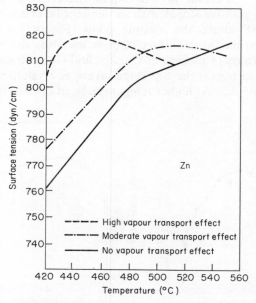

Figure 5.4.2 Effect on the $\gamma(T)$ characteristic for liquid Zn as a function of vapour transport

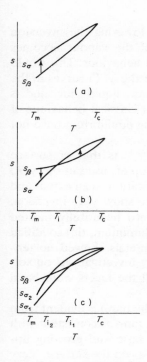

Figure 5.4.3 Schematic variation of the per particle bulk (β) and surface (σ) entropies as a function of temperature. (a) Conventional curves for which ($S_\sigma > S_\beta$) throughout. (b) An inversion arising from the existence of low entropy states just above the melting point. Conventional behaviour is regained at $T > T_i$. (c) Effect of enhanced vapour transport is to shift the surface entropy curve up the entropy axis, thereby lowering the inversion temperature T_i.

vapour transport. The equilibrium curve shows a strong inversion. Under progressively non-equilibrium conditions the familiar monotonic decreasing characteristic is regained.

It is instructive to construct schematic plots for the 'classical' entropy–temperature relations, and compare them with the 'inverting' systems. The classical curve (Figure 5.4.3a) shows the convergence of the bulk and surface entropies; initially the surface excess entropy $(S_\sigma - S_\beta) > 0$, at the triple point, and finally $(S_\sigma - S_\beta) = 0$ at the critical point when the liquid–vapour boundary disappears. Such a $\Delta S(T)$ relation would generate a monotonic decreasing $\gamma(T)$ curve as is generally reported in the literature for most systems. In the case of inversion of the surface tension, however, the bulk and surface entropies at the triple point may themselves be inverted corresponding to low entropy states at the surface, perhaps attributable to surface structuring. In such a case the $S(T)$ curve must have some such form as shown in Figure 5.4.3b. Provided the rate of entropy production at the surface exceeds that of the bulk, which at low temperatures seems an acceptable proposition, the gradient of the $\gamma(T)$ curve inverts, and remains largely constant over quite a wide temperature range subsequently. Systems showing almost coincidental bulk and surface entropy curves are presumably just those which have almost horizontal $\gamma(T)$ characteristics.

Under conditions of enhanced vaporization as White investigated for Zn (Figure 5.4.2), little change would be expected to occur to the $S_\beta(T)$ curve, but presumably $S_\sigma(T)$ is shifted to a higher value at each point along the temperature axis. This can only have the effect of lowering the inversion temperature T_i (Figure 5.4.3c), and if streaming of the vapour becomes excessive, S_β and S_σ will reinvert to give the classical behaviour. This would seem to account for the behavioural characteristics of the $\gamma(T)$ curves in the deliberately non-equilibrium experiments of White on liquid zinc, and is possibly the mechanism responsible for the apparent classical behaviour of many liquid metals when full precautions for ensuring equilibrium have not been taken.

A particularly interesting observation of White[34, 36] is that of surface faceting. It was found that the liquid metal sessile droplets used in the $\gamma(T)$ investigations developed surface faceting upon solidification to an extent and perfection directly related to the system's tendency to show $\gamma(T)$ inversion. Thus, zinc and cadmium, for example, solidified into beautifully faceted droplets (Figure 5.4.4), whilst other metal systems (aluminium, tin) solidified into apparently smooth sessile drops. These latter metals showed no tendency towards inversion. Laue back-reflection X-ray investigations on solidified drops of cadmium and zinc indicated that all the facets were basal planes, i.e. planes of densest packing.

That faceting initiates at the surface (as well as at the base) is evident from cross-sectioning the droplet.[34] The configuration of grains is consistent with solidification starting from several locations at the base and growing upwards, having begun independently at several locations on the surface. Very relevant is White's observation that the shrinkage cavity in more than thirty drops sectioned, though usually located just below the surface, in no case broke out into the surface. Given the high thermal conductivity of the metals and slow cooling, thermal gradients within the droplet cannot account for this behaviour.

Bearing in mind the mechanism for γ-inversion proposed above, it is

Figure 5.4.4 Sessile droplet of solidified zinc. The 'geodesic' facets are all basal planes. Sectioning reveals that the grains initiated at surface sites, quite independently from those at the base

Table 5.4.3.

Purity %	$\gamma_{Zn}(420\,°C)$ dyn cm^{-1}
99.99+	757 ± 5
99.999+	761 ± 5
99.9999	767.5 ± 5

difficult to escape the conclusion that surface faceting is a direct consequence of the development of stable density oscillations at the liquid metal
surface just above the triple point serving as nucleation centres in the
solidification process. (The minimimum wavelength of density fluctuations in
a liquid just above the solidification temperature with respect to spinodal
decomposition has recently been determined by Croxton:[38] investigations as
to whether an oscillatory density profile at the liquid surface can initiate
surface faceting is currently under investigation.)

It is possible that effects of surface impurity migration with temperature
could account for the inversion. Indeed, we have already discussed this
possibility at some length in Section 4.5. Regarding the system as a dilute
binary mixture of a contaminant solute in the metallic solvent, positive
surface adsorption may, at low temperatures, serve to lower the surface
tension. Indeed, in the limit of the impurity mole fraction $x_i \to 0$, Zadumkin
and Zvyagina[39] found that the condition

$$\left(\frac{\partial \gamma}{\partial x_i}\right)_{x_i \to 0} < 0 \quad \text{for} \quad \gamma_i - \gamma_m < 0 \tag{5.4.4}$$

(where γ_m is the surface tension of the pure metal) was substantiated by
94% of nearly 200 binary systems whose surface tension data they examined. The effect of metal purity on the surface tension of Zn at 420 °C is
given in Table 5.4.3.[33] Whilst ostensibly such an impurity
absorption/desorption process could account for the inversions in the $\gamma(T)$
curves (Section 4.5), we do also have to explain why other metallic systems
of lower purity do not also exhibit inversion and faceting. A detailed
discussion of chemical effects and their bearing upon measured values of the
surface tension of liquid metals has been given by White,[40] and we refer the
reader to the literature for further information.

5.5 Phenomenological descriptions of the temperature dependence of surface tension

Waves in an extended bulk liquid or solid are completely delocalized, but
tend to localize in the vicinity of a free surface when they may be classified
as either Rayleigh, capillary, or gravity waves. Rayleigh[42] gave the first
description of surface elastic waves propagating over a semi-infinite

medium. In the case of a liquid, such nondispersive waves can only propagate above a certain minimum frequency inversely proportional to the relaxation time τ_0 of the bulk fluid. τ_0 is typically a fraction of the Debye frequency. Clearly such waves are not of comparable importance in solids and liquids: in the latter case it is only the high-frequency components which contribute dynamically to the quantities of interest here. More generally we may say that the frequency spectrum of the surface waves constituting the heat motion at the free surface of a liquid body may be separated into two components: (i) the high frequency or Rayleigh part $(\nu > 1/\tau_0)$ and (ii) the low frequency or capillary part $(\nu < 1/\tau_0)$. The latter component describes waves restored by surface tension which are dispersive since the excess pressure due to a curved surface is inversely proportional to its radius, and the restoring force will, therefore, be greater on the more highly curved surfaces arising at short wavelengths.

For present purposes, at relatively high temperatures, we shall be primarily concerned with the Rayleigh contribution: at low temperatures the emphasis shifts toward the capillary region as we shall see when we come to consider a phenomenological description of quantum fluids (Chapter 6). Gravity waves, associated with macroscopic bodily motions of fluid in the earth's gravitational field, clearly do not concern us here.

Frenkel[43] appears to have given the first estimate of surface entropy (or $\partial\gamma/\partial T$) in terms of Rayleigh waves. An approximate expression for the free energy of an elastic continuum due to thermal motion is[44]

$$A = n'kT \ln\left(\frac{kT}{h\bar{\nu}'}\right) \tag{5.5.1}$$

where n' is the number of atoms per unit area, which in turn determines the number of degrees of freedom of the surface layer. $\bar{\nu}'$ is the average frequency of the surface waves. The surface excess entropy follows immediately as

$$S_s = -n'k \ln\left(\frac{kT}{h\bar{\nu}'}\right) + nk \ln\left(\frac{kT}{h\bar{\nu}}\right) \tag{5.5.2}$$

where $n, \bar{\nu}$ are the corresponding bulk quantities. After considerable analysis Frenkel finally obtains

$$\frac{\bar{\nu}}{\bar{\nu}'} = \exp\left\{\frac{3}{4}\left(\frac{3\pi}{2}\right)^{1/3}\left(\frac{v_B}{v_s}\right)_T\left[2+\left(\frac{G}{E}\right)^{3/2}\right]^{-2/3}\right\} \tag{5.5.3}$$

where $(v_B/v_s)_T$ is the ratio of the bulk and surface transverse wave velocities, being typically ~ 0.8. E and G are the Young and the shearing moduli, respectively. This treatment of Frenkel's is essentially a modification of Born and von Kármán's treatment of an elastic continuum, but with a discontinuous surface.

Bohdansky[45] utilizes (5.5.2), but in place of Frenkel's elaborate expression for $\bar{\nu}/\bar{\nu}'$ assumes that the ratio of bulk to surface mean frequencies is constant for liquids of similar structure. Anharmonicity of the vibrations is incorporated by Zadumkin,[46] and expressions for the free energies of a bulk and surface atom determined and related by means of the different coordination numbers appropriate to the two locations. A more sophisticated version of the same analysis was subsequently developed by Zadumkin and Pugachevich[47] in which a continuous variation in coordination from bulk liquid to surface is incorporated.

As we observed in Section 1.4, the surface entropy may be conveniently resolved into its thermal and configurational components, at least for the purposes of discussion. The preceding analyses relate solely to the thermal contributions. Attempts have been made, mainly on the basis of lattice theories, to account at least partially for the configurational contributions to the surface entropy. While such contrived models cannot be usefully incorporated into any modern concept of the statistical thermodynamics of the liquid surface, they do nevertheless show that an anomalously dense surface layer together with a fairly open bulk structure could give a negative excess surface configurational entropy, and at relatively low temperatures may dominate the thermal contribution with the consequence of an anomalous temperature dependence of the kind reported by White,[33, 34] discussed earlier in this section. In the presence of continuous evaporation the disordering effect supplements the thermal entropy contribution, and the sign of $\partial \gamma / \partial T$ may well reverse, as reported by White for certain metallic systems.

5.6 Experimental investigations of the metallic interfacial profile

Much of the uncertainty which impedes the qualitative assessment of the various theoretical descriptions of the interfacial profile arises from difficulties both of investigation and interpretation which surround any direct experimental attack. Under these circumstances, perhaps only under these circumstances, the machine simulations, *faute de mieux*, provide some insight into the molecular processes operating at the liquid surface—although a reappraisal of simulated data now appears in question (Chapter 10). It does seem, however, that the nature of the metallic interaction at present prohibits the simulation of inhomogeneous systems, and that any structural determinations will, of necessity, be of an experimental nature.

Attempts to study the ionic distribution at the surface of a liquid metal by means of electron diffraction have been unsuccessful, primarily due to complications introduced by non-equilibrium and multiple scattering.[59] Conventional X-ray scattering studies, so useful in elucidating bulk structures, are clearly inappropriate since the beam cannot be restricted to the interfacial zone. However, Compton has noted that at X-ray frequencies the refractive index of condensed matter is less than unity, in which case there will be total external reflection for X-rays of angles of incidence less than some critical angle, θ_c, typically $\sim 5 \times 10^{-3}$ radian (Figure 5.6.1). At such

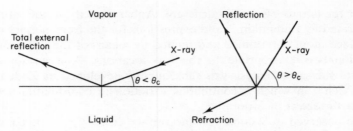

Figure 5.6.1 Conditions for total external reflection of X-rays
at a liquid surface

glancing incidence the X-rays penetrate only a small distance into the
surface. The reality of this effect was demonstrated in 1922,[60] the reflectivity
versus angle being interpreted on the basis of a sharp planar dividing surface
between an homogeneous condensed phase and a vacuum, the reflected and
refracted beams being described by Fresnel's equations.[61] A recent experimental investigation by Lu and Rice[58] exploits this property in an attempt to
determine the interfacial profile of liquid mercury.

In the X-ray region the refractive index is generally expressed in the form

$$n(\lambda) = 1 - \delta(\lambda) - i\beta(\lambda). \tag{5.6.1}$$

For λ in the range 1–10 Å the ratio $y = \beta/\delta$ is typically between 0.1 and
0.01, provided λ is far from the absorption edge. For a homogeneous
medium terminating at a planar interface the reflectivity for s- and p-
polarized X-rays is, very closely,

$$R_s R_s^* = \frac{A - (2x)^{1/2}(A-1)^{1/2}}{A + (2x)^{1/2}(A-1)^{1/2}} \tag{5.6.2}$$

where

$$A = x + [(1-x)^2 + y^2]^{1/2} \qquad x = \theta/\theta_c.$$

A realistic profile cannot, of course, be represented as a density discontinuity, but is a continuous variation over a range of $\sim 2\sigma$. Bloch and Rice[62]
subdivide the transition region into a subset of elemental strata, each of
which is deemed homogeneous, and this approach is adopted here. For
systems in the vicinity of the triple point the interfacial density varies by a
factor of $\sim 10^{-5}$ over a couple of atomic diameters, and the propriety of such
a simplification has been brought into question in earlier chapters.

However, Lu and Rice[58] find that the density profile of liquid Hg at
298 °K is of hyperbolic tangent form having a thickness estimated on the
basis of (2.11.3) to be 5.6±0.5 Å, including the effects of thermal (capillary)
broadening, which amounts to only ~ 1.4 Å at this temperature. The intrinsic width, then, arising from electron delocalization at the surface, accounts
almost entirely for the interfacial thickness—approximately two atomic
diameters, about that for argon at its triple point. What is surprising is that

the great qualitative difference in molecular potentials for the two systems yields such similar surface thicknesses. (It is an experimental fact, however, that the product of the surface tension and isothermal compressibility $\gamma\chi_T$ which is proportional to the interfacial thickness, is similar for a whole range of liquid systems, regardless of the nature of the interaction). The relaxed jellium calculations of Allen and Rice (Section 5.2) predict an interfacial thickness for liquid Hg which is approximately half that observed experimentally.

At present the dilemma remains. Experimental and theoretical uncertainties permit only speculative observation. Perhaps it is *structure* rather than interaction which is of importance here; after all, hard sphere structure factors for bulk liquid metals taken at the appropriate packing fraction, are known to account for the principal structural features of metallic systems,[63] despite the fundamental difference in the nature of the interaction and direct correlation. Lu and Rice[58] pursue this similarity in structure as a possible explanation of the similarity in interfacial thicknesses, but become discouraged by the known differences in the long range tail form of the direct correlation for insulators and metals, the latter being oscillatory. However, this need not be a cause for undue pessimism since structural details arise as *integrals* over the direct correlation functions, in which case much of the distinction between the oscillatory and monotonic tail forms may be washed out, and the resulting structures remain effectively similar.

References

1. M. Watabe and Maregawa, *2nd. Int. Conf. on Liquid Metals*, Taylor and Francis, London (1973), p. 133.
   J. Chihara, *2nd. Int. Conf. on Liquid Metals*, Taylor and Francis, London (1973), p. 137.
   N. H. March and M. P. Tosi, *Ann. of Physics*, **81**, 414 (1973).
   M. P. Tosi and N. H. March, *Nuovo Cimento*, **15B**, 308 (1973).
   M. P. Tosi, M. Parrinello and N. H. March, *Nuovo Cimento*, **23B**, 135 (1974).
2. V. K. Semenchenko, *Surface Phenomena in Metals and Alloys*, Pergamon, Oxford.
3. J. Bardeen, *Phys. Rev.*, **49**, 653 (1936).
   W. J. Suriatecki, *Proc. Phys. Soc.*, **A64**, 226 (1961).
   P. P. Ewald and H. Juretschke, In R. Gomer and C. S. Smith (Eds), *Structure and Properties of Solid Surfaces*, University of Chicago Press, Chicago (1953), p. 82.
   R. D. Berg and L. Wilets, *Proc. Phys. Soc.*, **A68**, 229 (1955).
4. A. J. Bennett and C. B. Duke, *Phys. Rev.*, **160**, 541 (1967).
   J. R. Smith, *Phys. Rev.*, **181**, 522 (1969).
5. N. D. Lang and W. Kohn, *Phys. Rev.*, **B1**, 4555 (1970).
   W. Kohn, Collective Excitations, In B. and S. Lundqvist, *Nobel Symposium*, Academic Press, New York (1974).
6. N. H. March, In N. H. March (Ed.), *Orbital Theories of Molecules and Solids*, Clarendon Press, Oxford (1974).
7. R. A. Craig, *Phys. Rev.*, **B6**, 1134 (1972).

8. J. Schmit and A. A. Lucas, *Solid State Commun,* **11,** 415 (1972).
   J. A. Appelbaum and D. R. Hamann, *Phys. Rev.,* **B6,** 2166 (1972).
9. M. Jonson and G. Srinivasan, *Phys. Lett.,* **43A,** 427 (1973).
   M. Jonson and G. Srinivasan, *Physica Scripta,* **10,** 262 (1974).
   P. J. Feibelman, *Solid State Commun.,* **13,** 319 (1973).
   P. J. Feibelman, *Phys. Rev.,* **176,** 551 (1968).
   W. Kohn, *Solid State Commun.,* **13,** 323 (1973).
   J. Heinrichs, *Solid State Commun.,* **13,** 1599 (1973).
10. N. W. Ashcroft, *Phys. Letters,* **23,** 48 (1966).
11. The importance of this approach for the calculation of surface energies has been emphasized by S. N. Zadumkin, *Fiz. Metal. i Metalloved.,* **11,** 3 (1961); **11,** 331 (1961). [*Phys. Metals Metallog.,* **11,** 3 (1961); **11,** 11 (1961)]. This conclusion may, of course, be extended to liquid metal systems.
12. G. D. Mahan, *Phys. Rev.,* **B12,** 5585 (1975).
13. M. D. Johnson, P. Hutchinson, and N. H. March, *Proc. Roy. Soc.,* **A282,** 283 (1964).
14. Y. Waseda and K. Suzuki, *Phys. Stat. Solidi,* **49,** 643 (1972).
   Y. Waseda and K. Suzuki, *Phys. Stat. Solidi,* **57,** 351 (1973).
15. R. Evans and R. Kumaradivel, *J. Phys. C.,* **9,** 1891 (1976).
16. J. A. Appelbaum and D. R. Hamann, *Phys. Rev.,* **B6,** 2166 (1972).
   G. P. Alldredge and L. Kleinman, *Phys. Rev.,* **B10,** 559 (1974).
   G. P. Alldredge and L. Kleinman, *Phys. Lett.,* **48A,** 337 (1974).
17. M. W. Finnis, *J. Phys. F.,* **4,** L37 (1975).
18. M. W. Finnis and V. Heine, *J. Phys. F.,* **5,** 2227 (1974).
19. R. F. Wallis, In S. G. Davison (Ed.), *Progress in Surface Science,* Vol. 4, Pergamon, Oxford (1975), p. 253.
20. R. Evans, *J. Phys. C.,* **7,** 2808 (1974).
21. R. Kumaradival and R. Evans, *J. Phys. C.,* **8,** 793 (1975).
22. B. C. Allen. In S. Z. Beer (Ed.), *Liquid Metals Chemistry and Physics,* Dekker, New York (1972), p. 161.
23. P. Vashista and K. S. Singwi, *Phys. Rev.,* **B6.,** 875 (1972).
24. L. Kleinmann, *Phys. Rev.,* **160,** 585 (1967).
25. N. W. Ashcroft and D. C. Langreth, *Phys. Rev.,* **155,** 682 (1967).
26. J. Frenkel, *Z. Phys.,* **51,** 232 (1928).
   K. Huang and G. Wyllie, *Proc. Phys. Soc.,* **A62,** 180 (1949).
   H. B. Huntingdon, *Phys. Rev.,* **31,** 1035 (1951).
   R. Stratton, *Phil. Mag.,* **44,** 1236 (1953).
27. See, for example, W. A. Harrison, *Pseudopotentials in the Theory of Metals,* Benjamin, New York (1966).
28. J. M. Ziman, *Adv. Phys.,* **13,** 89 (1964).
29. J. Schmit and A. A. Lucas, *Solid State Commun.,* **11,** 415 (1972).
   W. Kohn, *Solid State Commun.,* **13,** 323 (1973).
   R. A. Craig, *Solid State Commun.,* **13,** 1517 (1973).
30. C. A. Croxton, *Adv. Phys.,* **22,** 385 (1973).
   C. A. Croxton, *Liquid State Physics—A Statical Mechanical Introduction,* Cambridge University Press, London and New York, (1974), Ch. 4.
   C. A. Croxton. In C. A. Croxton (Ed.), *Progress in Liquid Physics,* Wiley, London (1978), Ch. 2.
31. T. E. Faber, *An Introduction to the Theory of Liquid Metals,* Cambridge University Press, London and New York (1972).
32. J. R. Wilson, *Metall. Rev.,* **10,** 381 (1965).
   D. Gerner and H. Mayer, *Z. Phys.,* **210,** 391 (1968).
33. D. W. G. White, *Trans. Metall. Soc.,* *A.I.M.E.,* **236,** 796 (1966).
34. D. W. G. White, *J. Inst. Metals,* **99,** 287 (1971).

35. C. Maze and G. Burnet, *Surf. Sci.*, **27**, 411 (1971).
36. R. T. Southin and G. A. Chadwick, *Scripta. Metall.*, **3**, 541 (1969).
37. J. Frenkel, *Phil. Mag.*, **33**, 297 (1917).
38. C. A. Croxton, *J. Phys. C.*, **12**, 2239 (1979).
39. S. N. Zadumkin and V. Ya. Zvyagina, *Russian Met. and Fuels*, **4**, 20 (1966).
    S. N. Zadumkin and V. Ya. Zvyagina, *Izvest. Akad. Nauk. S.S.S.R. (Metally)*, **4**, 58 (1966).
40. D. W. G. White, *Metals, Materials and Metallurgical Rev.*, July 1968.
41. See, for example, C. A. Croxton, *Liquid State Physics—A Statistical Mechanical Introduction* Cambridge University Press, London and New York (1974).
42. Lord Rayleigh, *Proc. London Math. Soc.*, **17**, 4 (1885).
43. J. Frenkel, *Kinetic Theory of Liquids*, Dover (1942), p. 308.
44. R. C. Brown and N. H. March, *Physics Reports*, **24**, 77 (1976).
45. J. Bohdansky, *J. Chem. Phys.*, **49**, 2982 (1968).
46. S. N. Zadumkin, *Russian J. Phys. Chem.*, **33**, 539 (1959).
47. S. N. Zadumkin and P. P. Pugachevich, *Proc. Acad. Sci. U.S.S.R., Physics and Chemistry Section*, **146**, 743 (1962).
48. R. Stratton, *Phil. Mag.*, **44**, 1236 (1953).
49. C. P. Flynn, *J. Appl. Phys.*, **35**, 1641 (1964).
50. N. H. March, *Liquid Metals*, Oxford, Pergamon Press (1968).
51. J. R. Smith, *Phys. Rev.*, **181**, 522 (1969).
    R. C. Brown and N. H. March, *J. Phys. C.*, **6**, L363 (1973).
52. W. A. Harrison, *Phys. Rev.*, **181**, 1036 (1969).
53. E. Wikborg and J. E. Inglesfield, *Sol. State. Commun.*, **16**, 335 (1975).
54. J. L. Morán–López, G. Kerker, and K. H. Bennemann, *J. Phys. F.*, **5**, 1277 (1975).
    F. Cyrot–Lackmann, M. C. Desjonquèrs, and J. P. Gaspard, *J. Phys. C.*, **7**, 925 (1974).
    D. Kalkstein and P. Soven, *Surf. Sci*, **26**, 85 (1971).
    R. Haydock and M. J. Kelly, *Surf. Sci.*, **38**, 139 (1973).
    M. J. Kelly, *Surf. Sci.*, **43**, 587 (1974).
55. J. W. Allen and S. A. Rice, *J. Chem. Phys.*, **67**, 5105 (1977).
56. J. C. Slater, *The Self-Consistent Field for Molecules and Solids: Quantum Theory of Molecules and Solids*, Vol. 4., McGraw-Hill, New York (1974).
    I. N. Levine, *Quantum Chemistry*, Vol. 1, (Allyn and Bacon, Boston (1970).
57. J. W. Allen and S. A. Rice, *J. Chem. Phys.*, **68**, 5053 (1978).
58. B. C. Lu and S. A. Rice (preprint, 1978): *J. Chem. Phys.*, **68**, 5558 (1978).
59. C. A. Croxton, *Ph.D. Thesis*, University of Cambridge (1969).
    R. M. Goodman and G. A. Samorjai, *J. Chem. Phys.*, **52**, 6331 (1970).
60. A. H. Compton, *Bull. Nat. Res. Council*, **20**, 48 (1922).
61. A. K. Compton and S. K. Allison, *X-rays in Theory and Experiment*, 2nd ed., Van Nostrand, Princeton, N.J.
62. A. Bloch and S. A. Rice, *Phys. Rev.*, **185**, 933 (1969).
63. N. Ashcroft and J. Lekner, *Phys. Rev.*, **145**, 83 (1966).

CHAPTER 6

The Quantum Liquid Surface

6.1 Introduction

The intervention of quantum mechanical effects in ^{4}He and ^{3}He is evident from a comparison of their respective phase diagrams with that predicted on the basis of the theory of corresponding states (Figure 6.1.1). The large zero-point energies, which unlike the other noble gases are comparable with the potential energy of the system, prohibit solidification at pressures less than 25–30 atm., the system retaining its fluidity even at absolute zero. A convenient measure of the quantum mechanical effect is given in terms of the parameter Λ^* which de Boer introduced into the theory of corresponding states:

$$\Lambda^* = (h^2/m\varepsilon\sigma^2)^{1/2}.$$

The quantum parameter Λ^* is a measure of the ratio of the de Broglie wavelength to the atomic diameter (Table 6.1.1). Of course, the distinction between ^{4}He and ^{3}He is beyond the scope of any quantum mechanical extension of the theory of corresponding states: the difference in mass alone cannot account for the differences in the phase diagrams of the two isotopes. It is also necessary to incorporate the statistics: ^{4}He atoms are bosons, while ^{3}He atoms are fermions. The most spectacular manifestation of these differences is the appearance of a fourth state of matter, the superfluid state. The intervention of statistics modifies the other three regions of the phase diagram also; ^{3}He must satisfy the Pauli exclusion principle, which raises its energy relative to a hypothetical ^{3}He boson fluid. The change is quite substantial in the case of the liquid and vapour phases being of the same order as the energy of a non-interacting Fermi gas of the same mass and density. In the solid phase, however, the system is virtually independent of the statistics since the particles are localized at separate sites with only a weak overlap. It follows, therefore, that the difference in energy between the solid and liquid phases should be much greater in a Fermi fluid than in a Bose fluid at the same density.[1]

According to the principle of corresponding states we should expect the reduced surface tension $\gamma^* = \gamma\sigma^2/\varepsilon$ to be a universal function of the reduced temperature $T^* = kT/\varepsilon$ for those systems which are described adequately in terms of a Lennard–Jones interaction. The deviation of the quantum liquids from the classical estimates of the superficial thermodynamic functions is already apparent in the case of neon (Table 6.1.1): experimental data available indicate that the reduced phenomenological relation describing the

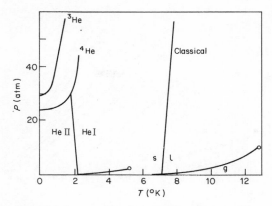

Figure 6.1.1 Schematic phase diagram of ^{4}He and ^{3}He compared with the prediction of the classical theory of corresponding states

temperature dependence may be written:

$$\gamma^* = \alpha(\Lambda^*)\left(1 - \frac{T^*}{T_c^*}\right)^\mu \qquad \mu = \frac{11}{9} \tag{6.1.1}$$

where the coefficient α introduces the Λ^*-dependence. In Figure 6.1.2a we show the concurrence of the experimental and 'corresponding state' characteristics determined on the basis of equation (6.1.1). Figure 6.1.2b shows the explicit dependence of α upon Λ^* which plays the role of γ_0 in the classical phenomenological $\gamma(T)$ relation.

The discrepancy between the classical predictions and the experimental observation of the surface tension and energy of quantum liquids may be understood in terms of the pair momentum current density tensor at the liquid surface, or what we might loosely call the 'kinetic temperature tensor'. Isotropy of the Maxwell distribution in momentum space for classical systems in thermodynamic equilibrium ensures that the temperature tensor is isotropic and identical to the thermodynamic temperature. But as soon as the kinetic component of the Hamiltonian $(-\hbar^2\sum_i \nabla_i^2/2m_i)$ becomes operationally dependent upon the density distributions, clearly the *kinetic* contribution to the pressure will be a function of the structural anisotropy, and

Table 6.1.1. De Boer parameter for the inert gases at the triple point

He	0.424
Ne	0.0939
Ar	0.0294
Kr	0.0161
Xe	0.0103

174

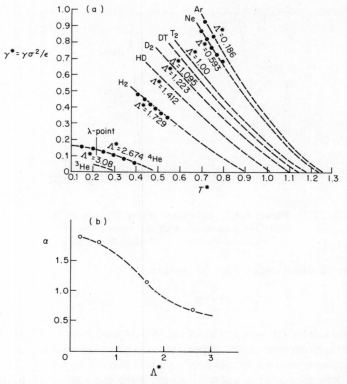

Figure 6.1.2 (a) Comparison of corresponding state and experimental determinations of surface tension with reduced temperature. (b) Functional form of $\alpha(\Lambda^*)$ in (6.1.1)

the kinetic temperature τ will generally be $\geqslant T$, the thermodynamic temperature. Perhaps this is most clearly understood in terms of the evident anisotropy in the zero-point contributions to the surface kinetics: obviously this is a characteristically quantum statistical phenomenon having no classical counterpart. The equation of state for a bulk quantum fluid may be written in classical form[2]

$$PV = Nk\tau - \frac{N\rho}{6} \int g_{(2)}(r)\nabla_1 \Phi(r) r \, d\mathbf{r}. \qquad (6.1.2)$$

For classical fluids τ is equal to the thermodynamic temperature T: all components τ_i of the τ-tensor are $\mathbf{1}_i T$, where $\mathbf{1}_i$ is the unit dyad in the configuration space of molecule i. For quantum mechanical fluids τ_i may depart widely from the classical value, and will in general be functions of position. Mazo and Kirkwood[2] observe that in the classical superposition approximation a quantum fluid may be described by the radial distribution function of a *classical* fluid at a temperature τ: all quantum effects, dynamical and statistical, are transferred to the determination of τ. Mazo and Kirkwood estimate $\tau = 12.6\,°\mathrm{K}$ at $T = 0$ for liquid ${}^4\mathrm{He}$, and this compares well with the semi-empirical results of Henshaw and Hurst.[3] We may therefore propose that in the pressure-tensor formulation of the surface tension $P(z)_{\text{quantum}} \geqslant P(z)_{\text{classical}}$, whereupon

$$\gamma = \int_{-\infty}^{\infty} \{P - P(z)\}\,\mathrm{d}z \tag{6.1.3}$$

is depressed below its classical value, and this is indeed found to be the case (Figure 6.1.1a).

In an attempt to avoid a direct quantum statistical description of the interphasal region, attention initially centred on the expression of the surface free energy in terms of elementary excitations. A proposition of Atkins,[4] for example, enables us to determine the limiting law of temperature dependence of ${}^4\mathrm{He}$ for temperatures sufficiently low that the whole liquid may be regarded as superfluid. Analysis is in terms of capillary waves as follows (c.f. Section 2.11). The surface part of the free energy per unit area may be written

$$A = \gamma_0 + 2\pi k_{\mathrm{B}} T \int \ln\left(1 - \exp\left(-\hbar\omega/k_{\mathrm{B}}T\right)\sigma(k)\,\mathrm{d}k \tag{6.1.4}$$

where $\sigma(k)$ is the distribution of capillary states per unit area in the range k to $k + \mathrm{d}k$,

$$\sigma(k)\,\mathrm{d}k = \frac{k\,\mathrm{d}k}{(2\pi)^2} \tag{6.1.5}$$

assuming a continuous distribution of k-states at the liquid surface. γ_0 represents the residuum of free energy at absolute zero which may be attributed to the surface modification of the zero point energy, plus the usual configurational excess: this may be identified as the surface tension at absolute zero. Combining equation (6.1.4), (6.1.5) we have

$$A = \gamma_0 - \frac{\hbar}{4\pi} \int \frac{k^2}{\exp\left(\hbar\omega/k_{\mathrm{B}}T\right) - 1}\,\mathrm{d}\omega. \tag{6.1.6}$$

At sufficiently low temperatures elastic Rayleigh waves are not excited, and only the long wavelength hydrodynamic capillary component contributes to the integral in (6.1.6). The shorter wavelength modes, having greater curvature, are more effectively restored under the action of the surface tension: the appropriate dispersion relation for capillary waves is[5]

$$\omega^2 = \frac{\gamma k^3}{\rho} \sim \frac{\gamma_0 k^3}{\rho} \tag{6.1.7}$$

where ρ is the liquid density. We have then, finally,

$$\gamma = \gamma_0 - \frac{\hbar}{4\pi} \left(\frac{\rho}{\gamma_0}\right)^{2/3} \int_0^{\infty} \frac{\omega^{4/3}}{\exp\left(\hbar\omega/k_{\mathrm{B}}T\right) - 1}\,\mathrm{d}\omega \tag{6.1.8}$$

where we have taken the upper limit as infinity since the integral converges rapidly. Integration gives

$$\gamma = \gamma_0 - \frac{(k_B T)^{7/3}}{4\pi\hbar^{4/3}} \left(\frac{\rho}{\gamma}\right)^{2/3} \Gamma(\tfrac{7}{3})\zeta(\tfrac{7}{3}) \tag{6.1.9}$$

where Γ and ζ are the gamma and zeta functions, respectively. We should point out that the intervention of viscous damping of the surface waves is explicitly excluded in restricting the analysis to superfluid ^{4}He ($<2.2\,^\circ$K): capillary waves of this kind do not exist in a Fermi liquid (^{3}He) since there the viscosity increases indefinitely as $T \to 0$, although there may be a tentative case to be made for the low temperature superfluid phase of ^{3}He ($T < 2\,^\circ$mK) (see Section 6.6).

It has been asserted[34] that one consequence of Atkins' capillary theory of surface tension is the divergence of the mean square displacement δ^2 of the surface from equilibrium. The surface displacement operator at a position $\mathbf{r}$ on a surface of area $\mathcal{A}$ is

$$Q(\mathbf{r}) = \frac{1}{\mathcal{A}^{1/2}} \sum_{\mathbf{k}} Q_{\mathbf{k}} \exp i\mathbf{k}\cdot\mathbf{r} \tag{6.1.10}$$

and the expectation value of Q^2 at temperature T is

$$\delta^2 = \langle |Q^2(\mathbf{r})|\rangle_T = \frac{1}{\mathcal{A}} \sum_{\mathbf{k}} \langle Q_{\mathbf{k}} Q_{-\mathbf{k}}\rangle_T \tag{6.1.11}$$

where $\langle\ \rangle_T$ denotes a quantum statistical average. Cole[35] observes that the full Hamiltonian for ripple waves on the surface of a semi-infinite fluid of mass density ρ and surface tension γ is of the form

$$\mathcal{H} = \sum_{\mathbf{k}} (v(k)Q_{\mathbf{k}}Q_{-\mathbf{k}} + t(k)P_{\mathbf{k}}P_{-\mathbf{k}}) \tag{6.1.12}$$

where

$$v(k) = \tfrac{1}{2}(\rho g + \gamma k^2) \tag{6.1.13}$$
$$t(k) = \tfrac{1}{2}k/\rho$$

giving the dispersion relation

$$\omega^2(k) = (\gamma/\rho)k^3 + gk \tag{6.1.14}$$

which, by comparison with (6.1.7), is seen to differ from Atkins' relation by the inclusion of the gravitational term which is unimportant except at long wavelengths.

It follows from (6.1.12) that

$$\langle Q_{\mathbf{k}}Q_{-\mathbf{k}}\rangle = \frac{\hbar\omega(k)}{4v(k)} \coth \tfrac{1}{2}\beta\hbar\omega(k) \qquad \beta = (k_B T)^{-1} \tag{6.1.15}$$

and for its long wavelength limit

$$\to \frac{k_B T}{\rho g} \qquad k \to 0, \qquad T \neq 0$$

$$\to \frac{\hbar}{2\rho} \left(\frac{k}{g}\right)^{1/2} \qquad k \to 0, \qquad T = 0. \tag{6.1.16}$$

Without the gravitational term in (6.1.12) and (6.1.14) a $1/k^2$ behaviour is obtained for (6.1.16) which gives rise to a logarithmic divergence to the sum for δ^2 (6.1.11). We can see from (6.1.11) and (6.1.15) that δ^2 is a sum of a zero-point amplitude δ_0^2 and a temperature-dependent term δ_T^2. Converting the sum to an integral, and recognizing that there will be an upper cut-off as k approaches the inverse interparticle distance, we take our upper limit as k_m such that the total number of surface modes equals the number of surface atoms:

$$\delta^2 = \delta_0^2 + \delta_T^2$$

$$\delta_0^2 = \frac{\hbar}{2\pi\rho} \int_0^{k_m} \frac{1}{2} \frac{k^2 \, dk}{\omega(k)} \sim \frac{\hbar k_m^{3/2}}{6\pi(\rho\gamma)^{1/2}} \tag{6.1.17a}$$

$$\delta_T^2 = \frac{\hbar}{2\pi\rho} \int_0^{k_m} \frac{k^2}{\omega(k)} \frac{dk}{\exp(\hbar\omega(k)/k_B T) - 1}. \tag{6.1.17b}$$

Over the range of interest $(10^{-4}\,{}^\circ K < T < 2\,{}^\circ K)$

$$\delta_T^2 = \frac{k_B T}{12\pi\gamma} \ln(\Gamma T^4) \qquad \Gamma = \frac{\gamma}{\rho g^3} \left(\frac{k_B}{\hbar}\right)^4. \tag{6.1.18}$$

We note that the inclusion of gravity ensured the convergence of δ_0^2, even though it does not appear explicitly in the final expression: a truly remarkable result. Cole does however make the observation that a planar surface would be unstable with respect to long wavelength fluctuations in the absence of a gravitational field, the stable configuration of a volume of liquid with surface tension being a spherical drop. In contrast, δ_T^2 depends explicitly on g, although this too diverges as $g \to 0$.[36]

The resulting values of δ are given in Figure 6.1.3 for liquid ^{4}He.

For classical fluids we have $\beta\hbar\omega(k_m) \ll 1$, and so from (6.1.17b)

$$\delta_T^2 = (4\pi\beta\gamma)^{-1} \ln(1 + (\gamma/\rho g)k_m^2) \gg \delta_0^2 \tag{6.1.19}$$

a result already obtained by Buff et al.[36]

An alternative approach is to regard the superfluid ^{4}He system as an ideal, degenerate assembly of non-interacting quasi-particles—phonons and rotons. Using equation (6.1.4) in conjunction with the dispersion relation $\omega = ck$ (c = velocity of sound in ^{4}He(II)), Singh[6] determines the excess free energy developed by the introduction of a bounding surface to be

$$\gamma = \gamma_0 - \pi m\zeta(2)kT^2/2h^2$$

$$= \gamma_0 - 7.5 \times 10^{-3}T^2 \text{ erg cm}^{-2}. \tag{6.1.20}$$

178

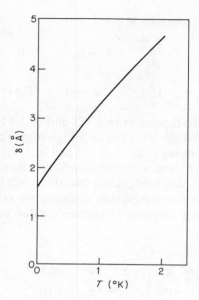

Figure 6.1.3 Dependence of the
mean square displacement of the
surface upon temperature

Of course, in a liquid the modes are not purely harmonic, so in addition to the phonon contributions we might also anticipate other elementary excitations to participate in the form of turbulence and vortex flow. Equations (6.1.9) and (6.1.20) are difficult to distinguish experimentally, although measurements of the surface tension[7,8] seem to suggest a temperature dependence which is larger than that predicted by Atkins' expression (6.1.9). Chen,[9] however, claims recently to have confirmed Atkins' law of surface tension–temperature dependence for superfluid ^{4}He, whilst Edwards et al.,[10] have subsequently improved Atkins' formulation by incorporating effects due to compressibility and phonon dispersion.

Let us point out, however, that the only experimental tests of the hypothesis that these surface excitations *are* quantized capillary waves, or *ripplons*,† have involved the measurement of the temperature dependence of surface tension, which does not exclude other possibilities.[7] As in Feynman's phonon theory of bulk ^{4}He, variational functions can be constructed to describe the surface excitations;[45] these functions are then varied to minimize the energy expectation value subject to the constraint that they be orthogonal to the ground state. It is found that two modes of excitation may arise,[46] each corresponding to a specific microscopic configuration—either ripplons[47] or the usual bulk phonons. The ripplon spectrum has been determined by Miller[46] and shown to possess the momentum dependence $k^{3/2}$ characteristic of classical capillary waves (6.1.7 *et seq.*) as initially

† Or perhaps even *surfons*.

proposed by Atkins.[4,7] An interesting development is that of Saam[47] who shows that if the angle of incidence or reflection of a neutron beam is less than the critical angle of reflection, then inelastic scattering can only be a consequence of absorption or emission of surface excitations. For the case where at least either the incident or the reflected beam is at less than the critical angle, the structure factor contains sharp peaks corresponding to the creation or annihilation of ripplons, separated from a broad background owing to the creation or annihilation of phonons near the surface of the liquid. If neither beam is below the critical angle, giant peaks corresponding to the creation or annihilation of bulk phonons appear in the background. Although Saam gives some numerical results for the predicted cross sections, no experimental confirmation has appeared as yet.

Interesting as these surface excitation approaches are, we have ultimately to confront the full quantum statistical problem if we require a microscopic description of the interfacial regions of quantum fluids which is capable of yielding both structural and thermodynamic information which can parallel that of the classical fluids developed in the preceding chapters. This we consider in the following section.

6.2 A formal quantum statistical description of the liquid surface

We consider a system of N identical particles confined to a cubic volume of side $l(l \gg \sigma)$ and resolved into a liquid and vapour phase by a weak gravitational field whose effect is otherwise negligible (Figure 6.2.1). The energy eigenvalues of the system are designated $E_1, E_2, \ldots, E_i, \ldots$ whereupon we may define the quantum-mechanical N-body partition function, or *zustandsumme*

$$Z_N(T, V, \mathscr{A}) = \sum_i \exp\left(-E_i/kT\right) \tag{6.2.1}$$

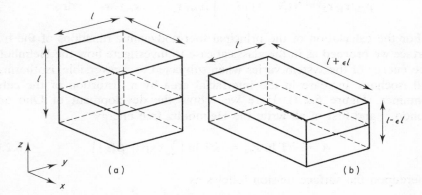

Figure 6.2.1 Geometry for the isothermal increase in interfacial area of a fluid at constant N, V

where the sum extends over all energy states, and we have included an explicit dependence upon surface area. (Note that if the sum were to run over energy *levels*, then we should have to multiply each level by its degeneracy.)

As in classical mechanics, the probability of finding a system of the canonical ensemble in the energy state E_i is $\exp(-E_i/kT)/\sum_i \exp(-E_i/kT)$, the partition function adopting its conventional role as a normalizing factor. The N-particle density matrix Γ_N is then given by[11]

$$\Gamma_N(\mathbf{r}_1, \mathbf{r}_2, \ldots, \mathbf{r}_N; \mathbf{r}_1', \mathbf{r}_N')$$

$$= \frac{1}{Z_N} \sum_i \exp(-\beta E_i) \Psi_i(\mathbf{r}_1, \ldots, \mathbf{r}_N) \Psi_i^*(\mathbf{r}_1', \ldots, \mathbf{r}_N') \quad (6.2.2)$$

at temperature $T = 1/k\beta$. Here the Ψ_i are a complete orthonormal set of symmetrized or antisymmetrized eigenfunctions: Ψ_i^* is the complex conjugate. The probability density corresponds to the diagonal part of Γ_N:

$$\rho_{(N)}(\mathbf{r}_1, \ldots, \mathbf{r}_N) = \frac{1}{Z_N} \sum_i \exp(-\beta E_i) \Psi_i(\mathbf{r}_1, \ldots, \mathbf{r}_N) \Psi_i^*(\mathbf{r}_1, \ldots, \mathbf{r}_N)$$

$$(6.2.3)$$

with reduced probability densities following as

$$\rho_{(n)}(\mathbf{r}_1, \ldots, \mathbf{r}_n) = \frac{N!}{(N-n)! \, n!} \int \rho_{(N)}(\mathbf{r}_1, \ldots, \mathbf{r}_N) \, d\mathbf{r}_{n+1} \cdots d\mathbf{r}_N. \quad (6.2.4)$$

For present purposes we shall be primarily interested in the one- and two-particle density distributions

$$\rho_{(1)}(\mathbf{r}_1) = N \int \cdots \int \rho_{(N)}(\mathbf{r}_1, \ldots, \mathbf{r}_N) \, d\mathbf{r}_2 \cdots d\mathbf{r}_N$$

$$(6.2.5)$$

$$\rho_{(2)}(\mathbf{r}_1, \mathbf{r}_2) = N(N-1) \int \cdots \int \rho_{(N)}(\mathbf{r}_1, \ldots, \mathbf{r}_N) \, d\mathbf{r}_3 \cdots d\mathbf{r}_N.$$

For the calculation of the principal thermodynamic functions of the free surface we proceed as in the classical case to investigate how the Helmholtz free energy of the system varies when subjected to a reversible, isothermal, and isochoric increase in the interfacial area by a distortion of the cubic container (Figure 6.2.1). Here we follow the development of Ono and Kondo,[12] and first of all write the Helmholtz free energy

$$A = -kT \ln Z_N = -kT \ln \left(\sum_i \exp(-\beta E_i) \right) \quad (6.2.6)$$

whereupon the surface tension follows as

$$\gamma = \left(\frac{\partial A}{\partial \mathscr{A}} \right)_{V,T} = \frac{kT}{Z_N} \sum_i \left(\frac{\partial E_i}{\partial \mathscr{A}} \right)_{V,T} \exp(-\beta E_i) \quad (6.2.7)$$

where, of course, E_i is a function of both volume V and interfacial area $\mathcal{A}$. It remains to calculate $(\partial E_i/\partial \mathcal{A})_{V,T}$.

The Hamiltonian operator of the system of N particles interacting through the pair potential $\Phi(r_{ij})$ is, as usual,

$$\mathcal{H} = -\frac{\hbar^2}{2m} \sum_{i=1}^{N} \nabla_i^2 + \sum_{i<j}^{N} \Phi(r_{ij}). \tag{6.2.8}$$

The wavefunctions Ψ must vanish at the walls of the containing vessel, and satisfy the time-independent Schrödinger equation

$$\mathcal{H}\Psi_N(\mathbf{r}_1, \ldots, \mathbf{r}_N) = E\Psi_N(\mathbf{r}_1, \ldots, \mathbf{r}_N) \tag{6.2.9}$$

where Ψ_N is an N-body or total wavefunction of the system having an energy eigenvalue E. We now perform two simultaneous, isothermal, distortions of the containing volume, whose net effect is simply to increase the interfacial area of the system, i.e. from l^2 to $l^2(1+\varepsilon)$ (Figure 6.2.1). The volume is compressed in the z-direction by an amount εl, and extended in the x-direction by the same amount. This change in the configuration will obviously modify both Ψ_N and E, so we write $E(\varepsilon)$ and $\Psi_N(\varepsilon)$. New boundary conditions are required for $\Psi_N(\varepsilon)$ to vanish on the displaced faces. Clearly,

$$\left(\frac{\partial E(\varepsilon)}{\partial \varepsilon}\right)_{\varepsilon=0} = l^2 \left(\frac{\partial E}{\partial \mathcal{A}}\right)_V$$

and so, provided we can evaluate $l^{-2}(\partial E/\partial \varepsilon)_{\varepsilon=0}$, we are able to determine the expression (6.2.7) for the surface tension.

Transforming to appropriately expanded variables:

$$\bar{x}_i = x_i/(1+\varepsilon) \qquad \bar{y}_i = y_i \qquad \bar{z}_i = z_i/(1-\varepsilon)$$

we may express the Hamiltonian operator (6.2.8) in these transformed Cartesian coordinates as follows:

$$\bar{\mathcal{H}}(\varepsilon) = -\frac{\hbar^2}{2m} \sum_i^N \left\{ \frac{\partial^2}{(1+\varepsilon)^2 \partial \bar{x}_i^2} + \frac{\partial^2}{\partial \bar{y}_i^2} + \frac{\partial^2}{(1-\varepsilon)^2 \partial \bar{z}_i^2} \right\} + \sum_{i<j}^N \phi(r_{ij}, \varepsilon). \tag{6.2.10}$$

The wavefunction adopts the form

$$\bar{\Psi}_N(\varepsilon) = \Psi_N((1+\varepsilon)\bar{x}_1, \bar{y}_1, (1-\varepsilon)\bar{z}_1, \ldots, (1+\varepsilon)\bar{x}_N, \bar{y}_N, (1-\varepsilon)\bar{z}_N)$$

which, when substituted in the Schrödinger equation

$$\bar{\mathcal{H}}(\varepsilon)\bar{\Psi}(\varepsilon) = E(\varepsilon)\bar{\Psi}(\varepsilon) \tag{6.2.11}$$

satisfies the original boundary conditions, irrespective of ε simply because we have scaled the system by a factor of ε in the x and z directions. Differentiating (6.2.11) with respect to ε, and evaluating at $\varepsilon=0$ we have, in an obvious notation,

$$\bar{\mathcal{H}}(0)\frac{\partial\bar{\Psi}_N(0)}{\partial\varepsilon}+\frac{\partial\bar{\mathcal{H}}(0)}{\partial\varepsilon}\bar{\Psi}_N(0)=E(0)\frac{\partial\bar{\Psi}_N(0)}{\partial\varepsilon}+\frac{\partial E(0)}{\partial\varepsilon}\bar{\Psi}_N(0). \quad (6.2.12)$$

Multiplying by $\bar{\Psi}_N^*(0)$, and integrating over configuration space throughout the vessel

$$\int \bar{\Psi}_N^*(0)\bar{\mathcal{H}}(0)\frac{\partial\bar{\Psi}_N(0)}{\partial\varepsilon}\,d\mathbf{r}^N + \int \bar{\Psi}_N^*(0)\frac{\partial\bar{\mathcal{H}}(0)}{\partial\varepsilon}\bar{\Psi}_N(0)\,d\mathbf{r}^N$$

$$= E(0)\int \bar{\Psi}_N^*(0)\frac{\partial\bar{\Psi}_N(0)}{\partial\varepsilon}\,d\mathbf{r}^N +\frac{\partial E(0)}{\partial\varepsilon}. \quad (6.2.13)$$

From the Hermitian nature of the operator $\mathcal{H}(0)$ the left-hand side of (6.2.13) may be rewritten as

$$\int \bar{\Psi}_N^*(0)\bar{\mathcal{H}}(0)\frac{\partial\bar{\Psi}_N(0)}{\partial\varepsilon}\,d\mathbf{r}^N = \int (\bar{\mathcal{H}}(0)\bar{\Psi}_N(0))^*\frac{\partial\bar{\Psi}_N(0)}{\partial\varepsilon}\,d\mathbf{r}^N$$

$$= E(0)\int \bar{\Psi}_N^*(0)\frac{\partial\bar{\Psi}_N(0)}{\partial\varepsilon}\,d\mathbf{r}^N \quad (6.2.14)$$

whereupon (6.2.13) reduces to

$$\frac{\partial E(0)}{\partial\varepsilon}=\int \bar{\Psi}_N^*(0)\frac{\partial\bar{\mathcal{H}}(0)}{\partial\varepsilon}\bar{\Psi}_N(0)\,d\mathbf{r}^N. \quad (6.2.15)$$

Recalling the expression for $\bar{\mathcal{H}}$ (6.2.10) we have

$$\frac{\partial E(0)}{\partial\varepsilon}=\int \Psi_N^*(0)\left[-\frac{\hbar^2}{m}\sum_i^N \left(\frac{\partial^2}{\partial z_i^2}-\frac{\partial^2}{\partial x_i^2}\right)\right.$$

$$\left.+\sum_{i<j}^N \left(\frac{x_{ij}^2-z_{ij}^2}{r_{ij}}\right)\frac{d\Phi(r_{ij})}{dr_{ij}}\right]\Psi_N(0)\,d\mathbf{r}^N \quad (6.2.16)$$

since $\bar{\Psi}_N(0)$ is nothing other than the original wavefunction Ψ_N! The expression for the surface tension (6.2.7) follows directly from (6.2.10) and (6.2.16):

$$\gamma = Z_N^{-1}l^{-2}\sum_i \exp(-E_i/kT)\int \Psi_i^*\left[-\frac{\hbar^2}{m}\sum_i^N \left(\frac{\partial^2}{\partial z_i^2}-\frac{\partial^2}{\partial x_i^2}\right)\right.$$

$$\left.+\sum_{i<j}^N \left(\frac{x_{ij}^2-z_{ij}^2}{r_{ij}}\right)\frac{d\Phi(r_{ij})}{dr_{ij}}\right]\Psi_i\,d\mathbf{r}^N \quad (6.2.17)$$

where E_i is the eigenvalue appropriate to eigenstate Ψ_i. The expression is now virtually complete; we note, however, that the average momentum of particle i is

$$\bar{\Phi}_i = Z_N^{-1}\sum_i \exp(-E_i/kT)\int \Psi_i^*\left(\frac{\hbar}{i}\nabla_i\right)\Psi_i\,d\mathbf{r}^N \quad (6.2.18)$$

in which case (6.2.17) simplifies to give

$$\gamma = \frac{N}{\mathscr{A}m}(\bar{p}_z^2 - \bar{p}_x^2) + \frac{1}{2\mathscr{A}} \int \int \left(\frac{x_{12}^2 - z_{12}^2}{r_{12}}\right) \frac{d\Phi(r_{12})}{dr_{12}} \rho_{(2)}(\mathbf{r}_1, \mathbf{r}_2) \, d\mathbf{r}_1 \, d\mathbf{r}_2$$

$$(6.2.19)$$

where we have used (6.2.3). We observe at once that the classical term in (6.2.19) is supplemented by what might be called the surface excess of the pair momentum current density tensor. This term is of purely quantum mechanical origin, having no classical counterpart, and arises from the configurational dependence of the momentum operator (6.2.18) applied in a region of structural inhomogeneity. A brief discussion was given in Section 6.1.

6.3 The surface structure of liquid ^{4}He

As we emphasised in Section 6.1, estimates of the principal thermodynamic functions of the liquid surface based on excitation analyses are incapable of yielding anything but the most general information on the single-particle density transition $\rho_{(1)}(z)$. Beyond these treatments, the structurally more specific calculations may be resolved into two main classes—the phenomenological and the variational, both of which we now proceed to consider in detail.

Representative of the former category are the treatments of Regge[13] and Padmore and Cole.[14] The basic method in the two approaches is to express the local free energy of the system as a functional $F[\rho_{(1)}(z)]$ of the single-particle density profile. Such a procedure has been discussed by Widom[15] and by Hohenberg and Kohn[16] for Bose and Fermi fluids, respectively. The optimum profile is then determined by the condition $\delta F/\delta \rho_{(1)}(z) = 0$, which is equivalent to the requirement of constancy of the chemical potential across the interface.

The Padmore–Cole functional is, essentially, of Cahn–Hilliard form (2.4.20), with its local density free energy and squared gradient terms supplemented by a quantum kinetic energy term:

$$F_{\text{kin}} = -\frac{\hbar^2}{2m} \int \sqrt{[\rho_{(1)}(\mathbf{r})]} \nabla^2 (\sqrt{[\rho_{(1)}(\mathbf{r})]}) \, d\mathbf{r}. \qquad (6.3.1)$$

The use of a Cahn–Hilliard functional away from the critical point is of questionable propriety, and this point was discussed at some length for classical fluids in Section 2.4. However, the kinetic component (6.3.1) assists the squared gradient term in favouring a delocalized interfacial zone, and so we may expect anisotropy in the zero point motions to broaden the density transition relative to a classical interface. We should point out that (6.3.1) is known to be exact for a non-interacting Bose system, or a system with weak point interactions.[15,17]

184

The resulting expression is solved numerically for the condition $\delta F/\delta\rho_{(1)}(z)$ $=0$, and the solution is shown in Figure 6.3.1, together with the results of a variational calculation[18] to be discussed in the next section. The monotonic profile is estimated to have a surface thickness of 5.9 Å on the basis of the 10–90% density criterion, and is substantially greater than its classical counterparts—presumably attributable to the delocalizing effect of the kinetic term. The asymmetry in the high and low density regions of the profile is principally a consequence of the asymmetry of the kinetic term with respect to the transformation of $\rho_{(1)}(z)/\rho_L \rightarrow 1-\rho_{(1)}(z)/\rho_L$.

The variational approaches may be subdivided as to whether a trial profile or a trial wavefunction is adopted in the variational minimization of surface energy.

Padmore and Cole,[14] for example, have obtained a profile graphically indistinguishable from that shown in Figure 6.3.1 by means of a variational adjustment of the trial profile

$$\rho_{(1)}(z) = \begin{cases} \rho_L\{1-[1+(z/\xi)^\gamma]\exp{(-z/\xi)^\gamma}\} & z>0 \\ 0 & z<0. \end{cases} \tag{6.3.2}$$

Of course, the adoption of such a profile excludes from the outset any possibility of a structured interface. However, the parameter ξ determines the thickness of the surface, whilst γ controls the asymmetry, and these authors find that the choice $\gamma=1.7$, $\xi=3.78$ Å yields a good representation of the numerical solution. Indeed, even a single-parameter trial function, with $\gamma=1$, yields a variational energy only 3% higher. However, it must be said that whilst it is easier to perform the variational calculation by prescribing the form of the density profile, it is also easy inadvertently to

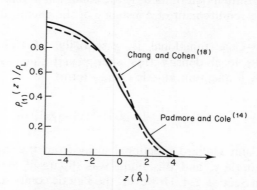

Figure 6.3.1 Functional minimization of the local free energy with respect to the density profile yields the optimum density transition $\rho_{(1)}(z)$. Also shown is Chang and Cohen's profile determined on the basis of equations (6.3.14a,b)

introduce unnecessary curvature into the single-particle factors in the wavefunction, and thereby raise the energy of the system.

Variational treatments based on a trial ground state wavefunction for ^{4}He generally centre on the Jastrow approximation, the motivation being that it is the simplest function which produces finite matrix elements for a potential energy which has short range divergences, such as helium.[1,22] The Jastrow function for a boson system (designated by the subscript B) is defined as

$$\Psi_B(\mathbf{r}_1, \ldots, \mathbf{r}_N) = \exp\left[\frac{1}{2}\sum_{i<j} u(r_{ij})\right] \tag{6.3.3}$$

where the function $u(r)$ is required to become increasingly negative at small r in order that the potential energy expectation value remain finite.

Nothing has been said as yet regarding the choice of $u(r)$ other than its short range behaviour. However, the similarity between the square of the Jastrow function and the Boltzmann factor establishes the mathematical equivalence of relating the radial distribution function to $(-\Phi(r)/kT)$ and the Jastrow radial distribution $g_J(r)$ to $u(r)$. Thus, formally,

$$u(r) = -\Phi(r)/kT \tag{6.3.4}$$

so that $u(r)$ operates as a fictitious pair potential at some arbitrary temperature T. The earliest and most widely used parametric representation of $u(r)$ involves two parameters,[32] p and a:

$$u(r) = -\left(\frac{a\sigma}{r}\right)^p \qquad \sigma = 2.556 \text{ Å}, \, a = 1.17, \, p = 5. \tag{6.3.5}$$

Somewhat surprisingly, this function is structureless and does not reflect the evident oscillations in $g_J(r)$. However, as Campbell[1] observes, it is the Laplacian acting on $u(r)$ which determines the kinetic energy, and increasing the structure will make the kinetic energy more positive. Certainly all calculations performed so far with realistic potentials have shown a monotonic $u(r)$ at all densities, including the lowest density of the solid.

The special boundary condition on the wavefunction Ψ_B arising from the presence of the free surface is that $\Psi_B \to 0$ as $z \to +\infty$, and the modification generally adopted is

$$\Psi_B^s(\mathbf{r}_1, \ldots, \mathbf{r}_N) = \exp\left[\frac{1}{2}\sum_{i>j} u(r_{ij}) + \frac{1}{2}\sum_i^N t_B(z_i)\right] \tag{6.3.6}$$

where the superscript s designates a system with a surface. Far outside the drop $t_B(z_i) \to +\infty$, whilst in the bulk $t_B(z_i) = 0$, this latter modulating function adopting a role analogous to a surface potential. In an earlier calculation by Lekner[31] the bulk ground state wavefunction for ^{4}He was used in place of the Jastrow factor. This, however, is quite incorrect since Lekner's function satisfies the wrong boundary conditions, although it does lead to tremendous simplifications! The result is, unfortunately, of no use since it yields a non-negative surface energy, giving rise to a minimum value of zero.

The Hamiltonian describing the boson system is

$$\mathcal{H} = -\frac{\hbar^2}{2m} \sum_{i=1}^{N} \nabla_i^2 + \frac{1}{2} \sum_{i,j \neq i} \Phi(r_{ij}) \tag{6.3.7}$$

and, using the variational wavefunction of the form (6.3.6), we obtain the expectation energy

$$\langle \mathcal{H} \rangle = -\frac{\hbar^2}{8m} \left\{ \iint \rho_{(2)}(\mathbf{r}_1, \mathbf{r}_2) \nabla_1^2 u(r_{12}) \, d\mathbf{r}_1 \, d\mathbf{r}_2 + \int \rho_{(1)}(\mathbf{r}_1) \nabla_1^2 t_B(\mathbf{r}_1) \, d\mathbf{r}_1 \right\}$$
$$+ \frac{1}{2} \iint \rho_{(2)}(\mathbf{r}_1, \mathbf{r}_2) \Phi(r_{12}) \, d\mathbf{r}_1 \, d\mathbf{r}_2 \tag{6.3.8}$$

where

$$\rho_{(2)}(\mathbf{r}_1, \mathbf{r}_2) = \frac{N(N-1) \int \cdots \int \Psi_B^2(\mathbf{r}_1, \ldots, \mathbf{r}_N) \, d\mathbf{r}_3 \cdots d\mathbf{r}_N}{\int \cdots \int \Psi_B^2(\mathbf{r}_1, \ldots, \mathbf{r}_N) \, d\mathbf{r}_1 \cdots d\mathbf{r}_N}$$
$$= \rho_{(1)}(\mathbf{r}_1) \rho_{(1)}(\mathbf{r}_2) g_{(2)}(\mathbf{r}_1, \mathbf{r}_2) \tag{6.3.9}$$

$$\rho_{(1)}(\mathbf{r}_1) = N \int \cdots \int \Psi_B^2(\mathbf{r}_1, \ldots, \mathbf{r}_N) \, d\mathbf{r}_2 \cdots d\mathbf{r}_N. \tag{6.3.10}$$

Differentiating (6.3.10) gives

$$\nabla_1 \rho_{(1)}(\mathbf{r}_1) = \rho_{(1)}(\mathbf{r}_1) \nabla_1 t(\mathbf{r}_1) + \int \rho_{(2)}(\mathbf{r}_1, \mathbf{r}_2) \nabla_1 u(r_{12}) \, d\mathbf{r}_2 \tag{6.3.11}$$

which enables us to eliminate $t_B(r)$ from (6.3.8):

$$\langle \mathcal{H} \rangle = \frac{\hbar^2}{8m} \left\{ \int \nabla_1 \rho_{(1)}(\mathbf{r}_1) \cdot \nabla_1 \ln \rho_{(1)}(\mathbf{r}_1) \, d\mathbf{r}_1 \right.$$
$$\left. - \iint [\nabla_1 u(r_{12}) \cdot \nabla_1 \rho_{(1)}(\mathbf{r}_1) + \rho_{(1)}(\mathbf{r}_1) \nabla_1^2 u(r_{12})] \rho_{(1)}(\mathbf{r}_2) g_{(2)}(\mathbf{r}_1, \mathbf{r}_2) \, d\mathbf{r}_1 \, d\mathbf{r}_2 \right\}$$
$$+ \frac{1}{2} \iint \rho_{(1)}(\mathbf{r}_1) \rho_{(1)}(\mathbf{r}_2) g_{(2)}(\mathbf{r}_1, \mathbf{r}_2) \Phi(r_{12}) \, d\mathbf{r}_1 \, d\mathbf{r}_2. \tag{6.3.12}$$

The surface energy is obtained by subtracting from the above expression the bulk energy of N particles, and dividing by the surface area:

$$u_s = -\frac{\hbar^2}{2m} \int_{-\infty}^{\infty} \sqrt{(\rho_{(1)}(z))} \frac{d^2}{dz^2} \sqrt{(\rho_{(1)}(z))} \, dz + \int_{-\infty}^{\infty} \rho_{(1)}(z) [\varepsilon(z) - \varepsilon_0] \, dz \tag{6.3.13}$$

where the first term represents the surface excess zero point energy (c.f.

(6.3.1)), and

$$\varepsilon(z) = -\frac{\hbar^2}{8m} \int g_{(2)}(\mathbf{r}_1, \mathbf{r}_2)[\nabla_2 u(r_{12}) \cdot \nabla_2 \rho_{(1)}(\mathbf{r}_2) + \rho_{(1)}(\mathbf{r}_2)\nabla_2^2 u(r_{12})]\, d\mathbf{r}_2$$

$$+ \frac{1}{2} \int \rho_{(2)}(\mathbf{r}_2)\Phi(r_{12})g_{(2)}(\mathbf{r}_1, \mathbf{r}_2)\, d\mathbf{r}_2$$

$$\varepsilon_0 = -\rho_L \int g_{(2)}(r)\left[\frac{\hbar^2}{8m} \nabla^2 u(r) - \frac{1}{2}\Phi(r)\right]d\mathbf{r}.$$

Before the surface energy (6.3.13) can be evaluated, we require some closure for $\rho_{(2)}(\mathbf{r}_1, \mathbf{r}_2)$. The approximation adopted is to write[18,21]

$$\rho_{(2)}(\mathbf{r}_1, \mathbf{r}_2) = \rho_{(1)}(z_1)\rho_{(1)}(z_2)g_{(2)}^L[r \mid \rho_{eff}(z_1, z_2)] \qquad (6.3.14)$$

where $g_{(2)}^L[r \mid \rho_{eff}]$ represents the local *isotropic* radial distribution function appropriate to a uniform density ρ_{eff}. A variety of specifications have been adopted:

$$\rho_{eff}(z_1, z_2) = [\rho_{(1)}(z_1)\rho_{(1)}(z_2)]^{1/2} \text{ [21]} \qquad (6.3.14a)$$

$$\rho_{eff}(z_1, z_2) = \tfrac{1}{2}[\rho_{(1)}(z_1) + \rho_{(1)}(z_2)] \text{ [18]} \qquad (6.3.14b)$$

$$\rho_{eff}(z_1, z_2) = \rho_{(1)}[\tfrac{1}{2}(z_1 + z_2)] \text{ [18]} \qquad (6.3.14c)$$

and these expressions have been critically evaluated in Section 2.4. Chang and Cohen[18] report that (6.3.14a) yields an inferior result to (b) and (c), both of which give essentially the same result (Table 6.3.1). The $g_{(2)}^L[r \mid \rho_{eff}]$ are determined on the basis of the BGY,[21] HNC,[18] and PY[18] equations.

A comparison of the various surface parameters is given in Table 6.3.1. It should be pointed out that the value of the surface energy near the minimum, corresponding to a surface thickness of 1.6–4.0 Å, is very flat,[19] and so the estimated thicknesses should not be taken too seriously.[20,56]

Buchan and Clark[23] have extended this analysis to the liquid ^{3}He surface: we shall consider their treatment in Section 6.4.

Liu, Kalos, and Chester[24] have tested the internal consistency of such an approach by adopting a trial distribution $\rho_{(1)}(z)$ and using (6.3.11) to determine $t(z)$ which, they point out, shows pronounced curvature accounting for an overestimate of the surface energy, in precisely the same manner

Table 6.3.1

Equation	Surface thickness ^{4}He (10%–90%)		Surface energy erg cm^{-2}	
(6.3.14a)[21]	2.4 Å		0.36	
(6.3.14b)[18]	4.6 Å*	4.4 Å†	0.42*	0.43†
(6.3.14c)[18]	4.6 Å*	4.4 Å†	0.42*	0.42†
Experiment[33]			0.378	

* using $\rho_{(1)}(z) = \tfrac{1}{2}e^{-\beta z}(z>0)$; $1 - \tfrac{1}{2}e^{\beta z}(z>0)$ $\quad (\beta = 0.7)$
† using $\rho_{(1)}(z) = (1 + e^{\beta z})^{-1}$ $\quad (\beta = 1.0)$

as the spurious structure in $u(r)$ discussed earlier. The wavefunction thus determined, (6.3.6) may then be substituted in

$$\rho_{(1)}(z) = \frac{N \int \cdots \int \Psi_B^{s2}(\mathbf{r}_1, \ldots, \mathbf{r}_N) \, d\mathbf{r}_2 \cdots d\mathbf{r}_N}{\langle \Psi_B^s \mid \Psi_B^s \rangle} \qquad (6.3.15)$$

and the distributions $\rho_{(1)}(z)$ compared. Discrepancy in the profiles may be attributed to the approximation introduced through equations (6.3.14a,b,c). Liu et al.[24] report an increase of 20% in the surface energy of the output distribution, together with a pronounced peak where none existed in the input function. Moreover, the specification of a monotonic trial profile does, of course, eliminate the possible development of structural features at the surface. Again, the assumption of locally isotropic distributions in regions of pronounced structural inhomogeneity is open to objection precisely as in the case of classical systems.

Liu et al. adopt a modified Jastrow wavefunction of the form (6.3.6), but discuss the surface structure of a thin ^{4}He film, in which case $t(z_i) \to \infty$ as $|z_i| \to \infty$. Writing the surface modulating function

$$\exp\left[-\frac{1}{2} \sum_{i=1}^{N} t_B(z_i) \right] = \prod_{i=1}^{N} h(z_i) \qquad (6.3.16)$$

two trial functions were adopted:

$$h(z) = [1 + \exp k(|z| - z_0)]^{-1} \qquad (6.3.16a)$$

and

$$h(z) \begin{cases} = 1 & |z| < z_0 \\ = 1 + \exp k(|z| - z_0)^q & |z| > z_0. \end{cases} \qquad (6.3.16b)$$

The variational parameters k and q control the width of the transition zone, whilst z_0 determines its location.

Using the trial functions (6.3.16a,b), the surface energy may be variationally determined with respect to the parameters k and q, and the transition profile thereby determined through (6.3.16). Liu et al. conclude that $\rho_{(1)}(z)$ shows a weak oscillatory structure. Moreover, these authors believe that an exact solution would reveal an *enhancement* of the oscillations, basing their conclusions on analogous calculations for ^{4}He in a channel.

In another variational determination of a trial wavefunction, Croxton[25] takes a symmetrical correlated Bijl–Dingle–Jastrow wavefunction which modifies in the vicinity of the ^{4}He surface:

$$\Psi_B^s(\mathbf{r}_1, \ldots, \mathbf{r}_N) = \prod_{i>j=1}^{N} \exp\left[\tfrac{1}{2} u(r_{ij}) \sum_{l=0}^{\infty} \sum_{m=0}^{l} A_{lm}(z_i) P_l^m(\cos \theta) \Phi(\pm im\phi) \right]. \qquad (6.3.17)$$

The coefficients of the harmonic expansion in the vicinity of the free surface specify the spatial dependence of the wavefunction, and as such, represents an improvement on the local isotropic representation. In the bulk liquid (6.3.17) reduces to the isotropic Jastrow function:

$$\prod_{i>j=1}^{N} \exp\left[\frac{1}{2} u(r_{ij})\Gamma(z_i, \theta, \phi)\right] \rightarrow \prod_{i>j=1}^{N} \exp\left[\frac{1}{2} u(r_{ij})\right] \qquad (6.3.18)$$

as $z_i \rightarrow -\infty$. In the present case $u(r_{ij})$ was determined by the Abe relation.[26,27]

The angular dependence of the trial function was restricted partly on symmetry grounds and partly on the grounds of computational expediency: obviously practical interest is in a trial function containing a relatively small number of components, beyond which the difficulties involved in solving the secular determinant become prohibitive. The trial function was, in fact, restricted to the first two zonal harmonics ($l = 0, 1; m = 0$)—the coefficients $A_{00}(z)$, $A_{10}(z)$ controlling the z-dependence of the wavefunction. The expectation value $\langle \mathscr{H} \rangle$ is then variationally minimized with respect to $A_{00}(z)$ and $A_{10}(z)$.

In a remarkable experiment, Edwards et al.[37] measured the elastic reflectivity of low-energy ($0.1\,°\text{K} \leqslant \hbar^2 k^2/2mk_B \leqslant 3\,°\text{K}$) ^{4}He atoms incident on the free surface of superfluid ^{4}He at $\sim 30\,°\text{mK}$, as a function of angle of incidence and momentum of the incident atoms. The results suggest that the ^{4}He atoms behave as if they were incident on a very smooth potential step. In a recent discussion by Echinique and Pendry,[38] it is assumed that the relevant interaction with the surface is via the van der Waals potential which extends beyond the (sharp) mass density cut-off, and which couples the incident atom to the quantized ripplons. Thus we imagine a ^{4}He atom approaching the surface to be scattered first by the weak van der Waals tail, and then to meet the ripplon coupling zone: scattering from the region close in to the surface without losing several quanta to ripplons is extremely unlikely, hence the low elastic reflectivity. However, it was found that such a model consistently yielded elastic reflectivities which were too high *unless* account was taken of the diffuse nature of the liquid ^{4}He surface. Echinique and Pendry adopt a linear density profile, achieving the bulk density in 5 Å, whereupon reasonably good agreement with experiment is obtained. Whilst these results do not necessarily suggest a linear profile on this scale, these authors have demonstrated that the results of Edwards et al. are inconsistent with a sharp density transition at the superfluid ^{4}He surface.

6.4 The surface structure of liquid ^{3}He

The N-body Hamiltonian for ^{3}He particles is identical to that of N ^{4}He particles, except that the ^{3}He mass is $\frac{3}{4}$ that of ^{4}He. In other words, if ^{3}He atoms were bosons, the results of the previous section would remain qualitatively unchanged: the lighter mass of the ^{3}He atom would increase the kinetic energy, reducing the binding energy and resulting in a lower equilibrium density than liquid ^{4}He. However, we are dealing with a system which obeys Fermi statistics and we take for the wavefunction of the ^{3}He system with a surface a suitably symmetrized wavefunction of position and spin coordinates of the N particles

$$\Psi_{F_s} \equiv \Psi_{F_s}(\mathbf{r}_1, \boldsymbol{\sigma}_1; \mathbf{r}_2; \boldsymbol{\sigma}_2; \cdots; \mathbf{r}_N, \boldsymbol{\sigma}_N). \tag{6.4.1}$$

The standard Jastrow approximation for a Fermi system is to take as a trial wavefunction

$$\Psi_{F_s}(\mathbf{r}_1, \boldsymbol{\sigma}_1; \mathbf{r}_2, \boldsymbol{\sigma}_2; \cdots; \mathbf{r}_N, \boldsymbol{\sigma}_N) = \mathcal{S} \Psi_{B_s}(\mathbf{r}_1, \mathbf{r}_2, \ldots, \mathbf{r}_N) M_F(\mathbf{r}_1, \mathbf{r}_2, \ldots, \mathbf{r}_N) \tag{6.4.2}$$

where $\mathcal{S}$ is the Slater determinant of plane-wave states and spin states appropriate to the bulk, $\Psi_{B_s}(\mathbf{r}_1, \mathbf{r}_2, \ldots, \mathbf{r}_N)$ is the mass three *boson* wavefunction with a surface, and $M_F(\mathbf{r}_1, \mathbf{r}_2, \ldots, \mathbf{r}_N)$ is a modulating factor which allows further variation of the wavefunction through the surface to accommodate the antisymmetry. The mass three boson problem is solved first, and then the statistics are incorporated. The modulating factor M_F is not 'state dependent'—all the symmetry is included in the Slater determinant—but it may be regarded as a product of position-dependent amplitudes of the plane wave states contained in $\mathcal{S}$. As Buchan and Clark observe, any dependence on position in $\mathcal{S}$ itself would have been excessively complicated, if treated variationally, and it would have been invalid to assume any average behaviour through the surface as this turns out to be sharp.

For the modulating factor M_F we write

$$M_F(\mathbf{r}_1, \mathbf{r}_2, \ldots, \mathbf{r}_N) = \prod_{i=1}^{N} \exp\{\tfrac{1}{2} t_F(z_i)\} \tag{6.4.3}$$

(c.f. 6.3.6, 6.3.16) where $t_F(z)$ is real and tends to zero or minus infinity as z tends to minus infinity or plus infinity.

The extension of the boson theory of Section 6.3 to a fermion system is complicated principally by the additional terms arising from the introduction of the Slater determinant. Using the wavefunction (6.4.2), the expectation value for a system with a surface is

$$\langle \mathcal{H} \rangle = E_B^s + E_1^s + E_2^s \tag{6.4.4}$$

where E_B^s is the eigenvalue satisfying $\mathcal{H} \Psi_B^s = E_B^s \Psi_B^s$ and

$$E_1^s = -\frac{\hbar^2}{2m} \eta^{-1} \sum_{n=1}^{N} \int \cdots \int \Psi_F^{s*} \Psi_B^s \nabla_n^2(\mathcal{S} M_F) \, d\mathbf{r}_1 \cdots d\mathbf{r}_N \tag{6.4.5}$$

$$E_2^s = -\frac{\hbar^2}{2m} \eta^{-1} \sum_{n=1}^{N} \int \cdots \int 2\Psi_F^{s*} \nabla_n \Psi_B^s \cdot \nabla_n(\mathcal{S} M_F) \, d\mathbf{r}_1 \cdots d\mathbf{r}_N \tag{6.4.6}$$

where $\eta = \int \cdots \int \Psi_F^{s*} \, d\mathbf{r}_1 \cdots d\mathbf{r}_N$ is the normalization integral. A little manipulation then reveals

$$E_1^s = \frac{\hbar^2}{2m} \sum_{n=1}^{N} k_n^2 - \frac{\hbar^2}{2m} \int \{ \tfrac{1}{2} t_F''(z) + \tfrac{1}{4}[t_F'(z)]^2 \} \rho_{(1)F}(z) \, d\mathbf{r}$$

$$- \frac{\hbar^2}{2m} N\eta^{-1} \int t_F'(z_1) \left\{ \hat{\mathbf{n}} \cdot \int \cdots \int \Psi_B^{s2} M_F^2 \mathscr{S}^* \nabla_1 \mathscr{S} \, d\tau_2 \cdots d\tau_N \right\} d\mathbf{r}_1 \quad (6.4.7)$$

where a prime denotes differentiation with respect to z and $\hat{\mathbf{n}}$ is a unit vector normal to the surface in the positive z-direction. Assuming that the particles completely fill a Fermi sphere $(T = 0)$ the first term in (6.4.7) is just $\tfrac{3}{5} N E_F$ (E_F = Fermi energy) which cancels when the energy of the bulk phase is subtracted to leave the surface excess energy of the fermion system.

$$E_2^s = -\frac{\hbar^2}{4m} \int t_F'(z) t_B'(z) \rho_{(1)F}(z) \, d\mathbf{r} - \frac{\hbar^2}{4m} \int t_F'(z) J_1(z) \, d\mathbf{r}$$

$$- \frac{\hbar^2}{2m} N\eta^{-1} \int t_B'(z_1) \left\{ \hat{\mathbf{n}} \cdot \int \cdots \int (\Psi_B^s M_F)^2 \mathscr{S}^* \nabla_1 \mathscr{S} \, d\tau_2 \cdots d\tau_N \right\} d\mathbf{r}_1 \quad (6.4.8)$$

where

$$J_1(\mathbf{r}_1) = \int \nabla_1 u(r_{12}) \rho_{(2)F}(\mathbf{r}_1, \mathbf{r}_2) \, d\mathbf{r}_2 = \hat{\mathbf{n}} J_1(z_1). \quad (6.4.9)$$

In the above expressions for E_1^s, E_2^s we have introduced the fermion one- and two-particle distributions defined analogously to (6.3.9, 6.3.10): the object of these manipulations is to express the surface energy of the fermion system in terms of known quantities, in particular the single-particle profile $\rho_{(1)F}(z)$ in terms of which variational adjustment may be made.

The surface energy per unit area may now be derived by subtracting from $\langle \mathscr{H} \rangle$ (6.4.4) the corresponding energy for a bulk liquid and dividing through by the surface area. After some manipulation we obtain

$$U_F^s = U_B^s - \frac{\hbar^2}{4m} \int \left\{ t_B' \rho_{(1)F}' - J_1 t_B' - \tfrac{1}{2} \rho_{(1)F}(t_B')^2 \right.$$

$$\left. + \frac{1}{2} \frac{J_1^2}{\rho_{(1)F}} - \frac{1}{2} \frac{(\rho_{(1)F}')^2}{\rho_{(1)F}} \right\} dz + \int [\Gamma_{\text{exch}}'(z) - \rho_{(1)F}(z) E_{\text{exch}}] \, dz \quad (6.4.10)$$

where

$$\Gamma_{\text{exch}}'(z_1) = \frac{\hbar^2}{4m} \int \rho_{(2)F}(\mathbf{r}_1, \mathbf{r}_2) \{ \nabla_1^2 u(r_{12}) + (\nabla_1 u(r_{12}))^2 \} \, d\mathbf{r}_{12}$$

$$+ \frac{\hbar^2}{4m} \int\!\!\int \nabla_1 u(r_{12}) \cdot \nabla_1 u(r_{13}) \rho_{(3)F}(\mathbf{r}_1, \mathbf{r}_2, \mathbf{r}_3) \, d\mathbf{r}_2 \, d\mathbf{r}_3 \quad (6.4.11)$$

and

$$E_{exch} = \frac{\hbar^2}{4m} N\rho \left[\int g_{(2)F}(r)\{\nabla^2 u(r) + (\nabla u(r))^2\} \, d\mathbf{r} \right.$$

$$\left. + \rho \int\int \nabla_1 u(r_{12}) \cdot \nabla_1 u(r_{13}) g_{(2)F}(r_{12}, r_{23}, r_{31}) \, d\mathbf{r}_1 \, d\mathbf{r}_2 \right]. \quad (6.4.12)$$

Using the Kirkwood superposition approximation for the three-particle distribution functions, Buchan and Clark have variationally minimized the expression (6.4.10) using a trial Fermi density profile for $\rho_{(1)F}(z)$:

$$\rho_{(1)F}(z) = [1 + \exp \beta(z - z_0)]^{-1} \quad (6.4.13)$$

where z_0 locates the transition and β governs its width.

It is found in all cases that the Fermi system has a considerably thinner surface than the equivalent mass-three boson system: the 90–10% thickness for mass-three bosons is 6.3 Å, whilst for fermions it is 3.7 Å, which is less than an interparticle distance for the densities investigated (Figure 6.4.1). This somewhat surprising result does not appear to be an artifact of the mathematical technique—in the density functional model the effect of statistics is to reduce the surface thickness, and it is concluded that a sharper surface is a feature of the proper inclusion of Fermi statistics, although no entirely convincing explanation is offered. However, Buchan and Clark find that the calculated bulk ^{3}He fluid binding energy is greater than the experimental value, and so the surface energy is too large, even though it is less than the experimental result: 0.147 ± 0.02 erg cm^{-2}, against the experimental value 0.512 erg cm^{-2}, both at a density 0.0164 Å^{-3}. The high value is attributed to the restricted form of the variational function (6.4.13). A more flexible trial function would undoubtedly yield some improvement, but becomes computationally prohibitive.

The assumption that the Fermi energy E_F remains constant across the

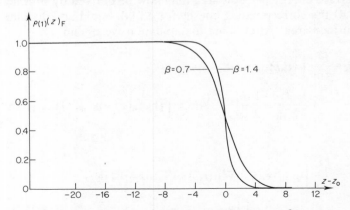

Figure 6.4.1 Comparison of density profiles for ^{3}He and a mass three boson fluid illustrating the effect of statistics

surface deserves further comment. The Thomas–Fermi assumption that E_F is proportional to $\rho_F^{2/3}$ would lead to an additional contribution to the surface energy

$$U_{TF} = \frac{3}{5} \int_{-\infty}^{\infty} \rho_{(1)F}(z)(E_F(z) - E_F)\, dz \qquad (6.4.14)$$

which, for the Fermi profile (6.4.13), yields $U_{TF} = -1.355 \; \rho_F^L E_F / \beta$. As Buchan and Clark[23] point out, this is a large term, and its inclusion would prevent the variational minimization of (6.4.13) with respect to β. However, in a calculation in which the particle wavefunction Ψ_F is explicitly considered at the surface, the treatment of E_F in isolation cannot be justified.

Finally, the present calculation does not include the zero-point energy of long-wavelength surface excitations,[28] which in fact amounts to a very small correction to the bulk energy.[29] The point is treated fully by Chang and Cohen,[18] who discuss the modification of the wavefunction to give the correct long wavelength behaviour.

6.5 Surface states of ^{3}He in dilute ^{3}He–^{4}He solutions

Recently, interest has been aroused by experiments of Esel'son and co-workers[30] in which it was found that the surface tension of superfluid ^{4}He was lowered by the presence of ^{3}He as an impurity by an amount greater than that expected from the usual rule of additivity. The results were interpreted phenomenologically by Andreev[39] in terms of the development of bound ^{3}He states at the ^{4}He surface—an extra binding energy which causes the ^{3}He atoms to accumulate at the superfluid ^{4}He surface. Alternatively and equivalently, he proposed that there exist surface states of ^{3}He with a minimum energy below that in the bulk. Clearly, any analytic description will depend essentially upon the detailed nature of the ^{4}He surface.[40,41] Indeed, the theoretical justification for Andreev's proposal has prompted several of the recent attempts to specify the ^{4}He density profile.[18,40]

Shih and Woo[40] have made a detailed determination of the surface states of ^{3}He in dilute ^{3}He–^{4}He solutions on the basis of their ^{4}He calculations (Section 6.3) in an attempt to provide a microscopic justification of Andreev's phenomenological model. In their treatment they replace the Nth ^{4}He atom by a ^{3}He atom which changes the ^{4}He Hamiltonian ($\mathcal{H}_4$) (6.3.7) as follows:

$$\mathcal{H} = \mathcal{H}_4 + \left(1 - \frac{m_4}{m_3}\right)\left(\frac{\hbar^2}{2m_4}\right)\nabla_N^2 \qquad (6.5.1)$$

where, of course,

$$\mathcal{H}_4 = \sum_{i=1}^{N} -\frac{\hbar^2}{2m_4} \nabla_i^2 + \sum_{j>i=1}^{N} \Phi(r_{ij}).$$

Adopting a trial wavefunction of Jastrow form

$$\Psi(\mathbf{r}_1, \ldots, \mathbf{r}_N) = \Psi_4 \exp\left[\tfrac{1}{2}\phi(z_N)\right] \qquad (6.5.2)$$

where, as usual (6.3.6),

$$\Psi_4 = \prod_{j>i=1}^{N} \exp\left[\tfrac{1}{2}u(r_{ij})\right] \prod_{l=1}^{N} \exp\left[\tfrac{1}{2}t(z_l)\right]$$

the expectation energy $\langle \Psi | \mathcal{H} | \Psi \rangle / \langle \Psi | \Psi \rangle$ is variationally minimized for a variety of trial functions $\phi(z)$. We find

$$\langle \Psi | \mathcal{H} | \Psi \rangle / \langle \Psi | \Psi \rangle = E_4 + E_a + E_b + E_c$$

where

$$E_a = (-\hbar^2 \pi / 2m_4 I) \iint\!\! \int \rho_{(2)}(\mathbf{r}_1, \mathbf{r}_N) \exp\left[\phi(z_N)\right] [du(r_{1N})/dz_N]$$
$$\times \phi'(z_N) \hat{r}_{1N}\, dz_N\, dz_1\, d\hat{r}_{1N}$$

$$E_b = (-\hbar^2 / 2m_4)(1 - m_3/m_4)(\pi / 2I) \iint\!\! \int \rho_{(2)}(\mathbf{r}_1, \mathbf{r}_N) \exp\left[\phi(z_N)\right]$$
$$\times u''(r_{1N}) \hat{r}_{1N}\, dz_N\, dz_1\, d\hat{r}_{1N}$$

$$E_c = (-\hbar^2 i / 2m_4 4I) \int \rho_{(1)}(z) \exp\{\phi(z)\}\{\phi'^2(z) + \phi''(z) + t''(z) + 2t'(z)\phi'(z)$$
$$+ (m_4/m_3)[t''(z) + \phi''(z)]\}\, dz \qquad (6.5.3)$$

and where

$$I = \int \rho_{(1)}(z) \exp\{\phi(z)\}\, dz$$

$$\hat{r}_{1N} = (x_{1N}^2 + y_{1N}^2)^{1/2}.$$

Each trial function $\phi(z)$ resulted in a symmetrical distribution for the ^{3}He atom centred on the Gibbs surface. The lowest energy corresponded to a Gaussian distribution $\exp[-(z/\lambda)^2]$ of width $\lambda = 2.5$ Å. Shih and Woo found that the extra binding energy ε_0 driving the surface adsorption of ^{3}He atoms to be $1.6\,°$K, in good agreement with surface tension measurements by Zinov'eva and Boldarev[43] who determined $\varepsilon_0 = 1.7 \pm 0.2\,°$K. Guo et al.[44] have more recently determined ε_0 to be $1.95 \pm 0.1\,°$K, although at a lower temperature and in the fully degenerate region where the ^{3}He system behaves like a dilute two-dimensional Fermi gas.[33]

Shih and Woo go on to regard the ^{3}He particle as moving in a single-particle well, in which case the oscillator frequency is determined to be $\hbar\omega \sim 2.6\,°$K. In other words, from the binding energy ε_0 we conclude that the surface states of ^{3}He in dilute ^{3}He–^{4}He solutions consist of a single bound state: this conclusion is in agreement with the semi-phenomenological treatment of Saam[42] who obtained $\varepsilon_0 = 3.1\,°K$.

A number of interesting questions remain,[45] however, since the treatment of Shih and Woo considers only the limiting case of *one* ^{3}He atom: but how do the conclusions vary with coverage, will the presence of some ^{3}He atoms at the surface induce others to 'float' up, and when will ε_0 drop to zero, announcing the completion of the ^{3}He monolayer?

6.6 Superfluid transitions in ^{3}He monolayers

A *priori* considerations suggest that ^{3}He, being composed of fermions, cannot have more than one particle occupying a single-particle state, and therefore cannot have a macroscopic single-particle condensate. On the other hand, of course, an example of superfluidity is provided by the superconductivity of electrons in a metal, which is a collection of fermions. The explanation of superfluidity in macroscopic fermion systems was given by Bardeen, Cooper, and Schrieffer[48] in terms of the association of a macroscopic number of particles into the same pair state which then forms a two-body condensate. The condition for pair-condensation is that the fermions have an attractive interaction between single-particle states near the Fermi surface, and the question naturally presents itself as to whether pair-condensed states can develop in liquid ^{3}He.[49,50,51]

Whilst pairing in superconductors occurs in relative s-wave angular momentum states, it was pointed out that the strongly repulsive short-range interaction between ^{3}He atoms makes pairing more likely in states of higher relative angular momentum—possibly in relative d-wave states.[50] However, following the discovery of anomalous NMR behaviour[52] in liquid ^{3}He it does appear that superfluid ^{3}He has at last been found.

It is well known that solution of the two-particle Schrödinger equation for two interacting helium atoms in three dimensions yields no bound states: diatomic He molecules do not exist. In *two* dimensions, however, Bagchi[53] has shown that there is a very weak attraction—binding energies $\sim 10^{-2}\,k$ for ^{4}He and $\sim 10^{-7}k$ for ^{3}He. As we have seen, two-dimensional ^{3}He systems develop spontaneously in dilute ^{3}He–^{4}He mixtures (Section 6.5), surface migration effectively constraining the ^{3}He atoms to a monolayer. Moreover, it may be shown that surface excitation exchange in the form of ripplons induce effective attractions between the fermions, and that in ^{3}He–^{4}He solutions such attractions completely cancel out the repulsion between bare ^{3}He atoms.[54] These circumstances, fortuitously, can only improve the experimental conditions for the observation of superfluidity in two dimensions: moreover, the increased pairing will also increase the transition temperature.[55] And although there has not as yet been any experimental report of such a phenomenon, such discontinuous behaviour at the liquid surface should be readily identifiable, for example, through observations of third sound, surface tension, and flow properties.

196

References

1. For an excellent description of the bulk structure of quantum fluids see
   C. E. Campbell. In C. A. Croxton (Ed.), *Progress in Liquid Physics*, Wiley, London (1978), Ch. 6.
   K. H. Bennemann and J. B. Ketterson (Eds), *The Physics of Liquid and Solid Helium, Part I*, Wiley, New York (1976).
2. R. M. Mazo and J. G. Kirkwood, *Proc. Nat. Acad. Sci. U.S.* **41,** 204 (1955); *J. Chem. Phys.*, **28,** 644 (1958).
3. D. G. Henshaw and D. Hurst, *Can. J. Phys.*, **33,** 797 (1955).
4. K. R. Atkins, *Can. J. Phys.*, **31,** 1165 (1953).
   L. D. Landau and E. M. Lifshits, *Statistical Physics*, Pergamon, Oxford (1969), p. 457.
5. J. Frenkel, *Kinetic Theory of Liquids*, Dover, New York (1955), Ch. 6.
6. A. D. Singh, *Phys. Rev.*, **125,** 802 (1962).
7. K. R. Atkins and Y. Narahara, *Phys. Rev.*, **138,** A437 (1965).
   J. R. Ekardt, *Ph.D. Dissertation*, Ohio State University, (1972).
   F. M. Gasparini, J. R. Ekardt, D. O. Edwards, and S. Y. Shen, *J. Low Temp. Phys.*, **13,** 437 (1973).
8. S. T. Boldarev and V. P. Peshkov, *Physica*, **69,** 141 (1973).
9. S. L. Chan, *Can. J. Phys.*, **50,** 1139 (1972).
10. D. O. Edwards, J. R. Ekardt, and F. M. Gasparini, *Phys. Rev.*, **A9,** 2070 (1974).
11. See, for example, D. ter Haar, *Elements of Statistical Mechanics*, Rinehard, New York (1954), pp. 147–155.
12. S. Ono and S. Kondo, *Hand. Phys.*, **10,** 237 (1960).
13. T. Regge, *J. Low Temp. Phys.*, **9,** 123 (1972).
14. T. C. Padmore and M. W. Cole, *Phys. Rev.*, **A9,** 802 (1974).
15. A. Widom, *Ph.D. Thesis*, Cornell University (1968).
16. P. C. Hohenberg and W. Kohn, *Phys. Rev.*, **136,** B864 (1964).
17. D. J. Amit, *Phys. Lett.*, **23,** 665 (1966); *J. Low Temp. Phys.*, **3,** 645 (1970).
18. C. C. Chang and M. Cohen, *Phys. Rev.*, **A8,** 1930 (1973).
19. R. Brout and M. Nauenberg, *Phys. Rev.*, **112,** 1451 (1958).
20. D. D. Fitts, *Physica*, **42,** 205 (1969).
21. Y. M. Shih and C.-W. Woo, *Phys. Rev. Lett.*, **30,** 478 (1973).
22. R. Jastrow, *Phys. Rev.*, **98,** 1479 (1955).
23. G. D. Buchan and R. C. Clark, *J. Phys. C.*, **10,** 3081 (1977).
24. K. S. Liu, M. H. Kalos, and G. V. Chester, *Phys. Rev.*, **B12,** 1715 (1975).
25. C. A. Croxton, *J. Phys. C.*, **6,** 411 (1973).
   C. A. Croxton, *Phys. Lett.*, A, **41,** 413 (1972).
26. R. Abe, *Prog. Theor. Phys. (Kyoto)*, **19,** 57 (1958); *Prog. Theor. Phys. (Kyoto)*, **19,** 407 (1958).
27. C. A. Croxton, *Liquid State Physics—A Statistical Mechanical Introduction*, Cambridge University Press, London (1974), pp. 184–189.
28. G. D. Buchan, *Phys. Lett.*, **59,** A35 (1976).
29. L. Reatto and G. V. Chester, *Phys. Rev.*, **155,** 88 (1967).
30. B. N. Esel'son, V. G. Ivantsov, and A. D. Shvets, *Zh. Eksp. Teor. Fiz.* **44,** 483 (1963) (*Sov. Phys. JETP*, **17,** 330 (1963)).
   B. N. Esel'son and N. G. Bereznyak, *Dokl. Akad. Nauk SSSR*, **98,** 564 (1954).
31. J. Lekner, *Prog. Theor. Phys.*, **45,** 36 (1971).
32. D. Schiff and L. Verlet, *Phys. Rev.*, **160,** 208 (1967).
33. H. M. Guo, D. O. Edwards, R. E. Sarwinski, and J. T. Tough, *Phys. Rev. Lett.*, **27,** 1259 (1971).
34. A. Widom, *Phys. Rev.*, **A1,** 216 (1970).
35. M. W. Cole, *Phys. Rev.*, **A1,** 1838 (1970).

36. F. P. Buff, R. A. Lovett, and F. H. Stillinger, Jr., *Phys. Rev. Lett.*, **14,** 491 (1965).
37. D. O. Edwards, P. Fatouros, G. G. Ihas, P. Mrozinski, S. Y. Shen, F. M. Gasparini, and C. P. Tam, *Phys. Rev. Lett.*, **34,** 1153 (1975).
38. P. M. Echinique and J. B. Pendry, *J. Phys. C.*, **9,** 3183 (1976).
39. A. F. Andreev, *Zh. Eksp. i Teor. Fiz.*, **50,** 1415 (1966) [*Sov. Phys. JETP*, **23,** 939 (1966)].
40. Y. M. Shih and C. W. Woo, *Phys. Rev. Lett.*, **30,** 478 (1973).
41. J. Lekner, *Phil. Mag.*, **22,** 669 (1970).
42. W. F. Saam, *Phys. Rev.* **A4,** 1278 (1971).
43. N. K. Zinov'eva and S. T. Boldarev, *Zh. Eksp. i Teor. Fiz.*, **56,** 1089 (1969) (*Sov. Phys. JETP*, **29,** 585 (1969)).
44. H. M. Guo, D. O. Edwards, R. E. Sarwinski, and J. T. Tough, *Phys. Rev. Lett.*, **27,** 1259 (1971).
45. C.-W. Woo In K. H. Bennemann and J. B. Ketterson (Eds), *The Physics of Liquid and Solid Helium. Part I*, Wiley, New York (1976), Ch. 5, p. 482.
46. M. D. Miller, *Ph.D. Thesis*, Northwestern University (1973).
47. W. F. Saam, *Phys. Rev.*, **A8,** 1048 (1973).
48. J. Bardeen, L. N. Cooper, and J. R. Schrieffer, *Phys. Rev.*, **108,** 1175 (1957).
49. D. J. Thouless, *Ann. Phys.* (*N.Y.*), **10,** 553 (1960).
50. K. A. Brueckner, T. Soda, P. W. Anderson, and P. Morel, *Phys. Rev.*, **118,** 1442 (1960).
51. V. J. Emery and A. M. Sessler, *Phys. Rev.*, **119,** 43 (1960).
52. D. D. Osheroff, R. C. Richardson, and D. M. Lee, *Phys. Rev. Lett.*, **28,** 885 (1972).
    D. D. Osheroff, W. J. Gully, R. C. Richardson, and D. M. Lee, *Phys. Rev. Lett.*, **29,** 920 (1972).
    See also the review by J. Wheatley, *Rev. Mod. Phys.*, **47,** 415 (1975).
53. A. Bagchi, *Phys. Rev.*, **A3,** 1133 (1971).
54. J. Bardeen, G. Baym, and D. Pines, *Phys. Rev.*, **156,** 207 (1967).
55. C.-W. Woo, *European Physical Society Topical Conference: Liquid and Solid Helium, Haifa, Israel* (1974).
56. J. Lekner and J. R. Henderson, *Physica*, **94A,** 545 (1978).

CHAPTER 7

The Surface of Liquid Water

7.1 Introduction

Polar fluids may be roughly resolved into two classes, simple and nonsimple, depending respectively upon the asymmetry of their force fields. Simple polar substances such as sulphur dioxide and the hydrogen and methyl halides have a direct electrostatic interaction between their permanent dipoles, or in the case of carbon dioxide and acetylene, between their quadrupoles, but are otherwise reasonably symmetrical. Water and ammonia, on the other hand, are electrically strongly asymmetric, or have a large polarity which is often located in one part of the molecule such as in the organic alcohols. Here, however, we shall confine our attention to the liquid water interface as being representative of a complex polar system and of intense theoretical and experimental interest. Its importance in the description of aqueous solutions and electrochemical processes ranging from the industrial to the biochemical has largely motivated the substantial progress which has been made in the understanding of bulk liquid water;[1-3] however, there have as yet been few attempts to describe the liquid–vapour interfacial region.

A careful study of the variation of surface tension with temperature shows that for a variety of nonpolar liquids the average molar surface entropy is ~24.0 joule deg^{-1}, whilst for water this value is only 9.8 joule deg^{-1}. The entropy deficit of about 14 joule deg^{-1} $mole^{-1}$, or about $1.7k$ per molecule has been attributed[4] to a preferred molecular orientation at the surface. Although orientational order is not the only agent responsible for the lowering of the slope of the $\gamma(T)$ characteristic, there are other reasons for expecting a higher degree of orientational order amongst the surface molecules with respect to those in the bulk. If the water molecule possessed only a centrally placed symmetrical dipole moment, the symmetry of the resulting dipolar field would render any orientation energetically equivalent to its opposite orientation, and so no specific surface polarization would develop.[5]

Frenkel[9] appears to have been the first to suggest that it is the permanent quadrupole moment of the water molecule which is responsible for the preferred orientation at the interfacial boundary. The quadrupole moment tends to displace field lines either toward the front or the back of the molecule, depending upon its sign, in which case the molecules will tend to orient themselves so as to place their electric fields as much as possible in the high dielectric–constant region of the anisotropic liquid rather than in the vapour, so as to minimize its free energy.[5]

Below such an oriented surface it has been suggested that there exists a transition zone extending for tens, or even hundreds, of molecular layers,[6] although this latter proposition has been contested.[7] In any case, there will exist a more or less well-developed polarized surface dipole layer with an associated surface potential; experimental measurements however do not agree even in sign, let alone magnitude.[8] We shall not involve ourselves here with the prevailing confusion which attends the definition, significance, and measurement of interfacial potentials, but merely refer the reader to the relevant literature,[10] although we shall make a brief comparison of various estimates of the surface potential in Section 7.5.

7.2 The water molecule

The existing body of theoretical and experimental evidence[11, 12] shows that the *isolated* water molecule forms hybrid bond orbitals with an HOH angle of about $104\frac{1}{2}°$, with the two lone pair electrons forming a near tetrahedral arrangement around the central oxygen: the geometry involved is shown in Figure 7.2.1. The positive charges are identified as partially shielded protons, and are rather deeply buried in the oxygen's electron cloud (at a distance of ~0.96 Å from the oxygen nucleus compared with 1.4 Å for the molecule's van der Waals radius). The two positive vertices of the molecule are of low polarizability, whilst the roughly tetrahedrally disposed negative vertices are of high polarizability. Consequently, in the study of condensed water it is insufficient to consider simply the 'bare' interaction of an isolated pair of molecules: substantial non-additivity is present accounting for almost 15% of the interaction energy in ice.[14, 15] Moreover, Barnes[2] has emphasized that the molecular dipole moment in condensed water tends to be

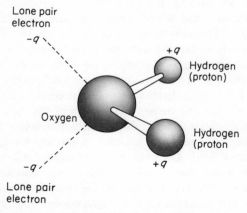

Figure 7.2.1 Geometry of the water molecule

enhanced by the mutual polarization of neighbours by 80–90%, in which case the isolated vapour dipole moment of $M = 1.83$ Debye is quite inappropriate. Similar observations apply to the higher multipoles, although the cooperative effect is rarely applied beyond the dipolar interaction and often then only in an effective pairwise manner;[16] the recent work of Barnes,[2] however, shows the importance of the induced dipole contribution—in particular it suggests that the dipole moment itself will be functionally dependent upon location within the anisotropic interphasal region. Fortunately for our purposes the variation across the transition zone is likely to be weak—the density and the associated dipole moment varying by factors of $\sim 10^3$ and ~ 1.5, respectively.

Generally, of course, a charge distribution may be expressed as a multipolar sum which supplements a central Lennard–Jones interaction. In dyadic notation[13]

$$4\pi\varepsilon U(\mathbf{r}_{12}) = \Phi(r_{12})_{\text{LJ}} + (q_2 + \mathbf{M}_2 \cdot \nabla_2 + \tfrac{1}{2}\mathbf{Q}_2 : \nabla_2\nabla_2 + \cdots)$$
$$\times (q_1 + \mathbf{M}_1 \cdot \nabla_1 + \tfrac{1}{2}\mathbf{Q}_1 : \nabla_1\nabla_1 + \cdots)(1/r_{12}) \qquad (7.2.1)$$

where q_i is the net electrostatic charge of particle i and $\mathbf{M}_i$ is the dipole moment vector:

$$\begin{bmatrix} M_x \\ M_y \\ M_z \end{bmatrix}_i \qquad (7.2.2)$$

and $\mathbf{Q}_i$ is the quadrupole moment tensor:

$$\begin{bmatrix} Q_{xx} & Q_{yx} & Q_{zx} \\ Q_{xy} & Q_{yy} & Q_{zy} \\ Q_{xz} & Q_{yz} & Q_{zz} \end{bmatrix}_i . \qquad (7.2.3)$$

The quantities q_i, $\mathbf{M}_i$, and $\mathbf{Q}_i$ have the physical significance of moments of the charge distribution, and are in many ways analogous to the moments of classical mechanics. The zeroth moment of the charge distribution is, for example, nothing other than the total integrated charge of the molecule and corresponds to the total mass of a density distribution except, of course, that in the case of electrostatics the total charge may be positive or negative. The first moments of the charge distribution are the dipole moments M_x, M_y, M_z and may be regarded as the three components of the dipole moment vector $\mathbf{M}$; in an external electric field these moments exert a torque on the molecule in a precisely analogous manner to the mechanical moment in the earth's gravitational field. Clearly, for a choice of Cartesian axes which aligns the dipole along the z axis we have the immediate simplification $M_x = M_y = 0$. Such axes are known as the *principal axes* of the charge distribution. The components of the quadrupole moment tensor may be regarded as the 'moments of inertia' of the charge distribution about the axes: if the distribution has an axis of symmetry, as in the case of the water molecule, then this choice of principal axes also diagonalizes the quadrupole tensor, leaving only the components Q_{xx}, Q_{yy}, Q_{zz} (Figure 7.2.2). We should mention that in the electrostatic case the 'moment of inertia' of the charge distribution may be negative or even zero owing to the opposed contributions from positive and negative charges, unlike its mechanical counterpart.

We could in principle proceed to higher and higher terms, the order being reflected in the number of differentiations in the operator. As we shall see, the relative ranges of the interaction forces rapidly fall off with increasing order of the multipolar structure: the dipole–dipole interaction falling off as r^{-3}, the dipole–quadrupole as r^{-4}, and the quadrupole–quadrupole term as r^{-5}. Clearly at liquid densities quadrupole–quadrupole contributions cannot be neglected, although they become subordinate to the dipolar and Coulombic terms in the vapour.

In multiplying out (7.2.1) $q_1 q_2 / r_{12}$ is the familiar Coulomb term, whilst $q_2 \mathbf{M}_1 \cdot \nabla_1 (1/r_{12})$ and $q_1 \mathbf{M}_2 \cdot \nabla_2 (1/r_{12})$ are the two charge–dipole contributions showing an inverse square dependence upon intermolecular separation. Since the water molecule is electrically neutral the Coulombic and charge–dipole terms do not arise, and consequently we shall consider them no further here. The presence of free ions will, however, be discussed in Chapter 11.

The first term of interest in (7.2.1) is the dipole–dipole interaction $\mathbf{M}_1 \mathbf{M}_2 : \nabla_1 \nabla_2 (1/r_{12})$. Writing $r_{12}^{-1} = [(x_1 - x_2)^2 + (y_1 - y_2)^2 + (z_1 - z_2)^2]^{-1/2}$ and performing the double differentiation we obtain

$$\mathbf{M}_2 \mathbf{M}_1 : \nabla_1 \nabla_2 (1/r_{12}) = \mathbf{M}_1 \cdot \overset{\leftrightarrow}{\mathbf{T}} \cdot \mathbf{M}_2 \qquad (7.2.4)$$

where ∇_i is composed of the partial components $\partial/\partial x_i$, $\partial/\partial y_i$, $\partial/\partial z_i$, and where the dyad

$$\overset{\leftrightarrow}{\mathbf{T}} = \begin{bmatrix} 1 & 0 & 0 \\ 0 & 1 & 0 \\ 0 & 0 & 1 \end{bmatrix} - 3 \underbrace{\begin{bmatrix} R_x R_x & R_y R_x & R_z R_x \\ R_x R_y & R_y R_y & R_z R_y \\ R_x R_z & R_y R_z & R_z R_z \end{bmatrix}}_{R^2} = \begin{bmatrix} T_{xx} & T_{yx} & T_{zx} \\ T_{xy} & T_{yy} & T_{zy} \\ T_{xz} & T_{yz} & T_{zz} \end{bmatrix}. \qquad (7.2.5)$$

The dipole vectors (7.2.2) associate with (7.2.5) to yield terms of the form $M_{x_1} T_{x_1 x_2} M_{x_2}$, $M_{z_1} T_{z_1 y_2} M_{y_2}$, etc. The representation (7.2.5) is readily obtained by evaluating all the differential components of $(1/r_{12})$ in (7.2.4).

A similar analysis enables us to express $\mathbf{M}_1 \mathbf{Q}_2 : \nabla_1 \nabla_2 \nabla_2 (1/r_{12})$ as $\mathbf{M}_1 \cdot \overset{\leftrightarrow}{\mathbf{U}} \cdot \mathbf{Q}_2$ and $\mathbf{M}_2 \mathbf{Q}_1 : \nabla_2 \nabla_1 \nabla_1 (1/r_{12})$ as $-\mathbf{M}_2 \cdot \overset{\leftrightarrow}{\mathbf{U}} \cdot \mathbf{Q}_1$ (since $\nabla_1 = -\nabla_2$), and $\mathbf{Q}_1 \mathbf{Q}_2 : \nabla_1 \nabla_1 \nabla_2 \nabla_2 (1/r_{12})$ as $\mathbf{Q}_1 \cdot \overset{\leftrightarrow}{\mathbf{W}} \cdot \mathbf{Q}_2$ where $\overset{\leftrightarrow}{\mathbf{U}}$ and $\overset{\leftrightarrow}{\mathbf{W}}$ represent the array of third and fourth order derivatives respectively (Table 7.2.1).

The resulting interaction energy $U(\mathbf{r}_{12})$ may, for symmetrical systems, simplify considerably by performing angular averages over the various molecular relative orientations, as we shall see in Section 7.3.

Finally, we should mention that in such a permanent (as opposed to induced) multipolar expansion of molecular charge distributions, the radial

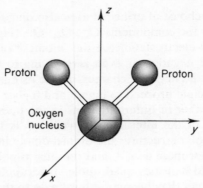

Figure 7.2.2 Choice of cartesian axes which coincides with the principal axes of the water molecule. This choice diagonalizes the quadrupole tensor, equation (7.2.3)

dependence depends sensitively upon the order of the multipole. Thus, the dipole–dipole interaction falls off as r^{-3}, the dipole–quadrupole as r^{-4}, the quadrupole–quadrupole as r^{-5}, and so on.[20] Induced multipoles, which we shall not consider in any detail here, fall off at the square of these rates. The radial dependences of the permanent multipolar interactions are illustrated schematically in Figure 7.2.3. Clearly the multipolar structure of the molecular structure, in as far as the intermolecular interaction is concerned,

Table 7.2.1. Evaluation of dipole–dipole, dipole–quadrupole, and quadrupole–quadrupole terms $\vec{\vec{T}}$, $\vec{\vec{U}}$, $\vec{\vec{W}}$ in the coordinate frame of Figure 7.2.2

(i) dipole–dipole

$$T_{zz} = \frac{1}{R^3}\left(1 - \frac{3z^2}{R^2}\right)$$

(ii) dipole–quadrupole†

$$U_{\alpha_1\beta_2\beta_2} = \frac{3\alpha}{R^5}\left(1 - \frac{5\beta^2}{R^2}\right)$$

$$U_{\alpha_1\alpha_2\alpha_2} = \frac{3\alpha}{R^5}\left(3 - \frac{5\alpha^2}{R^2}\right)$$

(iii) quadrupole–quadrupole†

$$W_{\alpha_1\alpha_1\beta_2\beta_2} = \frac{1}{R^5}\left(3 - \frac{15(\alpha^2+\beta^2)}{R^2} + \frac{105\alpha^2\beta^2}{R^4}\right)$$

$$W_{\alpha_1\alpha_1\alpha_2\alpha_2} = \frac{1}{R^5}\left(9 - \frac{90\alpha^2}{R^2} + \frac{105\alpha^4}{R^4}\right)$$

† where $\alpha_i = x_i,\ y_i,\ z_i;\ \beta_j = x_j,\ y_j,\ z_j;\ \alpha_i \neq \beta_j$.

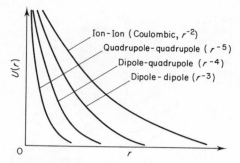

Figure 7.2.3 Schematic variation of the radial dependences of the permanent multipolar interactions

is apparent only at relatively small separations. Thus, the long range behaviour of correlation functions, etc, are dominated by the dipole–dipole term, working up to higher order terms as the intermolecular separation decreases.

Perhaps one further point is in order since we are discussing the radial distribution, and that concerns an additional modification of the surface function $g_{(2)}(z_1, \mathbf{r}_{12})$ due to the interaction of the field multipole with not only the origin multipole (as in the case of bulk correlation), but also with an *image* multipole located on the opposite side of the surface arising from the requirement that the interface is essentially an equipotential surface. This is, of course, in addition to any geometrical modification of the radial distribution which arises in the vicinity of the liquid surface. The 'direct' field gives the bulk correlation whilst the 'image' field gives the surface induced correlation. We shall return to this point in Section 7.4.

7.3 The interfacial structure and associated surface potential of liquid water

The principal and characteristic feature of the liquid water surface is the development of a surface polarization of the molecular dipole moments with an associated surface field and surface potential. As we outlined in Section 7.1, the surface molecules experience a torque tending to orient the dipole, although it is the higher order multipolar structure, in particular the quadrupole, which is responsible for the development of a mean torque in the inhomogeneous region: the spontaneous interfacial polarization and its associated surface field is simply a measure of the degree of surface dipolar orientation.

An analysis by Stillinger and Ben-Naim[18] visualizes the water molecule as a point dipole plus point quadrupole encased in a spherical exclusion envelope centred on the oxygen nucleus, Figure 7.3.1. In their treatment they establish the mean torque on the water molecule arising from its immersion in an inhomogeneous dielectric $\varepsilon(z)$.

204

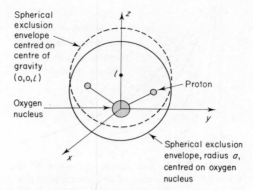

Spherical
exclusion
envelope
centred on
centre of
gravity
(o,o,l)

Oxygen
nucleus

Proton

z

l

y

x

Spherical exclusion
envelope, radius σ,
centred on oxygen
nucleus

Figure 7.3.1 Model of the water molecule:
a spherical exclusion envelope with a point
dipole and quadrupole

The electrostatic potential $\psi(r, \theta, \varphi)$ of the point multipole will satisfy Laplace's equation

$$\nabla^2 \psi(r, \theta, \varphi) = 0 \qquad (r > 0) \tag{7.3.1}$$

subject to certain boundary conditions. At $r = 0$ the potential will possess singularities characteristic of the dipolar and quadrupolar sources imputed to the fixed water molecule. The electrostatic potential will, or course, adopt two distinct forms inside and outside the spherical dielectric cavity, each of which may be expanded in spherical harmonics[20]

$$\psi_{\text{in}}^{(0)}(r, \theta, \varphi) = \sum_{l=0}^{\infty} \sum_{m=-l}^{+l} [A_{lm}^{(0)} r^l + B_{lm}^{(0)} r^{-l-1}] Y_{lm}(\theta, \varphi) \tag{7.3.2a}$$

$$\psi_{\text{out}}^{(0)}(r, \theta, \varphi) = \sum_{l=0}^{\infty} \sum_{m=-l}^{+l} C_{lm}^{(0)} r^{-l-1} Y_{lm}(\theta, \varphi) \tag{7.3.2b}$$

where $Y_{lm}(\theta, \varphi) = P_l^{(m)}(\cos \theta) \exp(im\varphi)$ are the unnormalized spherical harmonics which satisfy the associated Legendre equation

$$\nabla^2 Y_{lm}(\theta, \varphi) + l(l+1) Y_{lm}(\theta, \varphi) = 0 \tag{7.3.3}$$

and the superscripts (0) impute isotropy.

Stillinger and Ben-Naim determine the coefficients $A_{lm}^{(0)}$, $B_{lm}^{(0)}$, and $C_{lm}^{(0)}$ in the case of a homogeneous dielectric from the condition of continuity of the potential and radial component of the displacement vector over the surface of the exclusion envelope. Violation of the former condition would, of course, imply an infinite field at the surface of the envelope, whilst the existence of a tangential component of the displacement vector would initiate charge movements across the envelope surface. Isotropy of the dielectric constant ensures $\psi^{(0)}$ is entirely radially dependent and no net torque is exerted on the charge distribution within the molecule.

In the case of an inhomogeneous dielectric medium the same conditions apply, of course: indeed, continuity of the potential remains as before. However, the radial ψ derivative condition—continuity of the radial component of the displacement vector—becomes considerably more complicated when a spatial variation in the dielectric constant is introduced. Provided $\varepsilon(z)$ varies slowly some simplification of the problem ensues from linearization; however, this restricts the analysis to the discussion of critical interfaces. The isotropic $B_{lm}^{(0)}$ assignments remain as before, whilst the linearization enables us to write

$$A_{lm} \sim A_{lm}^{(0)} + |\nabla_z \varepsilon(z)| \, a A_{lm}^{(1)}. \tag{7.3.4}$$

There is an implicit use of local dielectric isotropy (whereby the usual inhomogeneous region dielectric tensor is replaced by a scalar), but this may be shown to be asymptotically correct as $T \to T_c$.

This done, then, the anisotropic part of the inner-region potential within the interfacial zone turns out to be

$$\psi_{in}(\mathbf{r}) = |\nabla_z \varepsilon(z)| \, a \sum_{l=0}^{\infty} \sum_{m=-l}^{+l} r^l A_{lm} Y_{lm}(\theta, \varphi) \tag{7.3.5}$$

where ψ_{in} is, of, of course, now explicitly a function of z. Such an angular-dependent potential will exert a torque on an anisotropic charge distribution, such as that of the water molecule. We may express the molecular charge distribution as a sum of the dipole and quadrupole components

$$q(\mathbf{r}) = q_d(\mathbf{r}) + q_q(\mathbf{r}) \tag{7.3.6}$$

whereupon the torque potential follows immediately as

$$V(\alpha, z) = \frac{1}{2} \int q(\mathbf{r}) \psi_{in}(\mathbf{r}) \, d\mathbf{r} \tag{7.3.7}$$

where the molecular dipole is oriented at an angle α to the z-axis. This orienting potential is supplemented by the field arising from the cooperative effect of the neighbouring polarized dipoles, $v(\alpha, z)$. The polarization density $\mathbf{P}(z)$ then follows from a direct evaluation of the relevant orientational average:

$$\mathbf{P}(z) = -\mathbf{E}(z)(4\pi)^{-1}$$

$$= \rho_{(1)}(z) M_z \frac{\int_0^\pi \cos\alpha \exp-[\{V(\alpha, z) + v(\alpha, z)\}/kT] \sin\alpha \, d\alpha}{\int_0^\pi \exp-[\{V(\alpha, z) + v(\alpha, z)\}/kT] \sin\alpha \, d\alpha}$$

$$\sim -\rho_{(1)}(z) M_z / 2kT \left\{ \int_0^\pi \cos\alpha [V(\alpha, z) + v(\alpha, z)] \sin\alpha \, d\alpha \right\} \tag{7.3.8}$$

where M_z is the z-component of the dipole vector, and the fact that the torques go to zero as $T \to T_c$ has been used to linearize the integrands.

Stillinger and Ben-Naim find

$$V(\alpha, z) = -\left\{5 \left|\frac{d\varepsilon(z)}{dz}\right| M_z Q_{zz}/a^3[1+2\varepsilon(z)][2+3\varepsilon(z)]\right\} \cos \alpha \quad (7.3.9)$$

$$v(\alpha, z) = -\{3\varepsilon(z)M_z E(z)/[1+2\varepsilon(z)]\} \cos \alpha \quad (7.3.10)$$

from which the potential difference between the two bulk phases follows by integration:

$$\bar{\psi}_L - \bar{\psi}_V = \frac{2\pi M_z^2 Q_{zz}}{3kTa^3} \int_{-\infty}^{\infty} \frac{\rho_{(1)}(z)\varepsilon'(z)\,dz}{[2+3\varepsilon(z)][1+2\varepsilon(z)+4\pi\beta M_z^2\rho_{(1)}(z)\varepsilon(z)]}$$

$$\beta = (kT)^{-1}. \quad (7.3.11)$$

The limiting behaviour near T_c is easily obtained since $\varepsilon'(z)$ contains the only important z dependence, $\rho_{(1)}(z)$ and $\varepsilon(z)$ in the remaining factors simply being replaced by their critical values ρ_c, ε_c. The critical potential drop then has the elementary form

$$\bar{\psi}_L - \bar{\psi}_V \sim \frac{20\pi\beta_c M_z^2 Q_{zz}\rho_c(\varepsilon_L - \varepsilon_V)}{3a^3(2+3\varepsilon_c)(1+2\varepsilon_c+4\pi\beta_c M_z^2\rho_c\varepsilon_c)} \quad (7.3.12)$$

from which it is seen that it is the difference in the bulk phase dielectric constants which primarily controls the rate at which $\bar{\psi}_L - \bar{\psi}_V$ vanishes as $T \to T_c$. Since all the factors in (7.3.12) other than Q_{zz} are invariably positive, it is the sign of the axial quadrupole moment which determines the direction of the preferential polarization at the liquid water surface. On the basis of equation (7.3.12) Stillinger and Ben-Naim obtain the subcritical surface potentials shown in Table 7.3.1, taking $Q_{zz} = 0.364 \times 10^{-26}$ esu cm^2. All surface potentials are consequently positive, in which case it is concluded that interfacial water molecules prefer to orient themselves with their protons pointing towards the liquid phase. However, as Stillinger and Ben-Naim observe, in this analysis the spherical dielectric cavity is centred on the oxygen nucleus; if the cavity were located on the molecular centre of gravity, a distance l along the axis of symmetry from the oxygen origin, then

Table 7.3.1. Values (in volts) for the surface potential $\bar{\psi}_L - \bar{\psi}_V$ calculated from equation (7.3.12)[18]

T °K	ε_L	ε_V	$\bar{\psi}_L - \bar{\psi}_V$(V)
643	9.74	6.55	1.003×10^{-3}
633	11.22	5.98	1.65×10^{-3}
623	12.61	5.65	2.20×10^{-3}
613	14.10	5.35	2.78×10^{-3}
603	15.51	5.24	3.28×10^{-3}
593	16.88	5.07	3.79×10^{-3}

$T_c = 674.15$ °K

the axial quadrupole moment would transform to $Q_{zz}^* = Q_{zz} - 4lM_z$ (effectively an application of the parallel axes theorem for moments of inertia), and, for a sufficiently large shift l, could conceivably reverse the sign of the surface potential, although the shift is unlikely to be of importance for the water molecule. However, it is now generally believed that the quadrupole moment for water is in fact, *negative*.

As Stillinger and Ben-Naim concede, the most serious source of error in their calculation is the spherical dielectric cavity assumption involving the application of macroscopic electrostatics at the molecular level which, however, is enforced by a lack of knowledge about local dielectric properties on a molecular scale. Again, the effect of the reaction field induced in the spherical cavity has not been considered. Its effect would be to enhance the dipole moment of the water molecule,[2, 19] and so would not be expected to modify the sign of the surface potential. The analogous effect of the reaction field on the quadrupole moment is entirely unknown.

Although the preceding analysis is strictly applicable only in the critical region, relying heavily upon simplifications which obtain near the critical point, Stillinger and Ben-Naim nevertheless estimate the surface potential of liquid water at 25 °C to be 0.029 V on the basis of equation (7.3.12).

A recent analysis of the problem by Croxton,[17] intended to be applicable to liquid water in the vicinity of the triple point, also regards the water molecule as a point dipole plus point quadrupole encased in a spherical exclusion envelope centred on the oxygen nucleus (Figure 7.3.1). In this treatment the introduction of a local dielectric constant is avoided and consequently difficulties arising from the specification of the local dielectric tensor at a molecular level do not arise. As we observed earlier, the Stillinger–Ben-Naim treatment relies heavily upon simplifications appropriate to the critical interface, and the analysis does not therefore bear legitimate extension to low temperature water in the vicinity of the triple point.

Croxton introduces a dipole order parameter $\eta_{(n)}(z)$ defined as

$$\eta_{(n)}(z_1) = \int_0^\pi f(\theta, z_1) \cos^n \theta \sin \theta \, d\theta \qquad (7.3.13)$$

where θ, ϕ represent the internal orientations of the dipole M_1 at depth z_1 below the origin of coordinates located on the Gibbs surface (Figure 7.3.2). $f(\theta, \phi)$ is its angular distribution which will vary with depth z_1 below the surface. It is assumed that all orientations ϕ are equally probable. It follows from (7.3.13) that the first moment $\eta_{(1)}(z_1)$ is zero in either bulk phase corresponding to the absence of any preferred dipolar orientation, whilst $\eta_{(1)} = +1$ (-1) for complete alignment of M along the positive (negative) z axis. It should be pointed out that although $\eta_{(1)} = 0$ in either bulk phase, we should generally expect $\eta_{(2)} \neq 0$ since there will nevertheless be local short range orientational order. Since it is anticipated on symmetry grounds that the surface dipole field will orient either parallel or antiparallel to the z-axis,

208

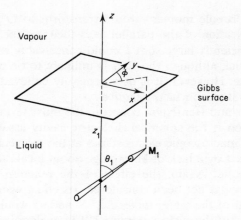

Vapour

Gibbs surface

Liquid

Figure 7.3.2 Geometry for the orientation of the water dipole moment with respect to the Gibbs surface

the sign of $\eta_{(1)}(z)$ should determine the molecular orientation, whilst its magnitude will indicate the degree of polarization.

The approach consists in the expression of the anisotropic (electrostatic) component of the surface excess free energy in terms of the $n = 1$ order parameter which is then variationally minimized with respect to $\eta_{(1)}(z_1)$

Davis[21] has developed the expression

$$\gamma = -\frac{1}{2} \int dz_1 \int d^3\mathbf{r}_{12}\, d^3\hat{\mathbf{e}}_1\, d^3\hat{\mathbf{e}}_2 \left[z_{12} \frac{\partial}{\partial z_{12}} - x_{12} \frac{\partial}{\partial x_{12}} \right] U(\mathbf{r}_{12}) \rho_{(2)}(\mathbf{r}_1, \mathbf{r}_2; \hat{\mathbf{e}}_1, \hat{\mathbf{e}}_2)$$

(7.3.14)

for the surface tension of a polyatomic fluid. $\hat{\mathbf{e}}_i$ denotes the Eulerian angles $(\theta_i, \phi_i, \psi_i)$ of molecule i and $d^3\hat{\mathbf{e}}_i = \sin \theta_i\, d\theta_i\, d\phi_i\, d\psi_i$ with $0 < \theta_i < \pi$, $0 < \phi_i < 2\pi$, and $0 < \psi_i < 2\pi$ (Figure 7.3.3—The angle ψ is redundant for water if the molecular axis is taken to be the axis of symmetry of the system). $\rho_{(2)}(\mathbf{r}_1, \mathbf{r}_2; \hat{\mathbf{e}}_1, \hat{\mathbf{e}}_2)$ represents the anisotropic pair correlation function

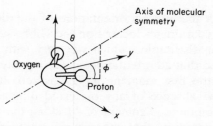

Axis of molecular symmetry

Oxygen

Proton

Figure 7.3.3 The set of Eulerian angles specifying the molecular orientation with respect to a cartesian reference frame

extended to include the internal angular configurations of particles 1 and 2. On the basis of a comparison of (7.3.14) with

$$\gamma = \int_{-\infty}^{\infty} \rho_{(1)}(z_1) a(z_1)\, dz_1 \qquad (7.3.15)$$

it follows that the surface excess free energy density is

$$\rho_{(1)}(z_1) a(z_1) = -\frac{1}{2} \int d^3 \mathbf{r}_{12}\, d^3 \hat{\mathbf{e}}_1\, d^3 \hat{\mathbf{e}}_2 \left[z_{12} \frac{\partial}{\partial z_{12}} - x_{12} \frac{\partial}{\partial x_{12}} \right] U(\mathbf{r}_{12}) \rho_{(2)}(\mathbf{r}_1, \mathbf{r}_2; \hat{\mathbf{e}}_1, \hat{\mathbf{e}}_2). \qquad (7.3.16)$$

Firstly the pair correlation function is decoupled:[24]

$$\rho_{(2)}(\mathbf{r}_1, \mathbf{r}_2; \hat{\mathbf{e}}_1, \hat{\mathbf{e}}_2)_{\hat{\mathbf{n}}} \sim \rho_{(1)}(z_1)_{\hat{\mathbf{n}}} \rho_{(1)}(z_2)_{\hat{\mathbf{n}}} f(\hat{\mathbf{e}}_1, z_1)_{\hat{\mathbf{n}}} f(\hat{\mathbf{e}}_2, z_2)_{\hat{\mathbf{n}}} g_{(2)}(\mathbf{r}_1, \mathbf{r}_2, z_1) \qquad (7.3.17)$$

where the subscripts $\hat{\mathbf{n}}$ refer to a specific orientation of the surface dipole vector field which, moreover, is assumed to be a slowly varying function on the scale of an intermolecular separation. We imagine the molecule as a point dipole plus a point quadrupole contained within a spherical envelope of the van der Waals radius ($\sigma = 1.4$ Å) and that angular correlation develops only through the multipolar interactions: angular correlation arising from the hard core structure of the molecule is neglected, whereupon (7.3.17) may be further decoupled

$$f(z_i, \hat{\mathbf{e}}_i)_{\hat{\mathbf{n}}} = f(z_i, \theta_i)_{\hat{\mathbf{n}}} f(z_i, \phi_i)_{\hat{\mathbf{n}}}, \qquad (7.3.18)$$

We need not specify $f(z_i, \phi_i)_{\hat{\mathbf{n}}}$ precisely, although we assume it is of the general form

$$f(z_i, \phi_i)_{\hat{\mathbf{n}}} = \sum_n A_n \cos^{2n} \phi_i + B_n \sin^{2n} \phi_i. \qquad (7.3.19)$$

The dipole–dipole contribution has the explicit form

$$U_{\mathrm{DD}} = \mathbf{M}_1 \cdot \overset{\leftrightarrow}{\mathbf{T}} \cdot \mathbf{M}_{12} = \begin{bmatrix} M_x \\ M_y \\ M_z \end{bmatrix}_1 \cdot \overset{\leftrightarrow}{\mathbf{T}}_{12} \cdot \begin{bmatrix} M_x \\ M_y \\ M_z \end{bmatrix}_2 \qquad (7.3.20)$$

$$= M^2 \begin{bmatrix} \sin \theta_1 \cos \phi_1 \\ \sin \theta_1 \sin \phi_1 \\ \cos \theta_1 \end{bmatrix} \begin{bmatrix} \sin \theta_2 \cos \phi_2 \\ \sin \theta_2 \sin \phi_2 \\ \cos \theta_2 \end{bmatrix} \overset{\leftrightarrow}{\mathbf{T}}_{12} \qquad (7.3.21)$$

although for our particular choice of coordinate frame, with the internal coordinates referred to the axis of molecular symmetry, considerable simplification ensues. We begin by performing the averages over $\hat{\mathbf{e}}_1$ and $\hat{\mathbf{e}}_2$ as required in (7.3.16). Averaging over ϕ_1 and ϕ_2, using the probability distribution (7.3.19), yields the reduced dipole vector

$$\begin{bmatrix} \cdot \\ \cdot \\ M_z \end{bmatrix}_i$$

whereupon

$$\hat{n}\langle U_{DD}\rangle_{\phi_1\phi_2} = 4\pi^2 M_z^2 \cos\theta_1 \cos\theta_2 T_{z_1z_2}. \tag{7.3.22}$$

Subsequent integration over θ_1, θ_2 in conjunction with the relevant angular distributions $f(z_i, \theta_i)_{\hat{n}}$ yields

$$\begin{aligned}\hat{n}\langle U_{DD}\rangle_{\hat{e}_1\hat{e}_2} &= 4\pi^2 M_z^2 \eta_{(1)}(z_1)\eta_{(1)}(z_2) T_{z_1z_2} \\ &= 4\pi^2 M_z^2 \eta_{(1)}(z_1)[\eta_{(1)}(z_1) + F(z_{12})]\end{aligned} \tag{7.3.23}$$

where (7.3.13) has been used and $\eta_{(1)}(z_2)$ is expressed in terms of $\eta_{(1)}(z_1)$ and their relative difference $F(z_{12})$.

Similar simplifications arise in the dipole–quadrupole and quadrupole–quadrupole terms for this particular choice of axes, the quadrupole tensor (7.2.3) reducing to diagonal form:

$$\begin{bmatrix} Q_{xx} & \cdot & \cdot \\ \cdot & Q_{yy} & \cdot \\ \cdot & \cdot & Q_{zz} \end{bmatrix}_i$$

$$\begin{aligned}\hat{n}\langle U_{DQ}\rangle_{\hat{e}_1\hat{e}_2} &= 8\pi^2 M_z \eta_{(1)}(z_1)[\eta_{(1)}(z_2)(Q_{xx}U_{z_1x_2x_2} + Q_{yy}U_{z_1y_2y_2}) \\ &\quad + (2\eta_{(1)}(z_2) - 1)Q_{zz}U_{z_1z_2z_2}]\end{aligned} \tag{7.3.24}$$

and

$$\begin{aligned}\hat{n}\langle U_{QQ}\rangle_{\hat{e}_1\hat{e}_2} &= 4\pi^2 \eta_{(1)}(z_1)\eta_{(1)}(z_2) \\ &\quad \times [Q_{xx}^2 W_{x_1x_1x_2x_2} + 2Q_{xx}Q_{yy}W_{x_1x_1y_2y_2} + Q_{yy}^2 W_{y_1y_1y_2y_2}] \\ &\quad + 4\pi\eta_{(1)}(z_1)(2\eta_{(1)}(z_2) - 1)[Q_{xx}Q_{zz}W_{x_1x_1z_2z_2} + Q_{yy}Q_{zz}W_{y_1y_1z_2z_2}] \\ &\quad + (2\eta_{(1)}(z_1) - 1)(2\eta_{(1)}(z_2) - 1)W_{z_1z_1z_2z_2}\end{aligned} \tag{7.3.25}$$

where, as before,

$$\eta_{(1)}(z_2) \sim \eta_{(1)}(z_1) + F(z_{12}) \tag{7.3.26}$$

From (7.2.1) the angle-averaged energy is, setting $\varepsilon = 1$ in the free space between molecules,

$$4\pi\hat{n}\langle U(\mathbf{r}_{12})\rangle_{\hat{e}_1\hat{e}_2} = 2\pi\Phi(r_{12})_{LJ} + \hat{n}\langle U_{DD}\rangle_{\hat{e}_1\hat{e}_2} + \tfrac{1}{2}\hat{n}\langle U_{DQ}\rangle_{\hat{e}_1\hat{e}_2} + \tfrac{1}{4}\hat{n}\langle U_{QQ}\rangle_{\hat{e}_1\hat{e}_2}. \tag{7.3.27}$$

In expressions (7.3.24), (7.3.25) use has been made of the fact that $U_{\alpha_i\beta_i\beta_i} = -U_{\alpha_i\beta_i\beta_i}$ and $W_{\alpha_i\alpha_i\beta_i\beta_i} = W_{\alpha_j\alpha_j\beta_i\beta_i}$. The second angular moment $\eta_{(2)}(z_i) = \int_0^\pi f(z_i, \theta_i) \cos^2\theta \sin\theta \, d\theta$ arises in the development of the angle-averaged dipole–quadrupole and quadrupole–quadrupole expressions. This may be related to the first moment $\eta_{(1)}(z_i)$ by forming small θ expansions about the orientations $\theta = 0$ and $\theta = \pi$, the most likely orientations of the surface dipole field, in which case we have

$$\eta_{(2)}(z_i) \sim 2\eta_{(1)}(z_i) - 1. \tag{7.3.28}$$

Such an approximation will be quite acceptable for low temperature water when thermal disorientation of the surface alignment is likely to be minimal.

Inserting (7.3.27) in (7.3.16) yields the local free energy density

$$
\begin{aligned}
\rho_{(1)}(z_1)\frac{a(z_1)}{4\pi} = {} & A_{LJ} + \eta_{(1)}(z_1)\{4\pi^2 M_z^2 \overline{T_{zz}\eta_{(1)}(z_2)} \\
& + 4\pi^2 M_z \overline{\eta(z_2)(Q_{xx}U_{z_1x_2x_2} + Q_{yy}U_{z_1y_2y_2})} + 8\pi^2 M_z Q_{zz}\overline{\eta_{(1)}(z_2)U_{z_1z_2z_2}} \\
& + \pi^2 \overline{\eta_{(1)}(z_2)[Q_{xx}^2 W_{x_1x_1x_2x_2} + 2Q_{xx}Q_{yy}W_{x_1x_1y_2y_2} + Q_{yy}^2 W_{y_1y_1y_2y_2}]} \\
& + 2\pi \overline{\eta_{(1)}(z_2)[Q_{xx}Q_{zz}W_{x_1x_1z_1z_1} + Q_{yy}Q_{zz}W_{y_1y_1z_2z_2}]} \\
& + \overline{\eta(z_2)W_{z_1z_1z_1z_1}}\} \\
& - 4\pi^2 \eta_{(1)}(z_1)M_z Q_{zz}\bar{U}_{z_1z_2z_2} \\
& - \pi \eta_{(1)}(z_1)[Q_{xx}Q_{zz}\bar{W}_{x_1x_1z_2z_2} + Q_{yy}Q_{zz}\bar{W}_{y_1y_1z_2z_2}] \\
& - \tfrac{1}{2}\eta_{(1)}(z_1)W_{z_1z_1z_2z_2}
\end{aligned}
\tag{7.3.29}
$$

where $\eta_{(1)}(z_2)$ is related to $\eta_{(1)}(z_1)$ by (7.3.26) and the bars represent the operation

$$
\bar{\Xi} = \int \left[z_{12}\frac{\partial}{\partial z_{12}} - x_{12}\frac{\partial}{\partial x_{12}} \right] \Xi \rho_{(2)}(\mathbf{r}_1, \mathbf{r}_1, z_1)\, d^3\mathbf{r}.
$$

Equation (7.3.29), which is quadratic in $\eta_{(1)}(z_1)$, may be variationally minimized with respect to $\eta_{(1)}(z_1)$ to yield a minimum orientational contribution to the interfacial free energy. In practice, an iterative minimization is made starting from some initial trial profile, the sign of the final profile determining the orientation of the dipolar field, parallel or antiparallel to the surface normal, whilst its magnitude $-1 < \eta_{(1)}(z_1) < +1$ establishes the degree of interfacial polarization. Close inspection of (7.3.29) reveals that $\eta_{(1)}(z_1) \to 0$ in either bulk phase as the stress tensor becomes isotropic and $\eta_{(1)}(z_2) \to \eta_{(1)}(z_1) \to 0$ as it does in either homogeneous phase.

The polarization density is readily determined as

$$
\mathbf{P}(z) = M_z \rho_{(1)}(z)\eta_{(1)}(z) \tag{7.3.30}
$$

whilst its associated mean interfacial electric field is

$$
\mathbf{E}(z) = -4\pi \mathbf{P}(z) \tag{7.3.31}
$$

(the negative sign obtains because the polarization is spontaneous, not induced by $\mathbf{E}$), in which case the surface potential becomes

$$
\bar{\psi}_L - \bar{\psi}_V = \int_{-\infty}^{\infty} \mathbf{E}(z)\, dz = -4\pi \int_{-\infty}^{\infty} \mathbf{P}(z)\, dz
$$

$$
= M_z \int_{-\infty}^{\infty} \rho_{(1)}(z)\eta_{(1)}(z)\, dz \tag{7.3.32}
$$

where $\bar{\psi}_L$ and $\bar{\psi}_V$ are the mean liquid and vapour potentials deep in each interior phase.

With a knowledge of $\eta_{(1)}(z)$, the electropolar component of the surface tension follows directly from the insertion of (7.3.29) in (7.3.15). Similarly, the surface potential follows from (7.3.32). The density profile $\eta_{(1)}(z)$ is determined on the basis of the BGY integral equation discussed in Chapter 2 using a Lennard–Jones interaction centred on the oxygen nucleus, the parameters (σ, ε) of which were determined by forming a Percus–Yevick fit to the X-ray radial distribution of Narten and Levy.[36] It is appropriate to mention that X-ray scattering reflects the *electronic* structure of the water molecule which is somewhat 'sphericalized' with respect to, say, a neutron scattering determination which is more sensitive to the *nuclear* structure of the molecule. The resulting density profile $\rho_{(1)}(z)$ is probably too sharp, neglecting as it does the covalent features of the interaction which enforce a substantially expanded structure with respect to the Lennard–Jones fluids.

The polarizability of the water molecule is considerable, and cooperative effects in the bulk liquid phase are known to *enhance* the dipole moment above its isolated or vapour value of 1.84 Debye to ~2.65 Debye. A similar, but unknown anisotropic polarization of the quadrupole tensor undoubtedly occurs; however, the enhancement is thought to be slight. In the absence of explicit knowledge of the octupolar and higher electropolar moments, the analysis necessarily terminates at the quadrupole–quadrupole term. The lower electropolar contributions to the surface tension, together with the Lennard–Jones component are shown in Table 7.3.2 for liquid water at 4 °C. These components account for approximately two thirds of the experimental value of ~75 dyn cm^{-1}: contributions arising from electrostatic image effects, higher electropolar and polarization effects, and the use of an anisotropic rather than the sphericalized X-ray radial distributions will all tend to increase the computed value of the surface tension. However, more elaborate assumptions are hardly justified at our present level of understanding.

The form of the orientation profile $\eta_{(1)}(z)$ at 4 °C is shown in Figure 7.3.4(a) from which we see that surface orientational order extends several molecular diameters into the bulk. Invariably the peak in the profile lies at ~10% density level which may be understood in terms of a competitive interplay between the strength of the orienting torque field and the effects of hindered rotation within the bulk liquid phase. It is clear from the $\eta_{(1)}(z)$ profile that the water molecules are oriented with their protons pointing

Table 7.3.2. Components of the surface tension of liquid water at 4 °C (dyn cm^{-1})

γ_{LJ}	γ_{DD}*	γ_{D^*Q}†	γ_{QQ}†	total	$\gamma_{expt.}$
37.4	1.78	8.48	2.54	50.20	75.6

* $M_Z = 2.65$ Debye

$$\dagger\,\mathbf{Q} = \begin{bmatrix} -6.56 & \cdot & \cdot \\ \cdot & -5.18 & \cdot \\ \cdot & \cdot & -5.51 \end{bmatrix} \text{Debye Å}$$

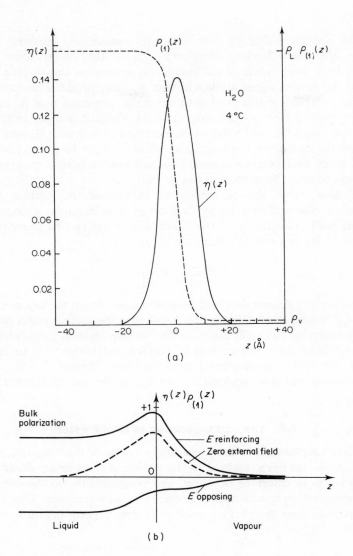

Figure 7.3.4 (a) Variation of the local order parameter $\eta(z)$ for liquid water at 4 °C.[17] Orientational order is seen to extend over ~15 molecular diameters. The density profile $\rho_{(1)}(z)$ is also shown. (b) Effect of an external electric field reinforcing and opposing the surface polarization

outwards: this is contrary to the conclusion of Stillinger and Ben-Naim[18] who, however, assume a *positive* quadrupole moment for the molecule. The orientation at the liquid surface is sensitively dependent upon the sign of **Q**. It is important that the components of the quadrupole tensor should be negative and almost equal: a consideration of the electronic distribution about the positively charged core structure immediately confirms that this must be the case. Various measurements of the liquid water surface potential have been made in an attempt to determine the surface orientation, but the results agree neither in sign nor magnitude. The spontaneous interfacial polarization will tend to broaden the transition zone if the surface field is reinforced by a normal external field, whilst if it tends to cancel the spontaneous field the width would be expected to decrease (Figure 7.3.4b). Ellipsometric studies of the interfacial width subject to the application of external fields have been proposed,[18] but unfortunately such experiments do not appear to have been carried out as yet.

There also exists the possibility of modifying the surface tension–temperature characteristic by application of a sufficiently strong external field. The bulk polarization of the water will, of course, be independent of the sense of the normal external field

$$\mathbf{P}(\underset{\text{bulk}}{\mathbf{E}}) = \mathbf{P}(\underset{\text{bulk}}{-\mathbf{E}}) \tag{7.3.33}$$

whilst the surface polarization will depend sensitively on the orientation of $\mathbf{E}$ (Figure 7.3.4b)—a reinforcing field will enhance the surface order parameter whilst an opposing field will disorient the surface dipoles. For a given value of external field the enhancement of surface order will tend to make the slope of the $\gamma(T)$ characteristic more positive (Figure 7.3.5). Such an investigation would also establish the net sign of the surface dipolar orientation.

7.4 The surface tension of liquid water

As we have emphasized in the preceding sections of this chapter, relatively little attention has been devoted to the statistical mechanical description of the liquid–vapour interface of water, and consequently no theoretical estimates, beyond largely qualitative speculations, are available. Davis,[21] however, has estimated the surface tension of low vapour pressure water for

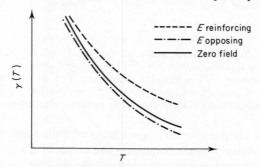

Figure 7.3.5 Effect of an external field upon the surface tension–temperature characteristic for liquid water

which, presumably, the long range dipole–dipole interactions represent the principal anisotropic contribution to the pair potential. In this dipolar approximation Davis adopts a central Lennard–Jones interaction supplemented beyond some cut-off radius a by a term of the form U_{DD} (7.3.20). Prior to insertion in equations (7.3.15), (7.3.16), the long range $(r \geqslant 5 \text{ Å})$ form of the *bulk* radial distribution is adopted, shown by Ben-Naim and Stillinger[22] to be

$$g_{(2)}^L \sim 1 - [9g_K(\varepsilon_0 - 1)/4\pi\rho_L(2\varepsilon_0 + 1)]U_{DD}/M^2 \qquad (7.4.1)$$

where ε_0 is the static dielectric constant of the liquid and g_K is the Kirkwood parameter arising from orientational correlations amongst dipoles, values of which have been estimated by Stillinger and Rahman.[16] It should be observed, as we did at the end of Section 7.2, that the requirement of an equipotential surface means account must be taken of the image field reflected in the Gibbs surface. This will, of course, modify the long range form of the radial correlation (7.4.1) which applies essentially to a bulk fluid.

On the basis of a Kirkwood–Buff step model approximation, insertion of (7.4.1) in (7.3.15) and (7.3.16) yields, after performing the angular averages

$$
\begin{aligned}
\gamma_{DD} &= -\frac{\rho_L^2 M^2}{6}\left[\frac{9g_K(\varepsilon_0 - 1)}{4\pi\rho_L(2\varepsilon_0 + 1)}\right]\int_{\substack{z>0 \\ r>a}} z\left[z\frac{\partial r^{-6}}{\partial z} - x\frac{\partial r^{-6}}{\partial x}\right]d\mathbf{r} \\
&= \frac{\pi\rho_L^2 M^2}{8a^2}\left[\frac{9g_K(\varepsilon_0 - 1)}{4\pi\rho_L(2\varepsilon_0 + 1)}\right]
\end{aligned}
\qquad (7.4.2)
$$

for the dipolar contribution to the surface tension of low pressure water. Although not mentioned by Davis, his analysis contains the implicit assumption that the order parameter is zero throughout the system, which it is almost certainly not in the vicinity of the surface. Consequently (7.4.2) represents a substantial underestimate of the dipolar contribution to the interfacial tension. Nevertheless, for water at $0\,°C$ Davis[21] takes $\rho_L = 3.33 \times 10^{22}$ molecules/cm^3, $\varepsilon_0 = 88$, $g_K = 3.66$, and $M = 2.35$ Debye and finds $\gamma_{DD} = 114$, 60, and 38 dyn/cm for cut-off radii of $a = 2.85$, 4, and 5 Å respectively (the nearest neighbour separation at this density being ~ 2.75 Å). The experimental value for the surface tension of water at $0\,°C$ is 76 dyn/cm, and whilst it is clear that long range dipolar contributions do make a substantial contribution to the final value, it is virtually impossible to disentangle and assess the effects of the various approximations arising from the choice of cut-off radius a, the use of the radial distribution (7.4.1), and the introduction of the discontinuous step model of the density profile.

Davis also neglects the effects of surface induced correlations in using (7.4.1) for the long range form of the radial distribution function: the requirement that the free surface of the liquid be an equipotential implies the existence of an effective image dipole on the opposite side of the surface. The magnitude of the latter for a fluid of bulk dielectric constant ε is $(\varepsilon - 1)/(\varepsilon + 1)$ times the original dipole. Thus a dipole in the vicinity of the surface experiences both a direct and an indirect orienting influence, yielding a modified radial distribution. Fulton[23] takes this into account, together with an alternative expression to (7.4.2) for the contribution of long range polarization fluctuations which does not involve the Kirkwood g_K factor. Thus, Fulton finds $\gamma_{DD} = 29$ dyn/cm, and the component arising from the development of surface induced correlation $\gamma_s = 10$ dyn/cm—a significant contribution. The sum of the two terms, representing the total anisotropic contribution to the surface tension, is 39 dyn/cm—fortuitously close to Davis' result of 38 dyn/cm.

7.5 The surface potential of liquid water

If the water molecule possessed only a centrally placed dipole moment, then positive and negative orientations of the dipolar field would be energetically equivalent. The effect of a permanent quadrupole moment is to redistribute the field lines about the molecule. As Frenkel[9] first observed, in an electrostatically inhomogeneous region such as the liquid surface, the symmetry breaking results in a spontaneous polarization of the water molecules with a consequent development of an interfacial dipolar field and associated surface potential. Clearly, the specification of the quadrupolar and higher electropolar features of the molecule is central to any quantitative description of the water surface structure. Although numerous experimental and theoretical attempts have been made to determine the surface potential of liquid water, they agree neither in sign nor magnitude, and the interfacial structure remains unresolved.

Frenkel[9] has given a simple order of magnitude estimate of the surface potential of liquid water, assuming a single interfacial layer of perfectly oriented dipoles of surface density σ, length d and polar charge $\pm q$:

$$\Delta V \sim 4\pi\sigma q d \tag{7.5.1}$$

Taking $\sigma \sim 10^{15}$ cm^{-2} and $\mu = qd \sim 1$ Debye yields $\Delta V \sim \pm 3$ volts, depending upon the orientation of the dipole field. In fact the numerical value of the surface potential may be expected to be substantially smaller.

That little can be said conclusively regarding the surface potential is neatly summarized in Table 7.5.1, and a relatively recent review of mainly dated data has been given by Llopis.[24] The specification of temperature in Table 7.5.1 would, under the circumstances, be superfluous, and indeed in most cases remains unquoted. Croxton's value of -3.69 volts warrants some comment, being an order of magnitude larger than the other determinations. He determines the surface potential on the basis of the expression

$$\Delta V = -4\pi M_z \int_{-\infty}^{\infty} \rho_{(1)}(z)\eta(z)\,dz \tag{7.5.2}$$

which is clearly dependent upon the form of the density profile $\rho_{(1)}(z)$. A somewhat more relaxed profile than the essentially Lennard–Jones estimate

Table 7.5.1. Surface potential of liquid water

	ΔV volts
Experimental	
Chalmers and Pasquill[25]	+0.26
Case and Parsons[37]	
Frumkin et al.[26]	+0.1
Kamienski[27]	+1
Theoretical	
Bernal and Fowler[12]	+0.4
Eley and Evans[28]	+0.4
Verwey[29]	−0.48
Hush[30]	−0.3
Strehlow[31]	−0.36
Passoth[32]	+0.28
Miscenko and Kwait[33]	−0.3
Stillinger and Ben-Naim[18]	+0.03
Croxton[34]	−3.69

of Croxton is appropriate to water with its relatively expanded structure. Such a profile would substantially reduce the integral (7.5.2) both directly, through $\rho_{(1)}(z)$, and indirectly through an overall reduction of the orientation profile $\eta(z)$. Although the *sign* of the surface potential would remain unmodified, Croxton's value of −3.69 volts for water at 4 °C must be regarded as an upper limit on the magnitude. However, the incorporation of free ion effects has not been discussed, which would undoubtedly lower the estimate (see Chapter 11).

References

1. F. Franks (Ed.), *Water—A Comprehensive Treatise*, Plenum Press, London and New York (1973–75), Vols 1–5.
2. P. Barnes, In C. A. Croxton (Ed.), *Progress in Liquid Physics*, Wiley, London (1978), Ch. 9.
3. A. Ben-Naim, In C. A. Croxton (Ed.), *Progress in Liquid Physics*, Wiley, London (1978), Ch. 10.
4. J. W. Good, *J. Phys. Chem.*, **61,** 810 (1957).
5. F. H. Stillinger and A. Ben-Naim, *J. Chem. Phys.*, **47,** 4431 (1967).
6. J. C. Henniker, *Rev. Mod. Phys.*, **21,** 322 (1949).
7. R. J. Watts–Tobin, *Phil. Mag.*, **8,** 333 (1963).
8. R. Parsons. In J. O'M. Bockris and B. E. Conway (Eds), *Modern Aspects of Electrochemistry*, Vol. 1, Academic Press, New York (1954), pp. 123–124.
9. J. Frenkel, *Kinetic Theory of Liquids*, Dover, New York (1955), p. 356.
10. R. Aveyard and D. A. Haydon, *An Introduction to the Principles of Surface Chemistry*, Cambridge Chemistry Texts, Cambridge University Press (1973), Ch. 2.
11. D. Eisenberg and W. Kauzmann, *The Structure and Properties of Water*, Oxford University Press, New York (1969), p. 44.

218

12. J. D. Bernal and R. H. Fowler, *J. Chem. Phys.*, **1**, 515 (1933).
13. W. Band, *Introduction to Mathematical Physic*, van Nostrand, Princeton (1959), pp. 191–203.
14. D. Hankins, J. W. Moskowitz, and F. H. Stillinger, *J. Chem. Phys.*, **53**, 4544 (1970). See also Erratum, *J. Chem. Phys.*, **59**, 995 (1973).
15. B. R. Lenz and H. A. Scheraga, *J. Chem. Phys.*, **58**, 5296 (1973).
16. F. H. Stillinger and A. Rahman, *J. Chem Phys.*, **60**, 1545 (1974).
17. C. A. Croxton, *J. Phys. C.*, (in press, 1980), *Phys. Lett.* **74** A, 325 (1979).
18. F. H. Stillinger and A. Ben-Naim, *J. Chem. Phys.*, **47**, 4431 (1967). ↲
19. L. Onsager, *J. Am. Chem. Soc.*, **58**, 1486 (1936).
20. C. A. Croxton, *Introductory Eigenphysics*, Wiley, London (1975). 140.
21. H. T. Davis, *J. Chem. Phys.*, **62**, 3412 (1975).
22. A. Ben-Naim and F. J. Stillinger, In R. A. Horne (Ed.), *Structure and Transport Processes in Water and Aqueous Solutions*, Wiley-Interscience, New York (1972).
23. R. L. Fulton, *J. Chem. Phys.* **64**, 1857 (1976).
24. J. Llopis in *Modern Aspects of Electrochemistry* (Eds. J. O'M. Bockris and B. E. Conway). Plenum Press, New York, (1971). Ch. 2.
25. J. A. Chalmers and F. Pasquill, *Phil. Mag.*, **23**, 88 (1937).
26. A. N. Frumkin, Z. A. Jofa and M. A. Gervich, *Zh. Fiz. Khim.*, **30**, 1455 (1956).
27. B. Kamienski, *Electrochim. Acta*, **3**, 208 (1960).
28. D. D. Eley and M. G. Evans, *Trans. Faraday Soc.*, **34**, 1093 (1938).
29. E. J. W. Verwey, *Rec. trav. Chim. Pays Bas*, **61**, 564 (1942).
30. N. S. Hush, *Austr. J. Sci. Res.*, **A1**, 482 (1948).
31. M. Strehlow, *Z. Elektrochem.*, **56**, 119 (1952).
32. G. Passoth, *Z. Phys. Chem.*, **203**, 275 (1954).
33. K. P. Miscenko and E. I. Kwait, *Zh. Fiz. Khim.*, **28**, 1451 (1954).
34. C. A. Croxton, *Phys. Lett.* A, (in press); *J. Phys. C*, (in press). (1980).
35. J. D. Parsons, *Phys. Rev.* **A19**, 1225 (1979) has recently effected a decoupling of the orientational and translational degrees of freedom for pair correlations which scale as $g_{(2)}(r/\sigma(\mathbf{e}_1, \mathbf{e}_2))$, where $\sigma(\mathbf{e}_1, \mathbf{e}_2)$ is an angle-dependent collision diameter. A specification of $\sigma(\mathbf{e}_1, \mathbf{e}_2)$ has been given for spheroidal molecules by B. J. Berne and P. Pechukas, *J. Chem. Phys.*, **56**, 4213 (1972).
36. A. Narten and H. Levy, *Science*, **165**, 450 (1969).
37. B. Case and R. Parsons, *Trans. Faraday Soc.*, **63**, 1224 (1967).

CHAPTER 8

Polymer Adsorption at a Solvent Surface

8.1 Introduction

A statistical theory of the conformational modification of polymer chains in the vicinity of a solvent boundary is almost entirely undeveloped; such determinations as there have been[1, 6] depend essentially upon one or other of a variety of assumptions—almost invariably the solvent surface is discontinuously sharp, and a quasi-crystalline model is adopted in which sites may be occupied either by a solvent molecule or segment of the polymer. We shall be largely preoccupied with such idealized systems, clearly exposing as they do the underlying energy and entropy contributions to the interfacial thermodynamics. Explicit effects of the surface in restricting the accessible macromolecular configurations have even been neglected, with some success, and intrachain interference (or the excluded volume effect) is generally neglected entirely, or treated in a rather crude way. If self-interference effects are ignored then, of course, the bulk conformation of the chain is Brownian, and the distribution of segments is Gaussian about the centre of mass.

Restricted walks in the vicinity of a boundary are characterized by a reduction in the number of accessible chain conformations and a consequential lowering of the configurational entropy. The suppression of degrees of freedom normal to the solvent boundary also implies a greater spreading or average molecular span as the boundary is approached. Such restricted systems have been examined in the case of zero excluded volume (random walk) by a number of workers,[1] and the adsorption isotherms investigated. The extension of the analysis of self-avoiding systems is not straightforward: indeed, the exact conformational specification of an isolated bulk self-avoiding chain molecule appears prohibitively difficult, and suggests that any detailed consideration of anisotropic interfacial systems in a molecularly coarse solvent is premature.

However, increased computing facilities have enabled *bulk* Monte Carlo studies of relatively short self-avoiding chains to be made in which averages are made over a representative sample of trial configurations.[2] Additionally, exact enumeration investigations have been made[3] in which a chain is taken through all possible conformations in forming a statistical average. Although the chains in the latter case are necessarily much shorter than those based on Monte Carlo sampling techniques, the numbers are nevertheless exact, and very reliable extrapolations may sometimes be made, as Domb and Sykes[4] have shown for bulk properties. The expedient of confining the chain to various two- and three-dimensional regular lattices—triangular, tetrahedral, cubic, etc.—dramatically reduces the number of accessible configurations, and at the same time introduces a self-avoiding feature into the lattice walk. If, for long chains, the configurational properties become independent of the lattice symmetry, then it is often supposed that similar behaviour will obtain in a continuum: the few off-lattice machine studies which have been made appear not inconsistent with this proposition.

As we have emphasized, extension of these investigations to incorporate interfacial effects is at a preliminary stage of development for anything other than the zero excluded volume condition. Although formal expressions exist for concentrated polymer solutions incorporating intrachain effects at realistic solvent boundaries, numerical results are confined to simple caricatures of the interface, but which, nevertheless, enable estimates of surface adsorption and modification of the solvent surface tension to be made.

8.2 Adsorption of a polymer chain at a solvent boundary

Much of the discussion relating to dilute binary liquid mixtures (Chapter 4) applies to the interfacial thermodynamics of polymer adsorption at the solvent surface. One may still speak of the relative adsorption $\Gamma_{(solv.)}^{(seg.)}$ of the polymer segments with respect to the solvent, arising as it does as a result of the competitive interplay between entropy and energy effects, both contriving to ensure a minimum excess Helmholtz free energy for the two-component system. Of course, there is in the present case the additional constraint which characterizes a polymeric system—that of preservation of sequential order within the chains—and it is this which distinguishes the system from that of a simple binary mixture. It is difficult to anticipate the nett effect of this additional constraint: we may, for example, regard each chain as a precipitated coil or globule, and treat the system as if it were a simple binary mixture of globules and solvent molecules. Such an approach may be reasonable for a poor solvent at low temperatures when it proves energetically advantageous for the chains to collapse into such coils. Of course, such an approach neglects the attrition of accessible conformations of the polymer as it approaches the interface, and these specifically polymeric contributions to the surface excess entropy are not taken into account. In fact the attrition of accessible states may amount to no more than a minor distortion of the coil as it approaches the surface and the approximation may be tolerable. Conversely, in a good solvent, and at high temperatures, the chain is widely deployed within the solvent and we anticipate that the attrition of accessible states as the chain approaches the surface will have important consequences for the surface excess configurational entropy, the system differing significantly from that of a simple binary mixture. Intrachain effects—that is the excluded volume problem—difficult to incorporate under any circumstances, serve to compound the difficulties.

Considerable qualitative insight is provided from a consideration of a crystalline lattice boundary in which each site may be occupied either by a solvent molecule or a segment of the polymer chain: undoubtedly renormalization group techniques will shortly be applied to the problem, although under circumstances somewhat idealized for present purposes. We assume for the sake of simplicity that the chains are sufficiently dilute as to behave independently, and that the system is athermal. The effects of excluded volume are achieved by confining the chain to a regular lattice. It is clear from Figure 8.2.1 that as the centre of gravity of the chain moves towards the boundary, the number of accessible chain conformations decreases. Indeed, for a random walk on a simple cubic lattice (i.e. a chain which is allowed to cross itself) one finds the fraction of configurations lost to the chain on account of the boundary to the total number of accessible configurations in the absence of a boundary to be, in the limit of large N,

$$\frac{A}{V}\left(\frac{2}{3\pi}\right)^{1/2} N^{1/2} \sim 0.46066 \frac{A}{V} N^{1/2} \tag{8.2.1}$$

where A and V represent the superficial number and total number of sites, respectively. For a similar, but self-avoiding, system Bellemans[5] finds the corresponding asymptotic form

$$(0.3003 \pm 0.0002)\frac{A}{V} N^{2/3} \tag{8.2.2}$$

showing a considerably stronger N dependence, and reflecting the greater average spreading of the chain. Of course, the decrease in the number of accessible internal configurations near the boundary has the immediate consequence of lowering the configurational entropy of the chain which attains its maximum value when the chain loses contact with the surface in its most extended conformations. If, additionally, we incorporate an energetic effect arising from the energy change ε resulting from the replacement of a solvent molecule by a polymer segment *in the surface layer*, then we may calculate the modification of the surface tension $\Delta\gamma$ relative to that of the

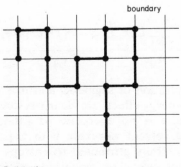

Figure 8.2.1 A self-avoiding walk on a regular square lattice

pure solvent as follows.[5] Let

c_N be the total number of configurations per unit volume allowed to a chain of N segments in the absence of boundary effects.

l_N be the number of configurations lost per unit area on account of the boundary

$g_N(\nu)$ be the number of configurations allowed to the chain per unit area when ν segments occupy superficial sites.

The modification of the solvent excess Helmholtz free energy per unit area due to the polymer is then

$$\Delta\gamma = kT \ln \frac{Z_N^{\sigma}}{Z_N^{\beta}} \tag{8.2.3}$$

where Z_N^{β}, Z_N^{σ} are the bulk and surface polymer partition functions (i.e. number of configurations). In the absence of energetic effects, we have quite simply

$$Z_N^{\beta} = c_N \qquad Z_N^{\sigma} = c_N - l_N. \tag{8.2.4}$$

Now, the total number of configurations allowed to the chain when ν elements occupy surface sites is $\sum_{\nu=1}^{N} g_N(\nu)$, whilst if the energy of replacement is ε this number is modified to $\sum_{\nu=1}^{N} g_N(\nu) \exp(\nu\varepsilon/kT)$ so that the surface *excess* of states (or deficit, depending upon the sign of ε) is $\sum_{\nu=1}^{N} g_N(\nu)[1-\exp(\nu\varepsilon/kT)]$. In this case, then, we have

$$Z_N^{\beta} = c_N$$
$$Z_N^{\sigma} = c_N - l_N - \sum_{\nu=1}^{N} g_N(\nu)[1-\exp(\nu\varepsilon/kT)]. \tag{8.2.5}$$

Substitution of this result in (8.2.3) and linearization of the logarithm yields

$$\Delta\gamma \sim \frac{CkT}{a} \frac{l_N + \sum_{\nu=1}^{N} g_N(\nu)[1-\exp(\nu\varepsilon/kT)]}{c_N} \tag{8.2.6}$$

where C is the concentration of segments and a is the area per superficial site. The entropy and energy effects are clearly resolved in this expression.

If $\varepsilon = 0$, that is there is no energetic advantage (or disadvantage) in replacing a superficial solvent molecule with a polymer segment, $\Delta\gamma$ is positive and may be entirely attributed to entropy effects associated with confinement of the polymer. The magnitude of the modification $\Delta\gamma \gtrless 0$ according as $\varepsilon \lessgtr 0$ depending upon the relative weighting of the superficial states $g_N(\nu)$.

A related quantity of interest is the fraction of superficially adsorbed elements of the chain:

$$\frac{\langle \nu_N \rangle}{N} = \frac{\sum\limits_{\nu=1}^{N} \nu g_N(\nu) \exp{(\nu \varepsilon / kT)}}{N \sum\limits_{\nu=1}^{N} g_N(\nu) \exp{(\nu \varepsilon / kT)}} \tag{8.2.7}$$

which is directly related to the heat of adsorption. If excluded volume effects are neglected, the system exhibits a transition at a critical temperature,[5] $\varepsilon / kT_c = 0.1823$. For $T > T_c$ the number of adsorbed segments is negligible compared with N, whilst for $T < T_c$ the chain is in an adsorbed state with the number of adsorbed segments $\sim N$ (Figure 8.2.2). Excluded volume effects at the surface may be expected to retard condensation of the chain on the superficial layer, as Bellemans has observed for relatively short self-avoiding walks ($N \leqslant 13$) on simple cubic lattices (Figure 8.2.2). As $N \to \infty$ one would ultimately expect to recover the infinite random chain behaviour, although the condensation temperature would be substantially lower on account of the excluded volume effects, as McCrackin[7] has already noted.

It is possible to define a superficial Flory temperature Θ at which entropy and energy effects exactly cancel in (8.2.6), whereupon $\Delta \gamma$ vanishes. In the asymptotic limit $N \to \infty$, Bellemans finds

$$\exp{(\varepsilon / k\Theta_\infty)} = 1.3323 \pm 0.0001. \tag{8.2.8}$$

Although Bellemans' simulation is but a simple caricature of a realistic system, it nevertheless illustrates the principal qualitative features of interfacial adsorption in a dilute polymer system incorporating excluded volume effects, and is embodied in the earlier Prigogine–Maréchal theory of interfacial polymer–solvent phenomena, to be developed in the following sections.

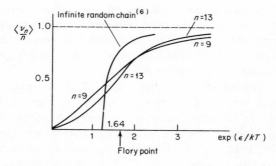

Figure 8.2.2 The fraction of superficially adsorbed chain segments as a function of temperature for self-avoiding walks on a simple cubic lattice ($N = 9$, 13), and for an infinite random chain

8.3 Surface thermodynamics of polymer solutions: Prigogine–Maréchal theory

As in the case of binary systems of molecules of differing size, we saw in the preceding section that the statistical mechanics of polymer–solvent solutions may be similarly resolved into entropy and energy contributions; the former relates to the system partition function as the number of ways the molecules may be arranged in space taking due account of their size, shape, and connectivity, whilst the configurational energy involves a knowledge of the number and nature of intermolecular contacts in terms of range and species. The two problems are, of course, closely interrelated, but provided the interaction energies between the various species are not too dissimilar they may be considered separately.

Two particular cases of surface adsorption arise both in the case of regular binary mixtures and polymer–solvent solutions according as, in the present case, the pure polymer has a greater or lower excess Helmholtz free energy per unit area γ_P greater than or less than that of the solvent γ_S. Provided there is no bulk segregation of the species it follows that the species of lower surface excess free energy will be preferentially adsorbed. If, for example, $\gamma_S < \gamma_P$ (e.g. polyisobutylene + n-heptane), corresponding to preferential solvent adsorption, the system would be expected to show a slight lowering of the solution surface tension below a simple volume fraction average over the entire concentration range

$$\gamma \leqslant \gamma_P \phi_P + \gamma_S \phi_S$$

$$\phi_S = \frac{N_S}{N_S + r N_P} \qquad \phi_P = \frac{N_P}{N_S + r N_P}$$

(8.3.1)

attaining one or other of the pure component values when either volume fraction ϕ_P or ϕ_S is zero. Very little qualitative difference from that of a simple mixture of spherical molecules would be expected. If, however, $\gamma_S > \gamma_P$ (e.g. polydimethyl-siloxane in toluene or tetrachloroethylene) corresponding to preferential polymer adsorption, the surface tension falls rapidly with concentration to that of the pure polymer, even for dilute systems. This is readily understood in terms of an interfacial zone composed almost entirely of preferentially adsorbed polymer, and is clearly demonstrated in Bellemans' determinations (Figure 8.2.2).

A simple theory which embodies most of these features in a semi-quantitative, and almost quantitative, fashion is that of Flory.[8] Whilst somewhat crude in comparison with the more sophisticated theories of Silberberg[9] and Hoeve,[10] Flory's treatment nevertheless clearly illustrates the interrelation between the entropy and energy contributions to the interfacial excess free energy.

We first consider a *bulk* liquid represented by a regular crystalline lattice on which molecules of the species P and S, corresponding to polymer and solvent respectively, are distributed. The components P occupy r connected sites and constitute linear unbranched r-mers in a solvent of S sites. The interaction energies between adjacent sites are then ε_{PP}, ε_{SS}, and ε_{PS} in an obvious notation. If the numbers of molecules in the solution are N_S, N_P respectively, then the total number of possible configurations is[11]

$$Z(N_S, N_P) = \frac{(N_S + rN_P)! \, \rho^{N_P}}{N_S! \, N_P! \, (N_S + rN_P)^{(r-1)N_P}}. \tag{8.3.2}$$

Of course connectivity and flexibility of the polymer would be expected to modify the number of accessible configurations. If $r = 1$, corresponding to a regular binary monomeric mixture of type P and S molecules, (8.3.2) reduces to that of a regular lattice solution, whilst flexibility is introduced through the factor ρ. For a perfectly flexible molecule $\rho = z(z-1)^{r-2}$ where z is the coordination of the lattice. Equation (8.3.2) represents the configurational entropy, whilst the configurational energy is simply[11]

$$U = U_S^0 + U_P^0 + \frac{\alpha}{N} \frac{rN_S N_P}{N_S + rN_P} \tag{8.3.3}$$

where the superscripts 0 designate the configurational energies of N_S and N_P *pure* S and P molecules, respectively. The second term represents the exchange energy developed between different species where $\alpha = N_A z\{\varepsilon_{SP} - \frac{1}{2}(\varepsilon_{SS} + \varepsilon_{PP})\}$, and N_A is Avagadro's number. The combination of equations (8.3.2) and (8.3.3) yields the bulk free energy of the system, and differentiation with respect to N_S and N_P yields the bulk chemical potentials μ_S, μ_P respectively:

$$\left. \begin{aligned} \mu_S &= \mu_S^0(T, p) + RT \ln \phi_S + RT \left(\frac{r-1}{r}\right) \phi_P + \alpha(\phi_P)^2 \\ \mu_P &= \mu_P^0(T, p) + RT \ln \phi_P - RT(r-1)\phi_S + r\alpha(\phi_S)^2 \end{aligned} \right\}. \tag{8.3.4}$$

The effects of flexibility, ρ, are contained within $\mu_P^0(T, p)$.

Modification of these results in the vicinity of a free surface were largely anticipated, at least qualitatively, in the preceding sections, and underlies the development of the surface excess functions. The difference in interaction energies ε_{SS}, ε_{PP}, and ε_{SP} is largely responsible for the preferential adsorption of one or other of the species, whilst the geometric suppression of many of the polymer configurations in the interfacial region implies a negative contribution to the surface entropy of the system, with a tendency for the chains to show a greater spreading as the interface is approached. Such an effect has been demonstrated by Bellemans[5] (8.2.2) on the basis of machine exact enumeration studies, and by Forsman and Hughes[19] for random flight chains. There have been a number of attempts to evaluate this difficult term. Initial studies[1] suggested that only a relatively small number of segments of the macromolecule were in contact with the surface, while more recent investigations[9,10] indicate that a large fraction (~ 0.7) of the segments contact the surface, which certainly seems to concur with Bellemans' findings (Section 8.2). Prigogine and Maréchal[12] concluded that reasonable predictions could be made on the basis of a *parallel-layer* model in which a Flory analysis is applied to polymer chains confined to a two-dimensional surface phase with all segments in contact with the surface, the thickness of the layer corresponding to one lattice site. For the present we restrict discussion to athermal solutions ($\alpha = 0$). In this case (8.3.4) reduces to

$$\left. \begin{aligned} \mu_S &= \mu_S^0(T, p) + RT\left[\ln \phi_S + \left(\frac{r-1}{r}\right)\phi_P\right] \\ \mu_P &= \mu_P^0(T, p) + RT[\ln \phi_P + (r-1)\phi_S] \end{aligned} \right\} \tag{8.3.5}$$

whilst the chemical potentials in the interfacial region are

$$\left. \begin{aligned} {}^\sigma\mu_S &= {}^\sigma\mu_S^0(T, p) + RT\left[\ln {}^\sigma\varphi_S + \left(\frac{r-1}{r}\right)^\sigma \varphi_P\right] - \gamma \mathscr{A} \\ {}^\sigma\mu_P &= {}^\sigma\mu_P^0(T, p) + RT[\ln {}^\sigma\varphi_P - (r-1)^\sigma\varphi_S] - \gamma r \mathscr{A} \end{aligned} \right\} \tag{8.3.6}$$

where γ is the solution surface tension and $\mathscr{A} = Na$ where a is the area associated with one lattice point. Now, at constant (T, p), thermodynamic stability enables us to equate the partial chemical potentials giving

$$\left. \begin{aligned} \gamma &= \frac{({}^\sigma\mu_S^0 - \mu_S^0)}{\mathscr{A}} + \frac{RT}{\mathscr{A}}\left[\ln\left(\frac{{}^\sigma\varphi_S}{\varphi_S}\right) + \left(\frac{r-1}{r}\right)({}^\sigma\varphi_P - \varphi_P)\right] \\ \gamma &= \frac{({}^\sigma\mu_P^0 - \mu_P^0)}{r\mathscr{A}} + \frac{RT}{r\mathscr{A}}\left[\ln\left(\frac{{}^\sigma\varphi_P}{\varphi_P}\right) + (r-1)({}^\sigma\varphi_S - \varphi_S)\right] \end{aligned} \right\} \tag{8.3.7}$$

where $\gamma_S = ({}^\sigma\mu_S^0 - \mu_S^0)/\mathscr{A}$, $\gamma_P = ({}^\sigma\mu_P^0 - \mu_P^0)/r\mathscr{A}$ are the surface tensions of the pure solvent and polymer, respectively.

Alternatively, the relative adsorption of each species may be expressed in terms of the difference $(\gamma_P - \gamma_S)$. Subtracting one equation from the other in (8.3.7) we have

$$\frac{{}^\sigma\phi_S}{\phi_S} = \left(\frac{{}^\sigma\phi_P}{\phi_P}\right)^{1/r} [\exp{(\gamma_P - \gamma_S)\mathscr{A}/RT}] \tag{8.3.8}$$

or

$$\ln\left[\left(\frac{{}^\sigma\phi_P}{\phi_P}\right)^{1/r} \bigg/ \left(\frac{{}^\sigma\phi_S}{\phi_S}\right)\right] = (\gamma_S - \gamma_P)\mathscr{A}/RT \tag{8.3.9}$$

where the l.h.s. of (8.3.9) is positive or negative according as there is preferential adsorption of the polymer or solvent, respectively, and reflects the well-known tendency for preferential adsorption of the component having the lower surface tension.

An alternative expression of the solution surface tension in terms of the component volume fractions follows by multiplying equations (8.3.7) by ϕ_S

and ϕ_P respectively, and adding:

$$\gamma = [\gamma_S \phi_S + \gamma_P \phi_P] + \frac{RT}{\mathscr{A}} \left[\phi_S \ln \frac{{}^\sigma \phi_S}{\phi_S} + \frac{1}{r} \phi_P \ln \frac{{}^\sigma \phi_P}{\phi_P} + \left(\frac{r-1}{r} \right) ({}^\sigma \phi_P - \phi_P) \right].$$

$$(8.3.10)$$

If the surface adsorption is small, that is, if ${}^\sigma \phi_S \sim \phi_S$, ${}^\sigma \phi_P \sim \phi_P$, (8.3.10) reduces to the crude linear volume fraction average

$$\gamma = \gamma_S \phi_S + \gamma_P \phi_P,$$

However, departures from this result embodied in (8.3.10) are not readily apparent. In fact, of course, preferential adsorption will depress γ below the volume fraction estimate. Segregation will enable energetically favourable interactions to develop, although at the expense of depressing the entropy and thereby raising the surface excess free energy. Nevertheless, if ${}^\sigma \phi_S \sim \phi_S$ and ${}^\sigma \phi_P \sim \phi_P$, then we require $(\gamma_P - \gamma_S)\mathscr{A}/RT \sim 0$ in (8.3.9) and we may write

$$\left. \begin{array}{l} {}^\sigma \phi_S = \phi_S + \phi_S \phi_P \Delta \\ {}^\sigma \phi_P = \phi_P - \phi_S \phi_P \Delta \end{array} \right\}$$

$$(8.3.11)$$

where Δ is a small quantity which tends to zero with $(\gamma_P - \gamma_S)\mathscr{A}/RT$, and may be shown from (8.3.9) to be

$$\Delta \sim \frac{(\gamma_P - \gamma_S)\mathscr{A}}{RT} \frac{1}{\phi_P + \phi_S/r}$$

$$(8.3.12)$$

to first order in $(\gamma_P - \gamma_S)\mathscr{A}/RT$. Equation (8.3.11) shows the correct limiting behaviour ${}^\sigma \phi_S = 0$ when $\phi_S = 0$ and ${}^\sigma \phi_S = 1$ when $\phi_S = 1$. Inserting (8.3.11) in (8.3.10) and expanding in powers of Δ up to the second:

$$\gamma = \gamma_S \phi_S + \gamma_P \phi_P - \frac{1}{2} \frac{RT}{\mathscr{A}} \phi_S \phi_P \left(\phi_P + \frac{\phi_S}{r} \right) \Delta^2 + \cdots$$

$$(8.3.13)$$

substitution of equation (8.3.12) in (8.3.13) yields the surface tension of the solution in terms of the bulk volume fractions

$$\gamma = \gamma_S \phi_S + \gamma_P \phi_P - \frac{1}{2} \frac{r \phi_S \phi_P}{[\phi_S + r \phi_P]} \frac{(\gamma_P - \gamma_S)^2 \mathscr{A}}{RT}.$$

$$(8.3.14)$$

Clearly, the surface tension is *depressed* below a simple linear volume fraction relationship, varying approximately as $(\gamma_P - \gamma_S)^2$ (Figure 8.3.1), and having a maximum absolute value when

$$\phi_S = \frac{\sqrt{r}}{1 + \sqrt{r}}.$$

$$(8.3.15)$$

Extension of the preceding analysis to non-athermal solutions ($\alpha \neq 0$) is straightforward, although we shall now be concerned with the number and nature of molecular contacts. The bulk liquid chemical potentials for each of

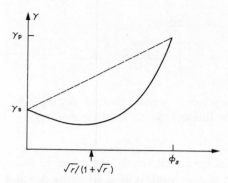

Figure 8.3.1 Surface tension of a binary mixture as a function of composition. The curve is seen to be depressed below the simple linear volume fraction relationship, shown by the broken curve

the species have already been given (8.3.4), whilst in the surface region we have

$$^\sigma\mu_S = {}^\sigma\mu_S^0(T, p) + RT\left[\ln {}^\sigma\phi_S + \left(\frac{r-1}{r}\right){}^\sigma\phi_P\right] + \alpha l({}^\sigma\phi_P)^2 + \alpha m(\phi_P)^2 - \gamma\mathscr{A}$$

$$^\sigma\mu_P = {}^\sigma\mu_P^0(T, p) + RT[\ln {}^\sigma\phi_P - (r-1){}^\sigma\phi_S] + r\alpha l({}^\sigma\phi_S)^2 + r\alpha m(\phi_S)^2 + \gamma r\mathscr{A}$$

$$(8.3.16)$$

where lz is the number of nearest neighbours within the plane and mz is the number in the neighbouring planes, so that $l + 2m = 1$. Incidentally, we note that the above expressions for chemical potential reduce to those for a regular solution, as they should, when $r = 1$.

Equating chemical potentials in the bulk and surface phases yields

$$\gamma = \gamma_S + \frac{RT}{\mathscr{A}}\left[\ln\left(\frac{{}^\sigma\phi_S}{\phi_S}\right) + \left(\frac{r-1}{r}\right)({}^\sigma\phi_P - \phi_P)\right] + \frac{\alpha}{\mathscr{A}}[l({}^\sigma\phi_P)^2 - (l+m)(\phi_P)^2]$$

$$\gamma = \gamma_P + \frac{RT}{r\mathscr{A}}\left[\ln\left(\frac{{}^\sigma\phi_P}{\phi_P}\right) + (r-1)({}^\sigma\phi_S - \phi_S)\right] + \frac{\alpha}{r\mathscr{A}}[l({}^\sigma\phi_S)^2 - (l+m)(\phi_S)^2]$$

$$(8.3.17)$$

which enables us to calculate both γ and surface composition as a function of bulk concentration. Provided the surface adsorption is not too large, i.e. if $(\gamma_P - \gamma_S)\mathscr{A}/RT$ and α/RT are both small compared with unity, we may proceed as for equations (8.3.11)–(8.3.14) and obtain an approximate solution as follows. Neglecting terms of order greater than the second in $(\gamma_P - \gamma_S)\mathscr{A}/RT$ and α/RT, we have

$$\gamma = \gamma_S\phi_S + \gamma_P\phi_P - \frac{\alpha m}{\mathscr{A}}\phi_S\phi_P - \frac{1}{2}\frac{r\phi_S\phi_P}{(\phi_S + \phi_P)}\left[\gamma_P - \gamma_S + \frac{\alpha m}{\mathscr{A}}(\phi_P - \phi_S)\right]^2\frac{\mathscr{A}}{RT}.$$

$$(8.3.18)$$

Alternatively, subtraction and rearrangement of equations (8.3.17) yields

$$\ln\left[\frac{(^\sigma\phi_P/\phi_P)^{1/r}}{(^\sigma\phi_S/\phi_S)}\right] = [(\gamma_S - \gamma_P)\mathscr{A}/RT] + \alpha(l+m)(\phi_S - \phi_P) - \alpha l(^\sigma\phi_S - ^\sigma\phi_P).$$

$$(8.3.19)$$

The second and third terms on the r.h.s. of (8.3.19) indicate the effect on the adsorption of the molecular interactions within the bulk and surface phases, respectively. Since α is almost invariably positive, corresponding to unfavourable polymer–solvent interactions, we see that for low polymer concentrations in both phases (i.e. $\phi_P < \phi_S$, $^\sigma\phi_P < ^\sigma\phi_S$) the second term favours polymer adsorption, whilst the third favours desorption corresponding to the well-known *azeotropy* effect whereby unfavourable interactions tend to eject from a phase whichever component is present in low concentration. Of course, the molecular interactions can reverse the sign of the adsorption with concentration.

8.4 Predictions of the Prigogine–Maréchal theory

Although an approximate analytic solution of (8.3.17) follows provided $(\gamma_P - \gamma_S)\mathscr{A}/RT \ll 1$ (8.3.18), few systems satisfy this requirement. Recent numerical solutions of (8.3.19) have been given by Siow and Patterson,[13] and the athermal ($\alpha = 0$) adsorption isotherms (8.3.19) $^\sigma\phi_P$ as a function of bulk concentration ϕ_P for various chain lengths r are shown in Figure 8.4.1. It is clear from Figure 8.4.1a that very high bulk polymer concentrations $\phi_P \sim 0.63$ are necessary before there is any significant surface adsorption, at least for long polymer chains: this system corresponds to preferential solvent adsorption. Conversely, in the case of preferential polymer adsorption, there is considerable surface adsorption even for very low bulk volume fractions.

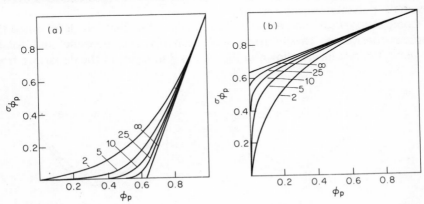

Figure 8.4.1 Athermal ($\alpha = 0$) surface adsorption isotherms $^\sigma\varphi_P$ as a function of bulk polymer concentration φ_P for various polymer lengths r. The values taken for $(\gamma_S - \gamma_P)\mathscr{A}/RT$ are (a) -1, corresponding to preferential solvent adsorption, and (b) $+1$, corresponding to preferential polymer adsorption[13]

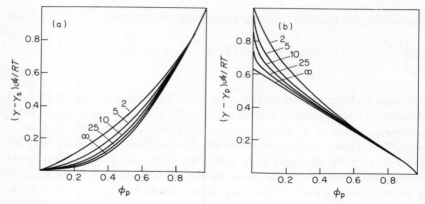

Figure 8.4.2 Athermal ($\alpha = 0$) surface tension increments as a function of bulk polymer concentration, φ_P. The values taken for $(\gamma_S - \gamma_P)\mathscr{A}/RT$ are (a) -1, corresponding to preferential solvent adsorption, and (b) $+1$, corresponding to preferential polymer adsorption[13]

In each case the amount adsorbed varies rapidly for lower molecular weights, attaining a limiting value as $r \to \infty$. These results appear to be in semi-quantitative agreement with Silberberg's[9] results, based on a considerably more elaborate model.

Solution of (8.3.19) for $^\sigma\phi_S$ may now be substituted into (8.3.17) to yield the surface tension of the system. In Figure 8.4.2 the surface tension increments $(\gamma - \gamma_S)\mathscr{A}/RT$ and $(\gamma - \gamma_P)\mathscr{A}/RT$ are given as functions of ϕ_P. Note how in Figure 8.4.2a, where $\gamma_S < \gamma_P$ corresponding to preferential solvent adsorption, the surface tension is insensitive to $^\sigma\phi_P$, particularly at low bulk polymer fractions. In the case of preferential polymer adsorption (Figure 8.4.2b), however, the surface tension shows a rapid decrease even at low bulk polymer concentrations.

Comparison with experiment is quite good. Maréchal has determined the surface tension of the benzene and diphenyl and benzene and tolane systems; the results are shown in Figure 8.4.3 in terms of the departure from

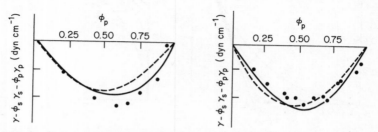

Figure 8.4.3 Departure of the surface tension from a simple linear volume fraction dependence. The points represent experimental data, whilst the full curve represents the numerical solution to (8.3.18) and the broken curve the approximation (8.3.17)

a simple linear volume fraction dependence. The parameter α/R was determined to be $\sim 77°$ for benzene and diphenyl at $70\,°C$ and $\sim 60°$ for benzene and tolane at $60\,°C$. ϕ_S = benzene in each case. The full curves represent a numerical solution to (8.3.18) whilst the broken curve represents the approximation (8.3.17). A more recent analysis by Siow and Patterson[13] comes to essentially the same conclusion.

8.5 Distribution of polymer segments at the solvent interface

In each of the preceding models (Bellemans, Section 8.2; Prigogine and Maréchal, Section 8.4) conformations of the polymer chain were confined to sequentially connected sites on a regular half lattice, the boundary of which represented the free interface, and the unoccupied sites a monomeric solvent. A more realistic description would necessarily involve the selfconsistent single-particle solvent $\rho_{(1;S)}(z)$ and polymer segment $\rho_{(1;P)}(z)$ distributions, which, on the basis of our experience with multicomponent fluids, we should expect to reflect the known preferential adsorption of the species yielding the lower surface excess free energy. The problem is considerably complicated, however, by the additional constraint of *connectivity* which characterizes the polymeric component. Moreover, not only do we require a knowledge of the self and cross two-particle correlations as in a binary mixture, but also the *internal* correlations developed within the polymer itself. Clearly the intersegment correlation will differ fundamentally depending whether the two segments in question are located on the same or on distinct chains: an immediate consequence of the connectivity of the macromolecule. Of course, the results should reduce to those of a simple binary mixture as the number of segments per chain $N \to 1$. Further complications will arise in cases where there is preferential polymer adsorption, for then interference effects, tangling, etc. will be substantial even at the lowest bulk volume fractions. A simple BGY analysis of the adsorption of an isolated polymer chain at the surface of a molecularly coarse solvent has been presented by Croxton[14] as follows.

The net force experienced by segment i of an N-mer in the vicinity of the solvent interface will arise from the non-adjacent polymer segments constituting the chain, and the surrounding solvent particles:

$$-\nabla_i \Phi(z_i) = \int_{\mathbf{p}} \sum_{p=1}^{N} \left\{ \nabla_i \Phi_{pp}(r_{ip}) - \nabla_i \Phi_{ps}(r_{ip}) Z(\mathbf{r}_{ip} \mid N) \rho_{(1;P)}(z/N) \, d\mathbf{p} \right.$$

$$\left. + \int_{\mathbf{s}} \nabla_i \Phi_{ps}(r_{is}) g_{(2;S)}(z_i, \mathbf{r}_{is}) \rho_{(1;S)}(z_S) \, d\mathbf{s}. \right. \tag{8.5.1}$$

p and s refer to polymer and solvent particles, respectively, whilst Φ_{pp} and Φ_{ss} refer to the polymer–polymer and polymer–solvent interactions. $Z(\mathbf{r}_{ip} \mid N)$ represents the anisotropic internal distribution of segments i and p within the N-mer, and $g_{(2;S)}(z_i, \mathbf{r}_{is})$ is the anisotropic distribution of solvent molecules about the ith segment. The $\int_s \mathrm{d}s$ term in (8.5.1) takes no account of the polymer configuration which obviously modifies the solvent distribution in the immediate vicinity of segment i. However, the negative term on the right hand side of (8.5.1) cancels ps configurations already adopted by the polymer, though with the implicit assumption that the solvent and segment particles are of almost identical size. The summation runs over all the internal polymer distributions about segment i within the N-mer. Further simplification may be made by adopting isotropic internal and external distributions $Z(r_{ip} \mid N)$, $g_{(2;S)}(r_{is})$ in the interfacial region, the expression of structural inhomogeneity being transferred to the single-particle distributions $\rho_{(1;S)}(z_s)$, $\rho_{(1;P)}(z_p \mid N)$ in what amounts to Green's closure.

Given the solvent profile $\rho_{(1;S)}(z_s)$, the response of the polymer is for the ith segment to adopt the distribution:[14]

$$
\rho_{(1;P)}(z_i \mid N) = \rho_P \exp \left\{ -\frac{1}{kT} \int_{-\infty}^{z_i} \left(\int_{\mathbf{p}} \sum_{\substack{p=1 \\ p \neq i-1,i,i+1}}^{N} \nabla_i [\Phi_{pp}(r_{ip}) - \Phi_{ps}(r_{ip})] \right. \right.
$$

$$
\left. \left. \times Z(r_{ip} \mid N) \rho_{(1;P)}(z_p \mid N)\, \mathrm{d}\mathbf{p} + \int_{\mathbf{s}} \nabla_i \Phi_{ps}(r_{is}) g_{(2;S)}(r_{is}) \rho_{(1;S)}(z_s)\, \mathrm{d}\mathbf{s} \right) \mathrm{d}z_i \right\} \quad (8.5.2)
$$

where ρ_P is the bulk number density of the polymer chains. This, of course, presumes that there is no interchain interference—an unrealistic assumption in the case of preferential adsorption of the polymer.

However, Croxton[15, 16] has determined the internal $Z(r_{ip} \mid N)$ and external distributions $Z(r_{1N} \mid N)$ on the basis of a diagrammatic convolution analysis both in the case of isolated and densely packed chains. Some typical distributions are shown in Figure 8.5.1; for details the reader is directed to the literature.

What is apparent is that the internal distributions differ depending both upon N and the location of segment i within the chain. Nor does an internal distribution of a subset of n segments have the same form as an end-to-end distribution of n segments. While this is hardly a surprising result, it does have the *a priori* implication in (8.5.2) that the segment distributions $\rho_{(1;P)}(z_i)$ and $\rho_{(1;P)}(z_j)$ $(i \neq j)$ will differ since each involves a different set of internal polymer distributions $Z(r_{ip} \mid N)$. And since there is no immediately apparent sum rule relating to these distributions it does appear that there may well be preferential adsorption of certain *regions* of the polymer chain, with obvious implications for the possible adsorption or desorption of the ends of the chains for which the overall effect of

$$
\int_{\mathbf{p}} \sum_{\substack{p=1 \\ p \neq 1-i,i,i+1}}^{1} \nabla_i [\Phi_{pp}(r_{ip}) - \Phi_{ps}(r_{ip})] Z(r_{ip} \mid N) \rho_{(1;P)}(z_p \mid N)\, \mathrm{d}\mathbf{p}
$$

is likely to be most pronounced.

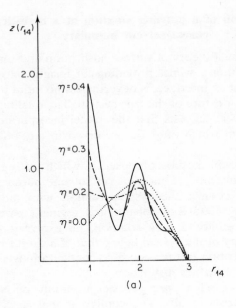

(a)

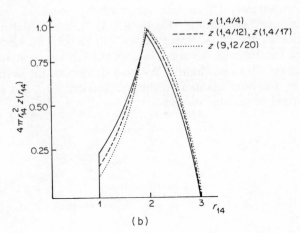

(b)

Figure 8.5.1 (a) Normalized end-to-end probability distribution functions for a 4-mer as a function of packing density. (b) Four segment distributions within chains of various lengths. The distributions are seen to depend sensitively upon location of the subset within the sequence

8.6 Surface tension of a polymer solution at a realistic, molecularly coarse solvent boundary

The Prigogine–Maréchal theory of surface adsorption and surface tension of polymer–solvent solutions, whilst providing an insight into the molecular processes operating at an interface, is nevertheless no more than a computationally convenient caricature of the true system. The qualitative correctness of the results, however, suggests that the model incorporates the essential features of the system which must be preserved in any refinement of the theory.

An alternative account may be proposed in which we regard the system rather as a binary mixture of monomeric solvent particles and flexible macromolecules. Of course, the system cannot, with any accuracy, be depicted simply as a mixture of solvent and segment particles—segment connectivity may, even for a bulk solution, be expected to depress the configurational entropy of the system below that of a simple binary mixture, whilst geometrical attrition of the accessible configurations as the surface is approached provides a further distinction.

We consider first of all a *pure* polymer assembly of N_c chains of N segments (NN_c segments in all). The configurational specification of the system is conveniently effected in terms of N_c 'external' vectors ${}^{\kappa}\mathbf{r}_i$ and $N_c(N-1)$ 'internal' vectors ${}^{\kappa}\mathbf{r}_{ij}$ (Figure 8.6.1). The external notation refers to the location of the ith segment of the κth chain with respect to a fixed origin, whilst the internal notation locates the jth segment with respect to the ith, again on the same chain.[17] We now determine the surface tension of the system in a similar manner to that of Section 3.1[18] but generalized to include nonrigid molecular systems.

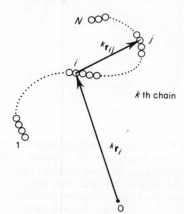

Figure 8.6.1 The 'internal' and 'external' vectors ${}^{\kappa}\mathbf{r}_{ij}$, ${}^{\kappa}\mathbf{r}_i$ respectively, involved in the configurational specification of the κth chain

The Hamiltonian of the system will, in addition to the kinetic term, contain components arising from inter- and intra-molecular interactions, the latter arising from internal degrees of freedom: segment–segment interactions within the molecule, connectivity energies, and energies arising from rotations, vibrations, plus electronic and nuclear contributions. Thus we may write

$$\mathcal{H} = \sum_{\kappa_i=1}^{NN_c} \frac{p_{\kappa_i}^2}{2m_{\kappa_i}} + \Phi_{N_c}\{\mathbf{r}^{NN_c}; \mathbf{I}^{NN_c}\} + \mathcal{H}^{\mathbf{I}} \tag{8.6.1}$$

where $\mathcal{H}^{\mathbf{I}}$ is the internal energy Hamiltonian, whilst the pairwise segment–segment interaction

$$\Phi_{N_c}\{\mathbf{r}^{NN_c}; \mathbf{I}^{NN_c}\} = \frac{1}{2} \sum_{\kappa_i=1}^{NN_c} \sum_{\lambda_i=1}^{NN_c} \Phi_{\kappa_i\lambda_i}. \tag{8.6.2}$$

It is understood that the sum in (8.6.2) excludes segment interactions on the same chain; moreover, Φ_{κ_i} will depend not only upon the segment separations, but also upon the internal states $^{\kappa}\mathbf{I}, {}^{\lambda}\mathbf{I}$ of chains κ, λ.

The configurational projection of the canonical ensemble partition function for the polymer assembly is, then,

$$Z_Q = \frac{1}{(NN_c)!} \int \exp\left(-\beta \mathcal{H}^{\mathbf{I}}\right) \int \exp\left(-\beta \Phi_N(\cdots {}^{\kappa}\mathbf{r}_i, {}^{\lambda}\mathbf{r}_j, \ldots; \ldots {}^{\kappa}\mathbf{I}, {}^{\lambda}\mathbf{I}, \ldots)\right)$$
$$\times (d^{\kappa}\mathbf{r}_j)^{N_c}(d\mathbf{I})^{N_c(N-1)!}. \tag{8.6.3}$$

We may now define the *external* segment–segment correlation function for segments i, j on chains κ, λ whose internal states are $^{\kappa}\mathbf{I}, {}^{\lambda}\mathbf{I}$, respectively:

$$\rho_{(2)}({}^{\kappa}\mathbf{r}_i, {}^{\lambda}\mathbf{r}_j; {}^{\kappa}\mathbf{I}, {}^{\lambda}\mathbf{I}) = \frac{1}{Z_Q(NN_c-2)!} \int \exp\left(-\beta\mathcal{H}^{\mathbf{I}}\right) \int \exp\left(-\beta\Phi_N\right)$$
$$\times (d^{\kappa}\mathbf{r}_j)^{N_c-2}(d\mathbf{I})^{N_c(N-1)!-2}. \tag{8.6.4}$$

In (8.6.4) all internal and external configurations are integrated over, the states $^{\kappa}\mathbf{r}_i, {}^{\lambda}\mathbf{r}_j; {}^{\kappa}\mathbf{I}, {}^{\lambda}\mathbf{I}$ being held fixed. Equation (8.6.4) therefore represents the probability of finding segments $^{\kappa}i, {}^{\lambda}j$ in the positions $^{\kappa}\mathbf{r}_i, {}^{\lambda}\mathbf{r}_j$, their respective chains being in the states $^{\kappa}\mathbf{I}, {}^{\lambda}\mathbf{I}$, all other states of the assembly being integrated over.

It is now straightforward, following the procedure of Section 3.1, to arrive at the result

$$\gamma = -\frac{1}{2} \int_{{}^{\kappa}\mathbf{I}} \int_{{}^{\lambda}\mathbf{I}} \int_{-\infty}^{\infty} \int \left\{ z_{\kappa_i\lambda_i} \frac{\partial \Phi_{\kappa_i\lambda_i}}{\partial z_{\kappa_i\lambda_i}} - x_{\kappa_i\lambda_i} \frac{\partial \Phi_{\kappa_i\lambda_i}}{\partial x_{\kappa_i\lambda_i}} \right\} \rho_{(2)}({}^{\kappa}\mathbf{r}_i, {}^{\lambda}\mathbf{r}_j; {}^{\kappa}\mathbf{I}, {}^{\lambda}\mathbf{I})$$
$$\times d\mathbf{r}_{\kappa_i\lambda_i} dz_{\kappa_i} d^{\kappa}\mathbf{I} d^{\lambda}\mathbf{I} \tag{8.6.5}$$

where $\mathbf{r}_{\kappa_i\lambda_i} = {}^{\kappa}\mathbf{r}_i - {}^{\lambda}\mathbf{r}_j$. Of course, there remains the specification of $\rho_{(2)}({}^{\kappa}\mathbf{r}_i, {}^{\lambda}\mathbf{r}_j; {}^{\kappa}\mathbf{I}, {}^{\lambda}\mathbf{I})$ before (8.6.5) cases to be anything other than a purely formal expression of the surface tension. In particular there is the problem of integration over the internal states $^{\kappa}\mathbf{I}$, $^{\lambda}\mathbf{I}$ of the two molecules. If we restrict ourselves to purely conformational internal states, that is neglect any rotational or vibrational contributions, and consider only the geometrically accessible configurations of the chain, then $\int d^{\kappa}\mathbf{I}$ reduces to a sum of integrals over the internal distribution of the remaining $N-1$ segments with respect to segment i. These we shall represent as

$$\int \sum_{j=1,\neq i}^{N} Z(^{\kappa}i^{\kappa}j \mid N)\, d^{\kappa}\mathbf{r}_{ij}$$

but which require a knowledge of the internal distributions within the N-mer, $Z(^{\kappa}i^{\kappa}j \mid N)$. These distributions, moreover, are for dense polymeric systems which will naturally differ from those of an isolated polymer chain. Such distributions have been determined by Croxton[15, 16] on the basis of a diagrammatic analysis incorporating a density-dependent pseudopotential which is taken to operate between the segments. A slightly more extended discussion has already been given in Section 8.5; we refer the reader to the literature for a detailed treatment, however.

It now remains to specify the pair distribution of segment centres, $\rho_{(2)}(z_i, \mathbf{r}_{ij})$, and also to extend the analysis to take account of the two-component nature of the solution—solvent and segment. For this we may proceed as in Section 4.4 (see equation 4.4.3), except that the segment–segment contribution has been determined as above, whilst we may approximate the solvent–solvent and the solvent–segment distributions by simple two-component pair functions or, alternatively, adopt the replacement procedure developed in Section 8.5 to determine the remaining contributions to the surface tension.

References

1. R. Simha, H. L. Frisch, and F. R. Eirich, *J. Phys. Chem.*, **57,** 584 (1953).
   A. Silberberg, *J. Chem. Phys.*, **48,** 2835 (1968).
2. F. L. McCrackin, J. Mazur, and C. L. Guttmann, *Macromolecules*, **6,** 859 (1973).
3. D. C. Rapaport, *J. Phys. A.*, **9,** 1521 (1976).
4. M. F. Sykes, *J. Chem. Phys.*, **39,** 410 (1963); C. Domb and M. F. Sykes, *J. Math Phys.*, **2,** 63 (1961).
5. A. Bellemans, *J. Polymer Sci.*, **C39,** 305 (1972).
6. R. J. Rubin, *J. Chem. Phys,* **43,** 2392 (1965).
7. F. L. McCrackin, *J. Chem. Phys.*, **47,** 1980 (1965).
8. P. J. Flory, *J. Chem, Phys.*, **9,** 660 (1941).
9. A. Silberberg, *J. Phys. Chem.*, **66,** 1872 (1962); **66,** 1884 (1962); *J. Chem. Phys.*, **46,** 1105 (1967); **48,** 2835 (1968).
   Z. Priel and A. Silberberg, *Polym. Prepr. Amer. Chem. Soc., Div. Polym. Chem.*, **11,** 1405 (1970).
10. C. A. J. Hoeve, *J. Polym. Sci.*, **C30,** 361 (1970); **C34,** 1 (1971).
11. R. Defay, I. Prigogine, A. Bellemans, and D. H. Everett, *Surface Tension and Adsorption*, Longmans, London (1966), Ch. 13.
12. I. Prigogine and J. Maréchal, *J. Colloid Sci.*, **7,** 122 (1952).
    L. Saroléa–Marhot, *Bull. Ac. Roy. Belg. (Cl. Sc.)*, **40,** 1120 (1954).

13. K. S. Siow and D. Patterson, *J. Phys. Chem.*, **77,** 356 (1973).
14. C. A. Croxton, *Phys. Lett.*, **59A,** 359 (1976).
    C. A. Croxton. In C. A. Croxton (Ed.), *Progress in Liquid Physics*, Wiley, London (1978), Ch. 2, p. 78 *et seq.*
15. C. A. Croxton, *J. Phys. A.* **12,** 2475 (1979).
16. C. A. Croxton, *J. Phys. A.* **12,** 2487 (1979); **12,** 2497 (1979).
17. J. G. Curro, *J. Chem. Phys.*, **64,** 2496 (1976).
18. H. T. Davis, *J. Chem. Phys.*, **62,** 3412 (1975).
19. W. C. Forsman and R. E. Hughes, *J. Chem. Phys.*, **38,** 2118 (1963); **38,** 2123 (1963).

CHAPTER 9

The Liquid Crystal Surface

9.1 Introduction

We have seen (equation (3.7.1) *et seq.*) that for molecular fluids having an anisotropic pair interaction, a torque field will develop in a region of density gradient, such as at the liquid–vapour interface. Whether surface-oriented states will develop depends upon the details of the density profile, the pair interaction, and the temperature. However, for strongly anisotropic molecules such as those which constitute the class of fluids known as liquid crystals (Figure 9.1.1), it would appear that there exists the possibility of enhanced molecular orientational order at the interface under the action of the surface torque field which would, of course, have immediate consequences for the surface excess quantities, in particular the surface tension and its temperature derivative, the surface excess entropy per unit area.

An approach which has proved useful in the discussion of the spontaneous development of long range orientational order which characterizes the liquid crystal systems is the mean field treatment which is closely analogous to the Weiss theory of ferromagnetism. Each molecule is assumed to experience an average orienting field due to its neighbours, but remains otherwise uncorrelated with their detailed configuration. Despite this neglect of local order, in particular the interrelation between spatial and angular correlation, it appears that many of the qualitative features of the nematic phase are well described by a mean field theory, although it must be said that in some respects the model fails badly, particularly in its estimates of the latent heat of the nematic–isotropic phase transition and of the specific heat.

9.2 A general theory of the liquid crystal interface[9]

The most widely used mean field approximation is that of Maier and Saupe[1] who postulate that in the absence of any permanent dipole moment the orientational energy of a molecule i is

$$u_i \propto (\tfrac{3}{2} \cos^2 \theta_i - \tfrac{1}{2})\eta \tag{9.2.1}$$

where θ_i is the angle which the long molecular axis makes with the preferred axis, the director, of the orientationally ordered system. The Tsvetkov order parameter η is given by

$$\eta = \tfrac{1}{2}\langle 3 \cos^2 \theta_i - 1 \rangle \tag{9.2.2}$$

where the angular brackets denote a statistical average taken over the

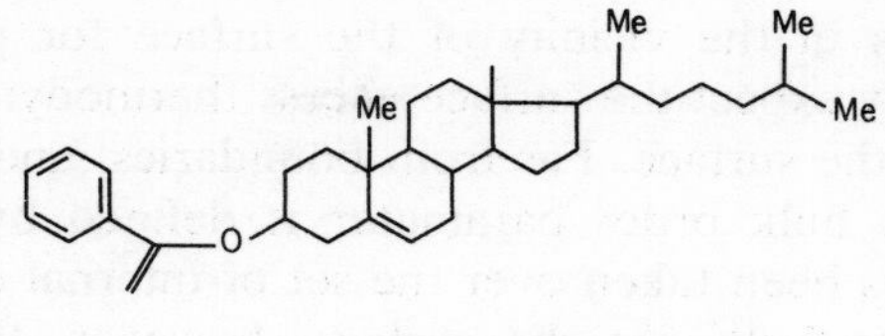

Me
Me
Me
Me
O
cholesteryl benzoate

(a)

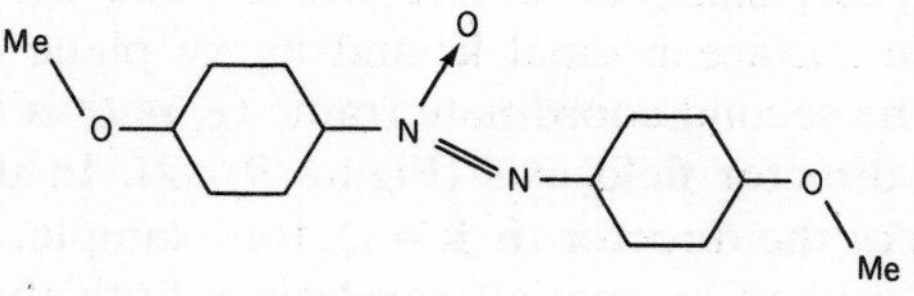

Me
O
N
O
N
O
Me
4-4'-dimethoxyazoxybenzene

(b)

Figure 9.1.1

distribution of molecular orientation with respect to the preferred axis, $f(\theta_i)$. For perfectly parallel alignment $\langle\cos^2\theta_i\rangle = 1$ $(\eta = 1)$, whilst for random orientations $\langle\cos^2\theta_i\rangle = \frac{1}{3}$ $(\eta = 0)$. In the nematic phase η has an intermediate value which is strongly temperature dependent (Figure 9.2.1), and exhibits a discontinuous change as the system switches from the nematic to the isotropic phase at T_{NI}.

Whilst the orientation of the director is arbitrary in the bulk fluid, the effect of a boundary—in particular a free surface—is to break the translational symmetry of the system whereupon both the local density and the orientational distribution depend upon the location of the molecular centre relative to the Gibbs dividing surface. Moreover, the direction of the

η
Nematic
Isotropic
T_{NI}
T

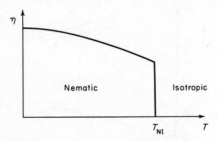

Figure 9.2.1 Schematic variation of the bulk order parameter with temperature. A discontinuous variation in the order parameter occurs at T_{NI} corresponding to the change in molecular structure from nematic to isotropic

director $\hat{\mathbf{n}}$ modifies in the vicinity of the surface for purely geometrical reasons and we may expect the surface excess thermodynamic properties to depend upon $\hat{\mathbf{n}}$ at the surface. Far from boundaries, and in the absence of external fields, the bulk order parameter is defined by (9.2.2), where a thermal average has been taken over the set of internal orientational states characteristic of the bulk. At the surface, however, the intervention of surface constraining and torque fields leads us to expect that a more general order parameter $\eta(z)_{\hat{\mathbf{n}}}$ will be necessary.

We specify two cartesian coordinate frames, the first having its z-axis coincident with the surface normal $\hat{\mathbf{k}}$, and its xy plane defining the Gibbs dividing surface. The second coordinate frame (ξ, η, ζ) is fixed with its ζ-axis along $\hat{\mathbf{n}}$, the local director field axis (Figure 9.2.2). In the case of perpendicular orientation of the director $(\hat{\mathbf{n}} \cdot \hat{\mathbf{k}} = 1)$, for example, the two coordinate frames are simply related by spatial translation: both the z and ζ axes are parallel.

The surface tension may be expressed in terms of the local *excess* Helmholtz free energy per particle $a_s(z_1)_{\hat{\mathbf{n}}}$ (c.f. 1.4.26)

$$\gamma = \int_{-\infty}^{\infty} a_s(z_1)_{\hat{\mathbf{n}}} \rho_{(1)}(z_1)\, dz_1 \tag{9.2.3}$$

where, for rigid molecules with pair interactions U_{12} depending only upon the centre of mass vector $\mathbf{r}_{12}$ and the set of molecular Euler angles (φ, θ, ψ),

$$\rho_{(1)}(z_1) a_s(z_1)_{\hat{\mathbf{n}}} = -\frac{1}{2} \int\int\int \left\{ z_{12} \frac{\partial U_{12}}{\partial z_{12}} - x_{12} \frac{\partial U_{12}}{\partial x_{12}} \right\}$$
$$\times \rho_{(2)}(z_1, \mathbf{r}_{12}; \mathbf{e}_1, \mathbf{e}_2)_{\hat{\mathbf{n}}}\, d\mathbf{e}_1\, d\mathbf{e}_2\, d\mathbf{r}_{12}. \tag{9.2.4}$$

x_{12} and z_{12} represent the components of the vector $\mathbf{r}_{12}$ linking the two

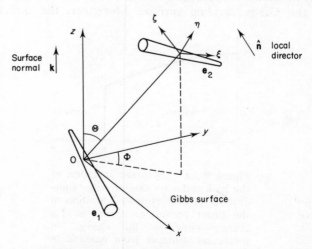

Figure 9.2.2 The molecular geometry

centres of mass, and $\mathbf{e}_1, \mathbf{e}_2$ denote the sets of Euler angles for the two molecules, $d\mathbf{e}_i = \sin\theta_i \, d\theta_i \, d\varphi_i \, d\psi_i$. Obviously the distribution amongst the states of the two interacting molecules will depend upon the location of their centres and the direction of the local director: $f(z_1; \mathbf{e}_1)_{\hat{\mathbf{n}}}$, $f(z_2; \mathbf{e}_2)_{\hat{\mathbf{n}}}$. We assume the orientation of the director remains constant throughout the interfacial region, which would appear to be a reasonable assumption since the bulk correlation range of the director is typically $>10^3$ Å.

To make further progress we must decouple the two-particle distribution $\rho_{(2)}(z_1, \mathbf{r}_{12}; \mathbf{e}_1, \mathbf{e}_2)_{\hat{\mathbf{n}}}$ which, in a region whose local director field is $\hat{\mathbf{n}}$, represents the probability of finding molecules 1 and 2 in the configuration $(z_1, \mathbf{r}_{12}; \mathbf{e}_1, \mathbf{e}_2)$. If, in the spirit of the mean field approximation, we assume that there is no angular correlation between neighbouring molecules, then we may write[10]

$$\rho_{(2)}(z_1, \mathbf{r}_{12}; \mathbf{e}_1, \mathbf{e}_2)_{\hat{\mathbf{n}}} = \rho_{(1)}(z_1)\rho_{(1)}(z_2)f(z_1; \mathbf{e}_1)_{\hat{\mathbf{n}}}f(z_2; \mathbf{e}_2)_{\hat{\mathbf{n}}}g_{(2)}(z_1, \mathbf{r}_{12}). \quad (9.2.5)$$

We point out that even in the bulk $g_{(2)}(-\infty, \mathbf{r}_{12})$ is a strongly anisotropic function. The pair interaction may generally be written

$$U_{12} = -\frac{v}{r^6}(\mathbf{e}_1 \cdot \overset{\leftrightarrow}{\mathbf{T}} \cdot \mathbf{e}_2)^2 \qquad v = \text{constant} > 0, \qquad r = |\mathbf{r}_{12}| \qquad (9.2.6)$$

where $\overset{\leftrightarrow}{\mathbf{T}}$ is the symmetric tensor $\overset{\leftrightarrow}{\mathbf{T}} = (\overset{\leftrightarrow}{\mathbf{1}} - 3\mathbf{rr}/r^2)$ and $\overset{\leftrightarrow}{\mathbf{1}}$ a unit tensor.

Now that the two-particle distribution has been decoupled (9.2.5) the angular averages over $f(z_1; \mathbf{e}_1)_{\hat{\mathbf{n}}}$ and $f(z_2; \mathbf{e}_2)_{\hat{\mathbf{n}}}$ may be performed. Clearly $f(z_2; \mathbf{e}_2)$ will depend upon $\hat{\mathbf{n}}$ being virtually z-independent for $\mathbf{n} \cdot \mathbf{k} = 0$ (parallel to surface) and strongly z-dependent for $\hat{\mathbf{n}} \cdot \hat{\mathbf{k}} = 1$ (perpendicular to surface). In a mean field approximation, however, we adopt the local midpoint value, and set

$$f(z_2; \mathbf{e}_2)_{\hat{\mathbf{n}}} = f(z_1; \mathbf{e}_1)_{\hat{\mathbf{n}}} \qquad (9.2.7)$$

which is likely to be a good approximation for $\hat{\mathbf{n}} \cdot \hat{\mathbf{k}} = 0$. Equation (9.2.6) then becomes

$$\langle U \rangle_{\mathbf{e}_1 \mathbf{e}_2} = \left(-\frac{v}{9r^6}\right)\{T_{\xi\xi}^2 + T_{\eta\eta}^2 + T_{\zeta\zeta}^2$$

$$+ \eta(z_1)_{\hat{\mathbf{n}}}[4T_{\zeta\zeta}^2 + 2T_{\xi\xi}^2 + 2T_{\eta\zeta}^2 - 2T_{\xi\xi}^2 - 2T_{\eta\eta}^2 - 4T_{\xi\eta}^2]$$

$$+ \eta^2(z_1)_{\hat{\mathbf{n}}}[T_{\xi\xi}^2 + T_{\eta\eta}^2 + 2T_{\xi\eta}^2 + 4T_{\zeta\zeta}^2 - 4T_{\xi\eta}^2 - 4T_{\eta\zeta}^2]\} \qquad (9.2.8)$$

where $T_{\kappa\lambda}$ are the cartesian components of $\overset{\leftrightarrow}{\mathbf{T}}$. The *local* order parameter is now defined as

$$\eta(z_1)_{\hat{\mathbf{n}}} = \frac{1}{2}\int_0^{\pi/2}(3\cos^2\theta - 1)f(z_1; \theta_1)_{\hat{\mathbf{n}}}\sin\theta \, d\theta \qquad (9.2.9)$$

where, as before, θ is the angle which the long molecular axis makes with respect to the local director (c.f. 9.2.2). Note that now $\langle U \rangle_{e_1 e_2}$ is a function of z_1 and $\hat{\mathbf{n}}$. We also point out that for mechanical stability the term $z_{12}(\partial U_{12}/\partial z_{12})\rho_{(2)}(z_1, \mathbf{r}_{12}; \mathbf{e}_1, \mathbf{e}_2)$ in (9.2.4) is independent of z_1 and, moreover, in either bulk isotropic phase we have

$$z_{12} \frac{\partial U_{12}}{\partial z_{12}} \rho_{(2)}(\pm\infty, \mathbf{r}_{12}; \mathbf{e}_1, \mathbf{e}_2) = x_{12} \frac{\partial U_{12}}{\partial x_{12}} \rho_{(2)}(\pm\infty, \mathbf{r}_{12}; \mathbf{e}_1, \mathbf{e}_2). \quad (9.2.10)$$

The general case of arbitrary orientation of the surface direct or $\hat{\mathbf{n}}$ is difficult to analyse. However, two specific cases, $\hat{\mathbf{n}} \cdot \hat{\mathbf{k}} = 1$ and $\hat{\mathbf{n}} \cdot \hat{\mathbf{k}} = 0$ corresponding to parallel and perpendicular orientation of the director relative to the surface, may be more readily discussed.

(a) $\hat{\mathbf{n}} \cdot \hat{\mathbf{k}} = 1$ (perpendicular orientation of the director)

In this case we make the transformation of axes $\xi \to x$, $\eta \to y$, $\zeta \to z$, and also transform to a cylindrical polar coordinate frame with $\hat{\mathbf{n}}$ along the polar axis. We then average over the Φ-distribution, whereupon (9.2.4) becomes (see Figure 9.2.2)

$$\rho_{(1)}(z_1) a_s(z_1)_\perp = -\frac{1}{2} \int_{-\infty}^{\infty} dz_{12} \int_0^{\pi/2} \sin\Theta \, d\Theta \int_{|z_{12}|}^{\infty} r_{12} \, dr_{12} \left(-\frac{v}{9 r_{12}^6} \right)$$

$$\times \{ \rho_{(2)}(-\infty, \mathbf{r}_{12} \mid \rho_L)[36 c_4(-\infty) \cos^4 \Theta - 12 c_2(-\infty) \cos^2 \Theta$$
$$+ \eta[36 c_2(-\infty) \cos^4 \Theta - 12 c_1(-\infty) \cos^2 \Theta - 72 c_4(-\infty) \cos^2 \Theta$$
$$+ 24 c_2(-\infty) \cos^2 \Theta - 72(c_2(-\infty) - c_4(-\infty)) \cos^4 \Theta$$
$$+ 24 c_2(-\infty) \cos^2 \Theta]$$
$$+ \eta^2[36 c_4(-\infty) \cos^4 \Theta - 12 c_2(-\infty) \cos^2 \Theta$$
$$+ 36(c_2(-\infty) - c_4(-\infty)) \cos^4 \Theta$$
$$- 12 c(-\infty) \cos^2 \Theta - 72 c_2(-\infty) \cos^4 \Theta + 24 c_1(-\infty) \cos^2 \Theta]]$$
$$- \rho_{(2)}(z_1, \mathbf{r}_{12})[36 c_4(z_1) \cos^4 \Theta - 12 c_2(z_1) \cos^2 \Theta$$
$$+ \eta(z_1)[36 c_2(z_1) \cos^4 \Theta - 12 c_1(z_1) \cos^2 \Theta - 72 c_4(z_1) \cos^2 \Theta$$
$$+ 24 c_2(z_1) \cos^2 \Theta - 72(c_2(z_1) - c_4(z_1)) \cos^4 \Theta$$
$$+ 24 c(z_1) \cos^2 \Theta]$$
$$+ \eta^2(z_1)[36 c_4(z_1) \cos^4 \Theta - 12 c_2(z_1) \cos^2 \Theta$$
$$+ 36(c_2(z_1) - c_4(z_1)) \cos^4 \Theta$$
$$- 12 c(z_1) \cos^2 \Theta - 72 c_2(z_1) \cos^4 \Theta + 24 c_1(z_1) \cos^2 \Theta]]\}_\perp$$
$$(9.2.11)$$

where we have used (9.2.10) and the decoupled form (9.2.5) for the two-particle density distribution.[10] The Φ-distribution has been taken to be

$f(z, \Phi)_{\hat{\mathbf{n}}}$, so that

$$\left.\begin{array}{ll} \int_0^{2\pi} f(z, \Phi)_{\hat{\mathbf{n}}} \cos \Phi \, d\Phi = c_1(z) & \int_0^{2\pi} f(z, \Phi)_{\hat{\mathbf{n}}} \cos \Phi \sin \Phi \, d\Phi = c(z) \\[2ex] \int_0^{2\pi} f(z, \Phi)_{\hat{\mathbf{n}}} \cos^2 \Phi \, d\Phi = c_2(z) & \int_0^{2\pi} f(z, \Phi)_{\hat{\mathbf{n}}} \sin \Phi \, d\Phi = s(z) \\[2ex] \int_0^{2\pi} f(z, \Phi)_{\hat{\mathbf{n}}} \cos^4 \Phi \, d\Phi = c_4(z). & \end{array}\right\} \quad (9.2.12)$$

Now, for the case $\hat{\mathbf{n}} \cdot \hat{\mathbf{k}} = 1$, $f(z, \Phi)_\perp = f(-\infty, \Phi)_\perp = 1.00$ we have from (9.2.12)

$$\left.\begin{array}{ll} c_1 = 0 & c = 0 \\ c_2 = \pi & s = 0 \\ c_4 = \pi & \end{array}\right\} \quad (9.2.13)$$

whereupon (9.2.11) simplifies to give

$$\rho_{(1)}(z_1) a_s(z_1)_\perp = -\frac{1}{2} \int_{-\infty}^{\infty} dz_{12} \int_0^{\pi/2} \sin \Theta \, d\Theta \int_{|z_{12}|}^{\infty} r_{12} \, dr_{12} \left(-\frac{v}{9r_{12}^6}\right)$$

$$\times \{\rho_{(2)}(-\infty, \mathbf{r}_{12} \mid \rho_L)[36\pi \cos^4 \Theta - 12\pi \cos^2 \Theta + \eta[-36\pi \cos^4 \Theta + 24 \cos^2 \Theta]$$

$$+ \eta^2[-36\pi \cos^4 \Theta - 12\pi \cos^2 \Theta]$$

$$- \rho_{(2)}(z_1, \mathbf{r}_{12})[36\pi \cos^4 \Theta - 12\pi \cos^2 \Theta + \eta(z_1)[-36\pi \cos^4 \Theta + 24 \cos^2 \Theta]$$

$$+ \eta^2(z_1)[-36\pi \cos^4 \Theta - 12\pi \cos^2 \Theta]]\}_\perp.$$

$$(9.2.14)$$

Formally,

$$\rho_{(2)}(z_1, \mathbf{r}_{12}) = \rho_{(1)}(z_1) \rho_{(1)}(z_2) g_{(2)}(z_1, \mathbf{r}_{12})$$

and it remains to specify $\rho_{(1)}(z_2)$ and $g_{(2)}(z_1, \mathbf{r}_{12})$. For near linear density profiles we may expand the single-particle density about $\rho_{(1)}(z_1)$, where to first order:

$$\rho_{(1)}(z_2) = \left(\rho_{(1)}(z_1) + z_{12}\left(\frac{\partial \rho_{(1)}(z_1)}{\partial z_1}\right)_{z_1}\right) \quad (9.2.15)$$

and for $g_{(2)}(z_1, \mathbf{r}_{12})$ we adopt the closure

$$g_{(2)}(z_1, \mathbf{r}_{12}) = 0.5 g_{(2)}[r_{12} \mid \rho_{(1)}(z_1)] + 0.5 g_{(2)}[r_{12} \mid \rho_{(1)}(z_2)] \quad (9.2.16)$$

which is invariant to particle exchange and is asymptotically correct for slowly varying profiles. The two functions $g_{(2)}[r_{12} \mid \rho_{(1)}(z_1)]$ and $g_{(2)}[r_{12} \mid \rho_{(1)}(z_2)]$ in (9.2.16) may be related by the Ornstein–Zernike equation, using (9.2.15):

$$g_{(2)}(z_1, \mathbf{r}_{12}) = 1 + c(r_{12}) + \left[\rho_{(1)}(z_1) + \frac{1}{2} z_{12}\left(\frac{\partial \rho_{(1)}(z_1)}{\partial z_1}\right)_{z_1}\right] F(\mathbf{r}_{12})$$

$$= g_{(2)}[r_{12} \mid \rho_{(1)}(z_1)] + \frac{1}{2} z_{12}\left(\frac{\partial \rho_{(1)}(z_1)}{\partial z_1}\right)_{z_1} F(\mathbf{r}_{12}) \quad (9.2.17)$$

where $F(\mathbf{r}) = \int c(\mathbf{r}_{13})h(\mathbf{r}_{23})\,d\mathbf{r}_3$; $F(\mathbf{r})$, $c(\mathbf{r})$, and $h(\mathbf{r})$ being the indirect, direct, and total correlation functions, respectively. Implicit in the approximation (9.2.17) is the assumption that $c(\mathbf{r})$ varies more slowly with density than does $\rho F(\mathbf{r})$. It is known that the direct correlation function is, in fact, a very strongly density-dependent function at small r. However, the principal contribution in (9.2.14) arises at intermediate to large r, where the approximation is likely to be reasonable. We may formally integrate (9.2.14) to yield the local excess Helmholtz free energy density:

$$
\begin{aligned}
\rho_{(1)}(z_1)a_s(z_1)_\perp = &-\rho_L^2[A_\perp + \eta B_\perp + \eta^2 C_\perp] \\
&+ \rho_{(1)}^2(z_1)[A_\perp(z_1) + \eta(z_1)B_\perp(z_1) + \eta^2(z_1)C_\perp(z_1)] \\
&+ \rho_{(1)}(z_1)\left(\frac{\partial\rho(z_1)}{\partial z_1}\right)_{z_1}^2[D_\perp(z_1) + \eta(z_1)E_\perp(z_1) + \eta^2(z_1)G_\perp(z_1)]
\end{aligned}
$$

$$(9.2.18)$$

where the coefficients $A_\perp(z_1), \ldots, G_\perp(z_1)$ represent the results of integrations (9.2.14) following the insertion of (9.2.16) and (9.2.17).

(b) $\hat{\mathbf{n}}\cdot\hat{\mathbf{k}} = 0$ (parallel orientation of the director)

Making the transformation $\zeta \to x$, $\eta \to y$, $\xi \to z$, we obtain the analogous expression to (9.2.11) for the surface excess Helmholtz free energy:

$$
\begin{aligned}
\rho_{(1)}(z_1)a_s(z_1)_\parallel = &-\frac{1}{2}\int_{-\infty}^{\infty} dz \int_0^{\pi/2} \sin\Theta\,d\Theta \int_{|z|}^{\infty} r\,dr\left(-\frac{v}{9r^6}\right)
\end{aligned}
$$

$$
\begin{aligned}
\times\{\rho_{(2)}(-\infty, r_{12}\,|\,\rho_L)[&36\cos^4\Theta - 12\cos^2\Theta \\
&+\eta[144\cos^4\Theta - 48\cos^2\Theta + 36c_2(-\infty)\cos^4\Theta - 12c_1(-\infty)\cos^2\Theta \\
&+36(1 - c_2(-\infty))\cos^4\Theta - 12s(-\infty)\cos^2\Theta] \\
&+\eta^2[144\cos^4\Theta - 48\cos^2\Theta - 72c_2(-\infty)\cos^4\Theta + 24c_1(-\infty)\cos^2\Theta \\
&-72(1 - c_2(-\infty))\cos^4\Theta + 24s(-\infty)\cos^2\Theta] \\
-\rho_{(2)}(z_1, \mathbf{r}_{12})[&36\cos^4\Theta - 12\cos^2\Theta \\
&+\eta(z_1)[144\cos^4\Theta - 48\cos^2\Theta + 36c_2(z_1)\cos^4\Theta - 12c_1(z_1)\cos^2\Theta \\
&+36(1 - c_2(z_1))\cos^4\Theta - 12s(z_1)\cos^2\Theta] \\
&+\eta^2(z_1)[144\cos^4\Theta - 48\cos^2\Theta - 72c_2(z_1)\cos^4\Theta + 24c_1(z_1)\cos^2\Theta \\
&-72(1 - c_2(z_1))\cos^4\Theta + 24s(z_1)\cos^2\Theta]\}_\parallel
\end{aligned}
$$

$$(9.2.19)$$

(c.f. (9.2.11)) where again c_1, c_2, etc. are defined by (9.2.12). Similar reasoning as for the perpendicular case yields the formal result

$$
\begin{aligned}
\rho_{(1)}(z_1)a_s(z_1)_\parallel = &-\rho_L^2[A_\parallel + \eta B_\parallel + \eta^2 C_\parallel] \\
&+ \rho_{(1)}^2(z_1)[A_\parallel(z_1) + \eta(z_1)B_\parallel(z_1) + \eta^2(z_1)C_\parallel(z_1)] \\
&+ \rho_{(1)}(z_1)\left(\frac{\partial\rho_{(1)}(z_1)}{\partial z_1}\right)_{z_1}^2[D_\parallel(z_1) + \eta(z_1)E_\parallel(z_1) + \eta^2(z_1)G_\parallel(z_1)]
\end{aligned}
$$

$$(9.2.20)$$

where the coefficients $A_\|(z_1), \ldots, G_\|(z_1)$ represent the result of the integrations (9.2.19) following the insertion of (9.2.15) and (9.2.17).

Although equations (9.2.18) and (9.2.20) are purely formal results, they do nevertheless show that the local free energy $a(z_1)$ has limiting Cahn–Hilliard 'square-gradient' form in the high temperature isotropic phase ($\eta = \eta(z_1) = 0$) when $\rho_{(1)}(z_1)$ is a slowly varying function of position, whilst in the nematic phase the Cahn–Hilliard form is modified by terms proportional to $\eta(z_1)$ and $\eta^2(z_1)$. Unfortunately, in the absence of any knowledge of the distributions $f(z, \Phi)$ and $g_{(2)}[\mathbf{r} \mid \rho_{(1)}(z_1)]$ it is difficult to be more specific regarding the coefficients $A, \ldots, G$ appearing in equations (9.2.18) and (9.2.20). However, physical considerations lead us to expect

$$A_\| = A_\perp > 0$$
$$A_\|(z_1) = A_\perp(z_1) > 0 \tag{9.2.21}$$
$$D_\|(z_1) = D_\perp(z_1) > 0$$

since in the high-temperature isotropic phase ($\eta = 0$) the Cahn–Hilliard result must be recovered regardless of the director $\hat{\mathbf{n}}$ in the nematic phase. Which of the two molecular orientations, parallel or perpendicular (or indeed some intermediate orientation), leads to a lower free energy depends in a subtle way upon the relative magnitudes of the coefficients $A, \ldots, G$, and unfortunately cannot be ascertained from these purely formal results.

Two further simplifications may be made, firstly by setting

$$g_{(2)}[\mathbf{r}_{12} \mid \rho_{(1)}(z_1)] = g_{(2)}^{L}(\mathbf{r}_{12}) \tag{9.2.22}$$

in (9.2.17), where $g_{(2)}^{L}$ is the bulk liquid crystal pair distribution. Such an approximation is likely to be reasonable for low temperature nematic systems possessing a relatively sharp surface density transition. Secondly, if we neglect the z dependence of the Φ distribution (which is valid for the perpendicular orientation), then we find, working to first order in the order parameter

$$\rho_{(1)}(z_1) a_s(z_1)_{\|,\perp} = A_{\|,\perp}[\rho_{(1)}^2(z_1) - \rho_L^2] + D_{\|,\perp}\rho_{(1)}(z_1) \left(\frac{\partial \rho_{(1)}(z_1)}{\partial z_1} \right)^2_{z_1}$$
$$+ B_{\|,\perp}[\rho_{(1)}^2(z_1)\eta(z_1) - \rho_L^2\eta] + E_{\|,\perp}\rho_{(1)}(z_1) \left(\frac{\partial \rho_{(1)}(z_1)}{\partial z_1} \right)^2_{z_1} \eta(z_1)$$
$$\tag{9.2.23}$$

where the $\|(\perp)$ subscripts are adopted throughout. The first two terms on the right-hand side of (9.2.23) represent the surface excess Cahn–Hilliard Helmholtz free energy of a disordered liquid ($\eta = 0$), whilst the subsequent terms represent the modification of the Cahn–Hilliard free energy due to surface orientational order. For a given bulk order parameter η, and a given density transition profile $\rho_{(1)}(z_1)$, we should determine the local order parameter $\eta(z_1)$ which minimizes the left-hand side of (9.2.23). If, however, we assume that $\eta(z_1)$ varies slowly across the interface, we may attribute the spatial variations of the orientational contributions to the surface excess Helmholtz free energy directly to the terms $\rho_{(1)}^2(z_1)$ and $\rho_{(1)}(z_1)(\partial\rho_{(1)}(z_1)/\partial z_1)_{z_1}^2$:

$$\rho_{(1)}(z_1)a_s(z_1)_{\|,\perp} \sim A_{\|,\perp}[\rho_{(1)}^2(z_1) - \rho_L^2] + D_{\|,\perp}\rho_{(1)}(z_1)\left(\frac{\partial\rho_{(1)}(z_1)}{\partial z_1}\right)_{z_1}^2$$

$$+ B_{\|,\perp}[\rho_{(1)}^2(z_1)\eta - \rho_L^2\eta] + E_{\|,\perp}\rho_{(1)}(z_1)\left(\frac{\partial\rho_{(1)}(z_1)}{\partial z_1}\right)_{z_1}^2\eta$$

$$(9.2.24)$$

where $A_{\|,\perp}$, $B_{\|,\perp}$, $D_{\|,\perp}$, $E_{\|,\perp} > 0$.

The expression (9.2.23) is very similar to the qualitative proposals for the surface variation of the order parameter proposed by Croxton and Chandrasekhar,[3] on the basis of which they predict possible features of the surface tension–temperature characteristics which may, in principle, arise in nematic–isotropic liquid crystal systems. We shall consider their discussion in more detail in Section 9.4.

9.3 The Kirkwood–Buff model of the nematic liquid crystal surface

Few conclusive results may be drawn on the basis of the purely formal analysis of the preceding section: further progress requires more drastic approximation before the expressions (9.2.11), (9.2.19) become amenable to discussion. Parsons[8] has considered what amounts to the equivalent of a Kirkwood–Buff analysis, but appropriate to nematic liquid crystal systems. In his treatment, Parsons assumes that the nematic liquid crystal retains its bulk density ρ_L right up to a planar surface of density discontinuity beyond which the fluid density is assumed to be zero.

Thus, as in the case of a simple fluid (Section 2.3) in the KBF approximation,

$$\rho_{(1)}(z) = \rho_L \qquad z < 0$$

$$\rho_{(1)}(z) = 0 \qquad z > 0$$

$$\rho_{(2)}(z_1, \mathbf{r}_{12}; \hat{\mathbf{e}}_1, \hat{\mathbf{e}}_2)_{\hat{\mathbf{n}}} = \rho_{(1)}(z_1)\rho_{(1)}(z_2)g_{(2)}(\mathbf{r}_2; \hat{\mathbf{e}}_1, \hat{\mathbf{e}}_2)_{\hat{\mathbf{n}}} \qquad (9.3.1)$$

where

$$g_{(2)}(\mathbf{r}_{12}; \hat{\mathbf{e}}_1, \hat{\mathbf{e}}_2)_{\hat{\mathbf{n}}} = g_{(2)}^L(r_{12})f(\theta_1)f(\theta_2)$$

it being assumed that the orientations θ_1 and θ_2 are uncorrelated and that the bulk two-particle distribution is isotropic. The mean field distribution of orientations is, of course, related to the bulk order parameter η as follows

$$\eta = \frac{1}{2}\int_0^{\pi/2}(3\cos^2\theta - 1)f(\theta)\sin\theta\,d\theta. \qquad (9.3.2)$$

The above assumptions are identical to the original KBF step model of the liquid surface, with the additional assumption that the uncorrelated orientational distributions $f(\theta_1)$, $f(\theta_2)$ are not modified in the vicinity of the transition zone which, of course, they should be.

Further, the bulk order parameter η is assumed to remain constant up to the dividing surface, and the radial distribution function retains a simplified bulk form $g_{(2)}(r)$ up to the interfacial plane. Clearly such an approximation which assumes that neighbouring molecules are spherically symmetrically distributed about the origin molecule is inappropriate to a liquid crystal system. Nevertheless, its adoption enables the integrals (9.2.11) and (9.2.14) to be evaluated with the final result

$$\left. \begin{array}{l} \gamma_\perp = \gamma_0 [1 + \tfrac{2}{3}\eta + \tfrac{1}{6}\eta^2] \\ \gamma_\parallel = \gamma_0 [1 - \tfrac{4}{9}\eta + \tfrac{8}{27}\eta^2] \end{array} \right\} \gamma_0 = \tfrac{3}{4}\pi\rho_L^2 v \int_0^\infty g_{(2)}(r) \frac{1}{r^2}\, dr \qquad (9.3.3)$$

corresponding to perpendicular and parallel orientation of the surface director. It is immediately apparent that $\gamma_\perp > \gamma_\parallel$, whereupon Parsons concludes that the molecules align parallel to the liquid surface, and that the surface tension for such a system will show a discontinuous increase across the nematic–isotropic phase transition, when $\eta \to 0$:

$$\frac{\Delta\gamma}{\gamma} = -\tfrac{4}{9}\Delta\eta + \tfrac{8}{27}(\Delta\eta)^2 \qquad (9.3.4)$$

where

$$\Delta\eta = \eta(T \leqslant T_{NI}) - \eta(T \geqslant T_{NI}).$$

Since $\eta = 0$ in the isotropic phase and $\sim\tfrac{1}{3}(T \leqslant T_{NI})$ in the bulk nematic we have

$$\frac{\Delta\gamma}{\gamma} \sim -10\%.$$

The small decrease in density which accompanies the transition is expected to be substantially less than that due to the discontinuous variation in the order parameter. Of course, as Parsons himself points out, the model is not self-consistent in as far as one of the initial assumptions of the model is that the director field does not modify in the vicinity of the surface, and yet one of the principal conclusions is that the surface free energy, and hence the surface tension, is minimized for $\hat{\mathbf{n}} \cdot \hat{\mathbf{k}} = 0$.

The difference in free energy between the parallel and perpendicular orientations

$$\frac{\gamma_\perp - \gamma_\parallel}{\gamma_0} = \frac{\eta}{9}\left(10 - \frac{7\eta}{6}\right) \qquad (9.3.5)$$

is, on the basis of this model, quite substantial; this suggests that the surface molecules are strongly anchored in the parallel configuration, and are unlikely to modify their orientation with temperature to any significant extent. Such a conclusion has been verified[6] for p-azoxyanisole (PAA) and n-p-methoxybenzylidene p'-n-butylaniline (MBBA) although in the latter case the long molecular axis is inclined at about 75° to the surface plane, which does not concur with Parsons' prediction. There are a number of possible explanations, however, the simplest being that the dipole axis of the molecule does not coincide with the geometrical axis. In many cases, of course, the molecules will have a permanent dipole moment, but Parsons concludes that the basic results of the Kirkwood–Buff analysis remain unmodified. However, the KB model implicitly excludes entropy contributions to the surface excess free energy which may substantially modify Parsons' conclusions regarding orientation; we shall consider this point in more detail in Section 9.5.

9.4 Surface tension of nematic and smectic-A liquid crystal systems

The Maier–Saupe nematic mean field analysis may be extended to smectic-A liquid crystal systems by assuming that in addition to being preferentially oriented along the z-axis (say), there is also layering in the z-direction, (Figure 9.4.1b). McMillan[2] has proposed that in this case the single-particle potential in the mean field of its neighbours is

$$u_i(z_i, \cos\theta_i) = \bar{u}\left\{1 + \alpha\cos\left(\frac{2\pi z_i}{d}\right)\right\}(\tfrac{3}{2}\cos^2\theta_i - 1)\eta \qquad (9.4.1)$$

where both translational and orientational effects are now included, the smectic layering being represented as the density wave $\alpha\cos(2\pi z_i/d)$. For $\alpha = 0$, (9.4.1) reduces to the Maier–Saupe nematic mean field potential. d is the length of the molecule. Now, although the minimum energy configuration corresponds to the molecular centre of mass lying on one of the planes with its long axis in the z-direction, there will be thermal fluctuations in the

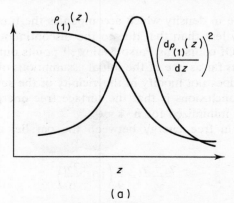

(a)

Figure 9.4.1a Schematic forms for the density transition profile $\rho_{(1)}(z)$ and the square of its gradient $(d\rho_{(1)}(z)/dz)^2$

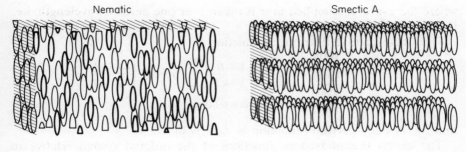

(b)

Figure 9.4.1b Molecular organization in the nematic and smectic-A liquid crystal phases

orientation and location of the molecule. Consequently a second order parameter σ is introduced into the single-particle mean field potential:

$$u_i(z_i, \cos\theta_i) = \bar{u}\left\{1 + \sigma\alpha\cos\left(\frac{2\pi z_i}{d}\right)\right\}(\tfrac{3}{2}\cos^2\theta_i - \tfrac{1}{2})\eta. \qquad (9.4.2)$$

The magnitude of α characterizes a particular molecular species, whilst σ is a measure of the amplitude of the density wave.

Now, the single-particle Boltzmann distribution will be of the form

$$f(z_i, \cos\theta_i) = \exp(-u_i/kT) \qquad (9.4.3)$$

and we may use this to determine selfconsistently the single-particle mean field potential. If we assume a pair interaction

$$V_{ij}(r_{ij}, \cos\theta_{ij}) = -(\bar{u}/Nr_0\pi^{3/2})\left\{\exp-\left(\frac{r_{ij}}{r_0}\right)^2\right\}(\tfrac{3}{2}\cos^2\theta_{ij} - \tfrac{1}{2}) \qquad (9.4.4)$$

where the exponential term reflects the short range character of the interaction, and r_0 is of the order of the length of the rigid part of the molecular length, then we have

$$u_i(z_i, \cos\theta_i) = \frac{\iint V_{ij}f(z_j, \cos\theta_j)\,d\mathbf{j}\,d\hat{\mathbf{j}}}{\iint f(z_j, \cos\theta_j)\,d\mathbf{j}\,d\hat{\mathbf{j}}} \qquad (9.4.5)$$

$$= \bar{u}\left\{(\tfrac{3}{2}\cos^2\theta_i - \tfrac{1}{2})(\tfrac{3}{2}\cos^2\theta_j - \tfrac{1}{2})\right.$$
$$\left. + \alpha\cos\left(\frac{2\pi z_i}{d}\right)(\tfrac{3}{2}\cos^2\theta_i - \tfrac{1}{2})\left\langle\cos\left(\frac{2\pi z_j}{d}\right)(\tfrac{3}{2}\cos^2\theta_j - \tfrac{1}{2})\right\rangle\right\}. $$
$$(9.4.6)$$

Selfconsistency between (9.4.2) and (9.4.6) requires that

$$\eta = \langle\tfrac{3}{2}\cos^2\theta_j - \tfrac{1}{2}\rangle \qquad (9.4.7)$$

$$\sigma = \left\langle\cos\left(\frac{2\pi z_j}{d}\right)(\tfrac{3}{2}\cos^2\theta_j - \tfrac{1}{2})\right\rangle \qquad (9.4.8)$$

where the z-average implied in σ is taken over one density wavelength, i.e. $\int_0^d dz$.

The single-particle distribution function is

$$Z_1 = \ln \left\{ d^{-1} \int_0^d \int_0^1 \exp \left\{ -\frac{\bar{u}}{kT} \left[\eta + \sigma\alpha \cos\left(\frac{2\pi z}{d}\right) \right] \right.\right.$$
$$\left.\left. \times (\tfrac{3}{2}\cos^2\theta - \tfrac{1}{2}) \right\} d(\cos\theta)\, dz \right\} \tag{9.4.9}$$

and the N-body partition function is $Z_N = N \ln Z_1$.

The excess thermodynamic functions of the ordered system relative to those of the disordered one can now be readily derived on the basis of equations (9.4.4), (9.4.5), and (9.4.6). For example, the orientational internal energy per particle with respect to a disordered fluid is

$$\frac{\hat{U}}{N} = -\frac{\int_0^d \int_0^1 u_i \exp(-u_i/kT)\, d(\cos\theta_i)\, dz_i}{\int_0^d \int_0^1 \exp(-u_i/kT)\, d(\cos\theta_i)\, dz_i} \tag{9.4.10}$$

$$= -N\bar{u}(\eta^2 + (\alpha\sigma)^2) \tag{9.4.11}$$

which we note is negative.

The (orientational) excess entropy with respect to a disordered fluid is

$$\hat{S} = kT \ln Z_N + \frac{\hat{U}}{T}$$

which, per particle,

$$= -k \left[\frac{\bar{u}(\eta^2 + (\alpha\sigma)^2)}{kT} \right.$$
$$\left. -\ln \left\{ d^{-1} \int_0^d \int_0^1 \exp\left(-\frac{\bar{u}}{kT}\left[\eta + \sigma\alpha \cos\left(\frac{2\pi z}{d}\right) \right] (\tfrac{3}{2}\cos^2\theta - \tfrac{1}{2}) \right) d(\cos\theta)\, dz \right\} \right]$$

$$\tag{9.4.12}$$

and the (orientational) excess Helmholtz free energy per particle

$$-\frac{1}{N} kT \ln Z_N = -kT \ln \left\{ d^{-1} \int_0^d \int_0^1 \exp\left\{ -\frac{\bar{u}}{kT}\left[\eta + \sigma\alpha \cos\left(\frac{2\pi z}{d}\right) \right] \right.\right.$$
$$\left.\left. \times (\tfrac{3}{2}\cos^2\theta - \tfrac{1}{2}) \right\} d(\cos\theta)\, dz \right\}. \tag{9.4.13}$$

The above analysis may now be extended to yield the *surface* excess functions arising from ordering relative to a disordered fluid by specifying *local* order parameters $\eta(z)_{\hat{n}}$ and $\sigma(z)_{\hat{n}}$, where we have explicitly indicated that these will depend upon the surface director field $\hat{n}$. It follows directly from (9.4.13) that the contribution to the surface tension arising from surface excess orientational order $\hat{\gamma}$ is

$$\hat{\gamma} = -\frac{kT}{\mathscr{A}} \int_{-\infty}^{\infty} \rho_{(1)}(z) \ln \left\{ d^{-1} \int_0^d \int_0^1 \exp\left(-\frac{\bar{u}}{kT} \left[\eta_{\hat{n}}(z) + \sigma_{\hat{n}}(z)\alpha \cos\left(\frac{2\pi z'}{d}\right) \right] \right. \right.$$

$$\left. \left. \times (\tfrac{3}{2}\cos^2\theta - \tfrac{1}{2}) \right) \, d(\cos\theta) \, dz' \right\} dz$$

$$+ \frac{kT}{\mathscr{A}} \int_{-\infty}^{\infty} \rho_L \ln \left\{ d^{-1} \int_0^d \int_0^1 \exp\left(-\frac{\bar{u}}{kT} \left[\eta + \sigma\alpha \cos\left(\frac{2\pi z'}{d}\right) \right] \right. \right.$$

$$\left. \left. \times (\tfrac{3}{2}\cos^2\theta - \tfrac{1}{2}) \right) \, d(\cos\theta) \, dz' \right\} dz \qquad (9.4.14)$$

(note the dummy variable z').

This somewhat cumbersome expression may be made physically more transparent by linearizing the functions whereupon we obtain, to second order

$$\hat{\gamma} \sim -\frac{\bar{u}^2}{10kT\mathscr{A}} \int_{-\infty}^{\infty} \left[(\rho_{(1)}(z)\eta_{\hat{n}}^2(z) - \rho_{L,V}\eta_{L,V}^2) + \frac{\alpha^2}{2}(\rho_{(1)}(z)\sigma_{\hat{n}}^2(z) - \rho_{L,V}\sigma_{L,V}^2) \right] dz$$

$$(9.4.15)$$

In obtaining the above expression we have assumed that the liquid–vapour density transition is accomplished within a distance of a molecular diameter, in which case $\rho_{(1)}(z) \sim \rho_L$ over the effective region of integration ($z < 0$).

The temperature derivative, or the orientational surface excess entropy, is, to the same degree of approximation,

$$-\frac{d\hat{\gamma}}{dT} = \hat{S}_s = -\frac{\rho_L}{\mathscr{A}} \int_{-\infty}^{\infty} \frac{\bar{u}}{T} (\eta_{\hat{n}}^2(z) - \eta^2 + \alpha(\sigma_{\hat{n}}^2(z) - \sigma^2)) \, dz$$

$$+ \rho_L \frac{kT}{\mathscr{A}} \int_{-\infty}^{\infty} d^{-1} \int_0^d \int_0^1 \left\{ -\frac{\bar{u}}{kT} \left[\eta_{\hat{n}}(z) - \eta + (\sigma_{\hat{n}}(z) - \sigma)\alpha \cos\left(\frac{2\pi z'}{d}\right) \right] \right\}$$

$$\times \exp\left(\tfrac{3}{2}\cos^2\theta - \tfrac{1}{2}\right) d(\cos\theta) \, dz' \, dz$$

$$= \frac{\bar{u}}{\mathscr{A}T}\left(\frac{\bar{u}}{10kT} - 1\right) \int_{-\infty}^{\infty} \left[(\rho_{(1)}(z)\eta_{\hat{n}}^2(z) - \rho_{L,V}\eta_{L,V}^2) + \frac{\alpha^2}{2}(\rho_{(1)}(z)\sigma_{\hat{n}}^2(z) - \rho_{L,V}\sigma_{L,V}^2) \right] dz$$

$$(9.4.16)$$

where we note that $\left(\dfrac{\bar{u}}{10kT} - 1\right)$ is invariably negative at the temperatures under consideration.

It does, of course, remain to specify the spatial variation of the local order parameters $\eta_{\hat{n}}(z)$ and $\sigma_{\hat{n}}(z)$. The z-dependence of these functions has already been discussed in Section 9.2, and is a generalization based on an analysis by Croxton and Chandrasekhar[3] for nematic liquid crystals. We shall continue their approach here.

(a) The nematic liquid crystal surface ($\sigma = 0$, $\eta > 0$)

It is proposed that the nematic order parameter in the vicinity of the liquid surface is determined solely on the basis of a competition between the disordering effect of the decreasing density across the liquid–vapour interface, and an ordering effect due to the development of a surface torque field. The analysis in Section 9.2 suggests the former is proportional to $\rho_{(1)}(z)$ and the latter to $(d\rho_{(1)}(z)/dz)^2$ (equation 9.2.23), (Figure 9.4.1).

On the basis of such a model a number of qualitative forms for the $\gamma(T)$ characteristic have been predicted.[3] At low temperatures, for example, just beyond ahe crystal–nematic transition temperature, the spatial delocalization of the surface may be sufficiently small and consequently the surface orientational field sufficiently high that there is a net *enhancement* of the local order parameter over its bulk value: that is $\eta_{\hat{n}}(z) > \eta$ in the vicinity of the surface. Setting $\sigma = 0$ in (9.4.16), corresponding to the nematic phase, we have

$$\hat{S}_{s} = -\frac{d\hat{\gamma}}{dT} = \frac{\bar{u}}{\mathscr{A}T}\left(\frac{\bar{u}}{10kT} - 1\right)\int_{-\infty}^{\infty} [\rho_{(1)}(z)\eta_{\hat{n}}^{2}(z) - \rho_{L,V}^{2}\eta_{L,V}^{2}]\,dz \qquad (9.4.17)$$

We see immediately that for a sufficiently large surface enhancement of orientational order the usual negative monotonic slope of the $\gamma(T)$ characteristic may actually reverse and increase with temperature, at least just above the melting point and just below the nematic–isotropic transition temperature T_{NI}, when the rapidly decreasing bulk order parameter is enhanced by the surface torque field (Figure 9.4.2). At higher temperatures, of course, spatial delocalization of the surface with a corresponding relaxation in the surface field will ensure the usual monotonic decrease of surface tension with temperature.

We have already observed (Section 1.4) that for spherically symmetrical interactions, minimization of the surface excess Helmholtz free energy is achieved as a competition between the surface excess energy term, which favours a sharp liquid–vapour density transition, and the surface excess entropy, which favours a spatially delocalized surface. In the case of strongly

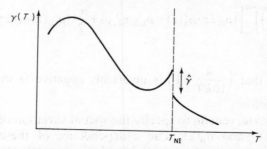

Figure 9.4.2 Schematic variation of the nematic surface tension showing a discontinuity at the nematic–isotropic transition temperature

anisotropic molecular systems such as we are considering here, *orientational* contributions to the excess Helmholtz free energy may be similarly regarded as a competitive interplay between the orientational energy term $\hat{U}_s$ which favours molecular orientation aligned with the surface director field, and the excess orientational entropy term $\hat{S}_s$ which favours angular dispersion:

$$\hat{A}_s = \hat{U}_s - T\hat{S}_s. \tag{9.4.18}$$

(Unless the spatial and orientational contributions to the free energy develop independently it is not legitimate to minimize them separately: the two components will generally be inextricably coupled. In the spirit of the preceding analysis, however, we assume here that such a decoupling is in order.) In terms of molecular quantities equation (9.4.18) becomes, for a nematic liquid crystal surface (c.f. 9.4.15),

$$\hat{\gamma} = \frac{1}{10\mathscr{A}} \int_{-\infty}^{\infty} \frac{\bar{u}^2}{kT} [\rho_{(1)}(z)\eta_{\hat{n}}^2(z) - \rho_{L,V}\eta_{L,V}^2] \, dz \tag{9.4.19}$$

whose contribution to the total (configurational + orientational) surface tension is positive or negative according as

$$\int_{-\infty}^{\infty} \rho_{(1)}(z)\eta_{\hat{n}}^2(z) \, dz \gtrless \int_{-\infty}^{\infty} \rho_{L,V}\eta_{L,V}^2 \, dz$$

the magnitude and sign of the contribution becoming apparent at the nematic–isotropic transition T_{NI} (Figure 9.4.2).

There is a complex interplay of the relative magnitudes of $\eta_{\hat{n}}(z)$ and η at T_{NI} resulting in a wide variety of pre- and post-transition phenomena. For example, it is possible that a very weak surface orientational (torque) field remains for a short thermal range beyond T_{NI} into the isotropic phase, resulting in a positive discontinuity $\hat{\gamma}$ in the surface tension, and reducing (and even inverting) the negative slope $d\gamma/dT$. Indeed, on the basis of equations (9.4.17) and (9.4.19), systems showing positive slopes just above T_{NI} must also show positive discontinuities $\hat{\gamma}$, and vice versa (Figures 9.4.2, 9.4.3). Ultimately, of course, all characteristics become monotonic, negatively-sloped functions of temperature.

Recent experimental observations on nematic liquid crystal systems[4,5] have confirmed the qualitative features predicted on the basis of the above analysis. For example, the surface tension of p-anisaldazine in the nematic and isotropic phases is shown in Figure 9.4.4. The development of a surface excess orientational order just before T_{NI} (corresponding to $\eta(z) > \eta$) with a discontinuous change in slope and magnitude of the $\gamma(T)$ characteristic is in qualitative agreement with Figure 9.4.2. On the other hand, a residuum of surface orientation just beyond T_{NI} in p-oxyanisole (Figure 9.4.5) agrees qualitatively with the curve in Figure (9.4.3). We observe that in this case $\hat{\gamma}$ will be positive (9.4.19) since in the isotropic phase $\eta_{\hat{n}}(z) > 0$ for $T \gtrsim T_{NI}$.

The question arises as to whether the 'anomaly' in the $\gamma(T)$ characteristic is due to a rather sudden reorientation of the molecules in the vicinity of

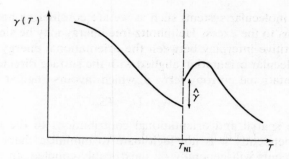

Figure 9.4.3 Schematic variation of the nematic surface tension showing a discontinuity and subsequent inversion in the vicinity of the nematic isotropic transition temperature

T_{NI}.[7] The alignment of the molecules at the surface of p-azoxyanisole (PAA) and n-p-methoxybenzylidene p′-n-butylaniline (MBBA) has been studied by light-scattering techniques.[6] In the case of PAA the molecules are found to be aligned parallel to the liquid surface, and the orientation appears to be independent of temperature, whilst for MBBA the molecules are tilted at 75° to the surface, the orientation being only slightly dependent upon temperature. It would therefore seem safe to say that the features of the $\gamma(T)$ curve are not attributable to any molecular reorientation at the surface, although of course, the specific surface orientation of the molecules does affect the magnitude of the surface tension.

The Kirkwood–Buff model of the nematic liquid surface (Section 9.3), in assuming that the bulk order extends up to the discontinuous liquid surface, forfeits those features of the preceding analysis which depend upon a local order parameter. Indeed, the model specifically neglects spatial and orientational contributions to the entropy component of the surface free energy. It

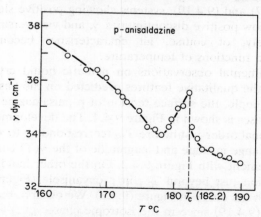

Figure 9.4.4 The experimental surface tension of p-anisaldazine (c.f. Figure 9.4.2)

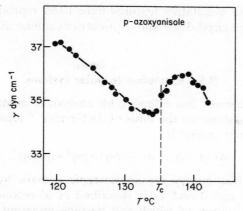

Figure 9.4.5 The experimental surface
tension of p-azoxyanisole (c.f. Figure 9.4.3)

may well be, therefore, that $\gamma_\parallel$ is *not* the lowest free energy as obtained by
Parsons—a lower value being achieved by inclusion of entropy contributions
resulting in some intermediate orientation of the surface director field.

(b) The smectic-A liquid crystal surface $(\sigma > 0, \eta = 0)$

A similar line of reasoning holds for the smectic-A liquid surface where, for
values of the local order parameter $\sigma_a(z) > \sigma$, corresponding to pronounced
layering in the vicinity of the surface, we again anticipate positive slopes in
the $\gamma(T)$ characteristic. Indeed, at the relatively low range of temperatures
over which the smectic phase is stable, the sharp density discontinuity at the
surface implies strong surface fields, with the result that a number of smectic
systems appear to show positive $\gamma(T)$ slopes throughout the entire phase
(Figure 9.4.6). Again, discontinuities in both the magnitude and the slope of
the surface tension may occur as for nematic fluids, and beyond the
smectic–nematic transition temperature T_{SN} we anticipate the usual nematic

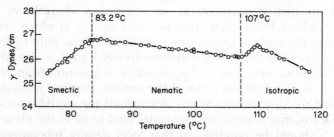

Figure 9.4.6 The experimental surface tension of
p-cyanobenzylidene p'-n-octyloxyaniline (CBOOA), showing
both smectic–nematic and nematic–isotropic transitions in
the surface tension

$\gamma(T)$ curves: these qualitative features have been reproduced experimentally[5] for p-cyanobenzylidene and p-n-octyloxyaniline (CBOOA) (Figure 9.4.6).

9.5   Extension to polar systems

More recently Parsons[11] has extended his analysis of nematic liquid crystals to include polar systems on the basis of de Gennes'[12] Landau treatment for which the *bulk* free energy is

$$A(\eta) = A_0 + a\eta^2 - b\eta^3 + c\eta^4 + d(\nabla\eta)^2. \qquad (9.5.1)$$

A_0 is the bulk free energy for an isotropic nematic system, whilst $a = a_0(T - T_c)$ where $a_0 > 0$ and T_c is described as a 'second-order transition temperature' the nature of which will become apparent below. The final term in (9.5.1) accounts for free energy contributions arising from distortions in the η field; $d > 0$. Similarly, the surface excess free energy may be written as

$$\gamma(\eta) = \gamma_0 - \gamma_1\eta + \gamma_2\eta^2 \qquad (9.5.2)$$

where γ_0 is the surface tension of the isotropic fluid, and γ_2 is taken to be positive for stability. Of course, there will be local variations in the order parameter in the vicinity of the surface, and it may be safely assumed that $\eta(z) \to 0$ in either bulk phase. Assuming $\eta(z)$ is nevertheless small across the interphasal region, we have from (9.5.1)

$$\frac{d^2\eta}{dz^2} - \left(\frac{a}{d}\right)^{1/2}\eta = 0 \qquad \eta(z) = \eta_0 \exp\{z(a/d)^{1/2}\}. \qquad (9.5.3)$$

η_0 is the value of the order parameter at the surface boundary condition

$$\left[\frac{\partial A}{\partial(\nabla\eta)}\right] \cdot \hat{\mathbf{k}} + \frac{\partial\gamma}{\partial\eta} = 0 \qquad (9.5.4)$$

where $\hat{\mathbf{k}}$ is the surface normal. Equation 9.5.4 represents the variational minimization of the total free energy with respect to the order parameter and the normal component of its distortion. It follows that $\eta_0 = \gamma_1/[(ad)^{1/2} + 2\gamma_2]$. For prolate uniaxial order $\eta_0 > 0$, in which case $\gamma_1 > 0$ in (9.5.2), and the surface excess order decays into the bulk over a distance $\sim(a/d)^{1/2}$. Typically $(ad)^{1/2} \lesssim 0.1$ dyn/cm, in which case η_0 should be observable for moderate values of $\gamma_1/2\gamma_2$. Again, it is apparent from (9.5.2) that γ is *lowered* by orientational order at the surface. Parsons' analysis in Section 9.3 showed that for van der Waals interactions the equilibrium state occurs when $\hat{\mathbf{n}} \cdot \hat{\mathbf{k}} = 0$, that is when the molecules tend to lie in the plane of the free surface. It should be pointed out that local *density* inhomogeneity is not involved in Parsons' treatment: the model is essentially KBF in nature. Density gradients exert additional torques on the surface molecules as we have described at some length earlier in this chapter, and to this extent the

Landau treatment provides an incomplete description of the interfacial structure.

For small deviations about Parsons' parallel alignment we may form a Taylor expansion and write

$$\gamma = \gamma_0 + \tfrac{1}{2}\gamma_Q(\hat{\mathbf{n}} \cdot \hat{\mathbf{k}})^2 \tag{9.5.5}$$

where $\gamma_Q > 0$: γ must be even in $(\hat{\mathbf{n}} \cdot \hat{\mathbf{k}})^2$ on account of the equivalence of the orientations $\hat{\mathbf{n}}$ and $-\hat{\mathbf{n}}$.

If, however, the two ends of the molecule are dissimilar, one end having a highly polar end group, for example, that end will tend to point into the more highly polar (bulk) medium. Moreover, the orientations $\hat{\mathbf{n}}$ and $-\hat{\mathbf{n}}$ are no longer equivalent, whereupon (9.5.5) is supplemented by a linear term:

$$\gamma = \dot{\gamma}_0 + \tfrac{1}{2}\gamma_Q(\hat{\mathbf{n}} \cdot \hat{\mathbf{k}})^2 - \gamma_P(\hat{\mathbf{n}} \cdot \hat{\mathbf{k}}). \tag{9.5.6}$$

Clearly there is a competitive interplay between the linear and quadratic terms—the former favouring a normal molecular orientation, whilst the latter favours a parallel alignment with respect to the surface. Minimization of the surface excess free energy in (9.5.6) with respect to $\hat{\mathbf{n}} \cdot \hat{\mathbf{k}} = \cos \theta$ yields

$$\cos \theta = \begin{cases} \gamma_P/\gamma_Q & \gamma_P < \gamma_Q \qquad (9.5.7a) \\ 1 & \gamma_P > \gamma_Q \qquad (9.5.7b) \end{cases}$$

from which it follows that $\gamma_P = \gamma_Q$ is a *critical point*: (9.5.7b) corresponds to uniaxial symmetry with $\hat{\mathbf{n}}$ along the surface normal $\hat{\mathbf{k}}$, whilst $\hat{\mathbf{n}}$ tilts with respect to $\hat{\mathbf{k}}$ when $\gamma_P < \gamma_Q$ (9.5.7a) and the surface symmetry is biaxial. Since γ_P and γ_Q will, in general, be functions of temperature, it follows that the tilt angle will in general also be temperature-dependent. Such variation of tilt angle with temperature, when it does occur, is, however, relatively weak.[6]

Strong fluctuations in the tilt angle also occur near the critical point $\gamma_P = \gamma_Q$ corresponding to a propogating mode involving oscillations of $\hat{\mathbf{n}}$ with respect to the surface normal $\hat{\mathbf{k}}$: equipartition suggests

$$|\theta|^2 = \frac{k_B T}{\gamma_P - \gamma_Q}. \tag{9.5.8}$$

At $\gamma_P = \gamma_Q$ the oscillation frequency of this mode must vanish, which Parsons interprets as a *soft mode* associated with the uniaxial–biaxial structural transition.

References

1. W. Maier and A. Saupe, *Z. Naturf.*, **13a,** 564 (1958); **14a,** 882 (1959); **15a,** 287 (1960).
2. W. L. McMillan, *Phys. Rev.*, **A4,** 1238 (1971).
3. C. A. Croxton and S. Chandrasekhar, *Proc. 1st. Intl. Conf. Liquid Crystals, Bangalore 1973*, Pramana Suppl. No. 1, p. 237.
4. S. Krishnaswamy and R. Shashidhar, *Mol. Cryst. Liq. Cryst.*, **35,** 253 (1976).

5. S. Krishnaswamy and R. Shashidhar, *Mol. Cryst. Liq. Cryst.*, **38,** 711 (1977).

6. M. A. Bouchiat and D. Langevin, *Phys. Lett.*, **34A,** 331 (1971).

7. F. Jahnig, *Proc. 1st. Intl. Conf. Liquid Crystals, Bangalore 1973*, Pramana Suppl. No. 1, p. 246.

   F. M. Leslie, *Proc. 1st. Intl. Conf. Liquid Crystals, Bangalore, 1973*, Pramana Suppl. No. 1, p. 252.

8. J. D. Parsons, *J. de Phys.*, **37,** 1187 (1976).

9. C. A. Croxton, *Mol. Cryst. Liq. Cryst.* (February, 1980).

10. For interactions whose radial distributions scale as $g_{(2)}(r/\sigma^*(e_1, e_2))$, where $\sigma^*(e_1, e_2)$ is an *effective* collision diameter for a particular relative orientational configuration of the molecules, Parsons (*Phys. Rev.* **A19,** 1225 (1979)) finds that the radial and angular features of the integrand decouple completely, affording a substantially more realistic representation than that of (9.2.5). Further discussion will be given in Chapter 11.

11. J. D. Parsons, *Phys. Rev. Lett.*, **41,** 877 (1978).

12. P. G. de Gennes, *Phys. Lett.*, **30A,** 454 (1969).

CHAPTER 10

Machine Simulations

10.1 Introduction

The difficulty of direct experimental investigation of the density transition profile at the liquid surface has been emphasized in the preceding chapters, and, indeed, underlies much of the inherent theoretical uncertainty which attends the microscopic description of the interphasal region. In the case of homogeneous systems, machine simulation played a central role between theory and experiment in yielding information representative of the structure and thermodynamics of infinite assemblies of particles interacting through analytically convenient caricatures of the realistic potential. It would seem that an extension of these techniques to the inhomogeneous surface region would finally resolve the form of the transition profile. Unfortunately, simulation of the interfacial zone proves substantially more difficult than for homogeneous systems and, in consequence, a degree of uncertainty prevails in the simulations reported in the literature regarding the development or otherwise of stable density oscillations at the liquid–vapour interface, in particular for the dielectric, inert gas systems. The very difficulties specific to the simulation of the liquid surface prove to be just those which are possibly responsible for the initiation of density oscillations in the transition zone.

Generally, simulation of an inhomogeneous system necessarily involves a greater number of particles than its homogeneous counterpart, if detailed information regarding the inhomogeneity is to be obtained. Persistence of long-lived fluctuations and long-ranged correlations require $\sim 10^7$ configurations for a Monte Carlo ensemble, or time averaging over $\sim 10^{-9}$ sec. for a molecular dynamics simulation for an assembly of $\sim 10^3$ particles if adequate phase sampling is to be achieved. Indeed, it has been suggested that some of the confusion surrounding the simulation of Lennard–Jones interfacial systems may be attributed to the extremely slow convergence of the profiles; such is the weakness of coupling between liquid and vapour.

In this chapter we propose not to review the actual simulation techniques *per se*, for this is adequately covered elsewhere,[1] but rather to discuss fundamental questions regarding the philosophy of approach which have recently arisen and which may ultimately result in a reappraisal of the status of the machine investigations. (Nevertheless, a brief description is given in Appendices 10.1 and 10.2.)

10.2 Suppression of long wavelength fluctuations

In all cases, whether in two or three dimensions, the particles are confined to a fundamental cell whose specification is of central importance in as far as an acceptable compromise between the conflicting demands of computational expediency and statistical significance of the simulated result has to be found—both of which are contingent upon the cell dimension. This is an appropriate point to draw together several seemingly unrelated aspects of the description of liquid surfaces, resulting in a resolution, at least partially, of apparently conflicting descriptions of the liquid surface, with important consequences for simulation studies.

The apparent conflict arises from the seemingly opposed descriptions of the surface which arise on the basis of the local free energy approach (Section 2.4) based on modern perturbation theories of liquids, and the capillary wave description. The former descriptions are known to be in good agreement with computer simulations, yielding a well-defined density profile whose width is independent of system size; this we shall call the *intrinsic interfacial width*. Any effects of an external gravitational field are generally ignored. The capillary wave description (Section 2.11), on the other hand, gives an explicit dependence of the interfacial width upon system size and strength of the external gravitational field g. Indeed, for an infinite system a slow divergence of the interfacial width as $g \rightarrow 0^+$ is predicted at all temperatures.[9]

We may resolve this apparent conflict by making an arbitrary subdivision of the capillary wave numbers into a long wavelength $(k < \pi/L)$ and a short wavelength $(k > \pi/L)$ regime where L is of the order of a correlation length in the bulk liquid. Following Weeks,[4] we make a schematic subdivision of such an infinite system in Figure 10.2.1a in which we illustrate the development of long ranged surface correlations due to long wavelength capillary fluctuations in the local height of the Gibbs surface. These fluctuations arise on account of density fluctuations in different regions of the interface, or in terms of Figure 10.2.1a, from changes in the occupation numbers of each of the columns.[4,7] The development of these long ranged fluctuations was recently predicted by Wertheim,[3] and although he restricted his analysis regarding the effects of an external field, these same fluctuations nevertheless can, in principle, cause the interfacial width to diverge as $g \rightarrow 0^+$. Only an experiment in a space laboratory can settle the question.

The subsystems in Figure 10.2.1a of cross-sectional area $\sim L^2$ are what we study in computer simulation, although of fixed occupation number, of course. Moreover, local free energy mean field theories have been shown by Davis[5] to have an implicit cut-off appropriate to the consideration of a finite system of cross-sectional area L^2—hence the illusory good agreement between simulation and the mean field theories. More specifically, these subsystems do have a finite intrinsic surface thickness. However, a precise and non-arbitrary definition of the local intrinsic width probably cannot, but for most practical considerations need not, be made.[4]

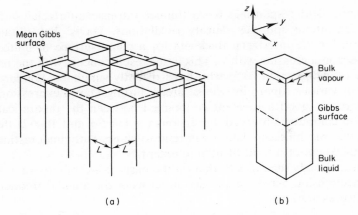

Figure 10.2.1 (a) Subdivision of an infinite fluid into an array of vertical columns whose horizontal width is bulk correlation length. Long wavelength capillary fluctuations may then be related to the local height of the Gibbs surface. (b) The fundamental cell in machine simulations of the liquid surface. Replication is generally in the xy plane, whilst z must extend from bulk liquid to bulk vapour, the Gibbs surface being located well away from the upper or lower boundaries

The obvious corollary regarding the dynamic behaviour of the fluctuations, those of increasing wavelength occurring on slower and slower time scales as $k \rightarrow 0$, suggests that low frequency modes are also suppressed in the course of finite simulations. Particularly, the long time tails in the horizontal time correlation functions will be suppressed, although the time scale leading to the slow divergence of the vertical correlations is probably so long that they need not concern us in practice.[4]

Provided we appreciate that it is the subsystems in Figure 10.2.1a which we are able to simulate (although of fixed occupation number), and that the results are subject to the reservations outlined above, we may now consider in detail the further restrictions which necessarily apply within one of the representative subsystems. Certain of the consequences of the imposition of periodic boundary conditions to eliminate partially the effects of rigid cell walls have already been mentioned. Stochastic boundary conditions have been proposed whereby particles lost from an unconstrained system are randomly returned, but this would appear simply to replace one problem by another (probably harder). The height of the cell, z, imposes no great restriction since it must in any case extend from the bulk liquid into the bulk vapour (Figure 10.2.1b). The lower boundary must not, however, initiate any density oscillations which might extend into the transition zone region. The x and y dimensions, on the other hand, in addition to suppressing much shorter wavelength fluctuations, also impose an artificial planarity on the surface[2] since the condition for periodic replication in the xy plane effectively constrains the transition region, and may act as a kind of surface constraining field responsible for a thinner intrinsic interfacial width, and may even initiate spurious density oscillations. Davis[5] has obtained an expression for the interfacial thickness for a plane surface of finite extent, the dispersion increasing with L. This latter result concurs with the results of Chapela *et al.*[2] (Section 10.5) who have recently simulated Lennard–Jones systems of various square interfacial areas. Fitting Davis' expression to the $L = 20\sigma$ result, good agreement is obtained between the theoretically predicted and computer observed dispersions for $L = 5\sigma$ and 10σ. It therefore appears that machine simulation will tend to consistently underestimate the surface thickness of a fluid of infinite extent.

We have already seen (6.1.8) that on the basis of a capillary wave analysis the surface tension may be expressed in terms of the angular frequencies of the surface excitations:

$$\gamma = \gamma_0 - \frac{\hbar}{4\pi}\left(\frac{\rho}{\gamma_0}\right)^{2/3}\int_0^\infty \frac{\omega^{4/3}}{\exp(\hbar\omega/kT)-1}\,d\omega \qquad (10.2.1)$$

where γ_0 is a positive constant and the nature of the integrand enables the upper limit to be extended to $+\infty$. Clearly, the suppression of low frequency contributions, replacing the lower limit by $\omega_{min} = \pi/L$, will result in a consistent overestimate of the surface tension of a simulated fluid of square interfacial area L^2 with respect to that of an infinite system—a conclusion reached by Weeks[4] on the basis of a more elaborate argument. Weeks estimates that for the two Lennard–Jones molecular dynamics simulations of Rao and Levesque[6] with cell parameters $L = 6.78\sigma$ and 13.15σ, the lowering of the surface tension in transforming to a system of macroscopic dimension would be about 8% and 3%, respectively.

No attempt will be made here to 'transform' the results of the finite simulations to their macroscopic thermodynamic limit, however, since our understanding of the problem remains incomplete as yet—a qualitative appreciation of the shortcomings and their consequences inherent in computer simulation is perhaps the best we can manage at present. Whether the interfacial thickness is approaching a finite limit or is weakly diverging as the cross-sectional area of the simulation is increased is not yet clear. The very important theoretical question is raised, however, that if we cannot attain the thermodynamic limit without inclusion of a weak gravitational field, then the development of a rigorous statistical mechanical description of interfacial phenomena is likely to be considerably more difficult than anticipated.

10.3 Effects of mean liquid density fluctuations within the subsystem

Machine simulation techniques have been widely employed in an attempt to resolve the controversy as to whether a Lennard–Jones fluid just above the triple point is capable of developing a partially layered structure rather than a monotonic transition profile. In particular we enquire as to whether the presence or absence of surface structure can be considered in any way an artifact of the simulation process. Long range undulations in the Gibbs dividing surface are expected to develop[3,4] and these, of course, are inconsistent with the xy replication scheme adopted in all simulations, and cannot be followed either spatially or temporally in the course of conventional Monte Carlo or molecular dynamics computations. Such fluctuations between regions of the interface separated by many bulk correlation lengths provides an *additional* contribution to the interfacial width[4] and depends upon the system size in the manner predicted by capillary wave theory (Section 10.2). What we are concerned with here, however, is the intrinsic interfacial structure which is subsequently subjected to long wavelength undulations. This we outlined in the previous section (Section 10.2).

Generally, the procedure adopted in such a simulation is as follows. After some 'settling down' period $(0 \rightarrow \sim 1.7 \times 10^6$ configurations for Monte Carlo, $0 \rightarrow \sim 4 \times 10^3$ time steps for molecular dynamics, depending upon the method used) the statistics are allowed to accrue. The particles are confined to a cell whose base generally serves as the location of the origin of coordinates, and the basic cell is then replicated in the x and y directions to form a relatively thick adsorbed layer of liquid, or, alternatively, a two-sided film with a median origin plane. The methods and cell parameters for some recent simulations, chronologically arranged, are given in Tables 10.3.1a,b. In each case, the origin of coordinates is located at distances from five to fifteen diameters below the Gibbs surface.

Osborn and Croxton[10] have investigated the effect of fluctuations in the mean bulk density, and hence the location of the Gibbs surface, for a model oscillatory transition profile. Obviously, if measurement is referenced to an origin many atomic diameters distant from the Gibbs surface, then density fluctuations will engender a smearing of any surface structure[11] with the inevitable conclusion that the profile is monotonic. Rao and Levesque[6] monitor the r.m.s. bulk density fluctuations in the course of their simulation of a Lennard–Jones liquid film, and find that the r.m.s. density fluctuations settle down to about 1.5% after about 10^{-10} sec., which is about the lifetime of the density fluctuations which continually form and disperse.[2] Of course, a significant movement of the Gibbs surface requires the cooperative effort of a large number of particles, and the period of such long wavelength fluctuations is correspondingly large—on a time scale $\sim 10^{-10}$ sec. which suggests that oscillations may be detected in simulations of shorter extent. Osborn and Croxton observe that r.m.s. density fluctuations of more than 3% entirely suppress the oscillatory structure of the model profile, whilst fluctuations ~ 1–2% suppress the structure to an extent where it may be indistinguishable from the statistical noise. They emphasize that it is not merely the net shift in the location of the Gibbs surface, but also the *distribution* or spread of locations arising in the course of the simulation which determines the degree of structural suppression. Few authors quote the shift of the Gibbs surface, although Rowlinson[12] reports a *net* shift $<0.1\sigma$ whilst Abraham *et al.*[13] report a shift $\sim 1.3\sigma$. Unless a cumulative

Table 10.3.1. a. Single component three-dimensional simulations of the liquid–vapour interface

	No. of particles/ method	T^*	Cell parameters $L \times h(\sigma)$	Approximate location of Gibbs surface (σ)	Settling down period (configs./ time steps)	Duration (configs./ time steps)	$\gamma\sigma^2/\varepsilon$	Remarks
Lee et al.[14] (1974)	1000 MC	0.7035	6.54×40	~±15	not quoted	*6×10⁶	1.13§	Two-sided film, strong 'gravitational' field providing constraint.
Liu[15] (1974)	129 MC	0.75	?×15	~10	1.7×10⁶	*5–10×10⁶	1.01§	Adsorbed layer on attractive wall.[†]
		0.90	?×15				0.6§	
		1.10	?×20				0.35§	
		1.30	?×20				0.18§	
Opitz[15] (1974)	300 MD	1.03	6×4	~6				
		1.127	6×24					
Abraham et al.[13] (1975)	1024 MC	0.701	26.5×4.4	~±9	1.7×10⁶	6.2×10⁶		Two-sided film[†]
Chapela et al.[2a] (1975)	255 MC	0.701	5×25	~13	1.5×10⁶	*5.6×10⁶	2.08	Adsorbed layer on attractive cell base[†]
	255 MC	0.918	5×25	~13	1.5×10⁶	5.6×10⁶	0.85§	
	255 MC	1.127	5×25	~13	1.5×10⁶	3.5×10⁶	0.25§	
Rao and Levesque[6] (1976)	1728 MD	0.704	13.15×39.45	~±2.5	not quoted	6400×1728	1.18±0.04	Two-sided film.[†] Kalos et al. have recently reanalysed this simulation in an assessment of the development of long range transverse correlation (Section 10.11)
	1024 MD	0.744	6.78×56.24	~±7	not quoted	39000×1024		
Chapela et al.[2b] (1977)	255 MC	0.701	5×25	~13	not quoted	15.2×10⁶	1.30	Adsorbed layer on attractive cell base
	1020 MD	0.699±0.001	10×25	~13	not quoted	40.0×10³	1.10	
	4080 MD	0.701±0.001	20×25	~13	not quoted	4×10³	1.10	
	255 MD	0.708±0.002	5×25	~13	not quoted	38.9×10³	1.07	
	255 MD	0.759±0.002	5×25	~13	not quoted	42.0×10³	0.90	
	1020 MD	0.785±0.001	10×25	~13	not quoted	38.5×10³	0.83	
	255 MD	0.823±0.002	5×25	~13	not quoted	45.0×10³	0.78	
	1020 MD	0.836±0.001	10×25	~13	not quoted	38.0×10³	0.74	
	255 MC	0.918	5×25	~13	not quoted	6.3×10⁶	0.67	
	255 MC	1.127	5×25	~13	not quoted	3.5×10⁶	0.35	

* indicates evidence of layering
† no tail corrections applied
§ read to graphical accuracy
? L not quoted

b. binary three-dimensional computer simulations of the liquid–vapour interface

No. of particles/method	T_a^*, T_b^*	Cell parameters $L \times h(\sigma)$	Approximate location of Gibbs surface (σ)	Settling down period (configs./time steps)	Duration (configs./time steps)	$\gamma\sigma^2/\varepsilon$	Remarks
Chapela et al.[2b] (1977) 255 MC	0.918, 0.701	5×25	~13	not quoted	15×10^6	1.21 ± 0.09	Equimolar mixture $m_a = m_b$, $\sigma_a = \sigma_b$, $\varepsilon_{aa}/\varepsilon_{bb} = 0.763$, $\varepsilon_{ab} = (\varepsilon_{aa}\varepsilon_{bb})^{1/2}$. More volatile mixture (a) preferentially adsorbed as a surface layer. Adsorbed layer on attractive base
255 MD	0.933±0.002, 0.713±0.002	5×25	~13	not quoted	8.68×10^4	0.94 ± 0.07	

profile is formed relative to the centre of mass of the film in the latter case, then, of course, the reported profile is entirely meaningless. No authors report the *variance*, however. An interesting feature which seems to characterize the simulation is the initial development of structured profiles which, in some cases at least, subsequently become indistinguishable from the statistical noise, Why oscillations should initially develop is not clear, although as we have mentioned, replication of the distribution in the x and y directions does effectively constrain the surface, and may be responsible for the development of spurious oscillations. Nevertheless, it is generally agreed that the slow equilibration of the surface layer, attributable to the relatively weak liquid–vapour coupling, necessitates longer runs, and presumably the accumulated effects of mean density fluctuations over such long, distantly-referenced runs must arise.

In the case of the simulation of double-sided liquid films, a common practice is to symmetrize the simulation by averaging the 1.h.s. and r.h.s. profiles:

$$\rho_{(1)}(z) = 0.5\{\rho_{(1)}(-z)_{1.h.s.} + \rho_{(1)}(z)_{r.h.s.}\}. \qquad (10.3.1)$$

Although individually both profiles may be highly structured, they do not and need not correspond in phase of the oscillatory structure,[14] which suggests that the procedure of summing profiles referenced to a distant origin in this way may be invalid. It does appear that the certitude with which the nature of the density profiles determined on the basis of simulations so far reported is called into some question. Osborn and Croxton go on to suggest that future machine simulations of the liquid surface should either reference the system from the (fluctuating) Gibbs surface, or at least record the distribution in location of the Gibbs surface, and the variance in the bulk liquid density in the course of the simulation.

10.4 Effects of truncation of the pair interaction

Before we consider some specific simulations of interfacial systems it is appropriate to mention an essential difference between the Monte Carlo and molecular dynamics techniques regarding the effect of truncation of the pair interaction. Termination of the interaction beyond some radial separation R is a necessary expedient significantly reducing the number of interacting pairs, and hence a substantial reduction in the computation time.

In a Monte Carlo simulation in which a large number of trial configurations of the assembly are tested, we work directly with the truncated potential shown in Figure 10.4.1. If the true interaction is of Lennard–Jones form, then the effective Monte Carlo interaction is:

$$\Phi_{MC}(r) = \Phi_{LJ}(r) \qquad (r < R)$$
$$= 0 \qquad (r > R). \qquad (10.4.1)$$

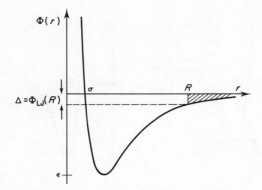

Figure 10.4.1 Introduction of 'tail effects' in truncating the pair interaction at R. Note that the pair potential becomes discontinuous at R

It is then relatively easy to correct for 'tail' effects, shown by the shaded region in Figure 10.4.1.

In a molecular dynamics computation, however, we solve the Newtonian equation of motion of the assembly and it is the *force* which we truncate beyond some arbitrary cut-off distance R. We cannot accept a truncated interaction of the form $\Phi_{MC}(r)$ since the discontinuity in the potential implies that an atom will experience an impulse as it crosses the spherical shell of radius R of another atom. To ensure continuity of the interaction the potential is often shifted by the amount $\Delta = \Phi_{LJ}(R)^{6,18}$ to yield the effective molecular dynamics interaction

$$\Phi_{MD}(r) = \Phi_{LJ}(r) - \Phi_{LJ}(R) \qquad (r < R)$$
$$= 0 \qquad (r > R). \qquad (10.4.2)$$

Thus, while the Monte Carlo results are for the true truncated Lennard–Jones potential, the molecular dynamics results are for a shifted potential. This has a number of effects which are not easily allowed for; for example, the coexisting liquid and vapour densities are slightly lower than those for a true Lennard–Jones fluid.[14] We may approximately regard the temperature of the molecular dynamics calculation as being scaled by a factor of $\varepsilon/[\varepsilon - \Phi(R)]$. However, in Figure 10.4.2a we show the nature of the correction by the shaded region which is required to recover the true Lennard–Jones result. Alternatively, the potential $\Phi_{MD}(r)$ (10.4.2) may be scaled by the factor $\varepsilon/[\varepsilon - \Phi(R)]$ to yield $\Phi^*_{MD}(r)$, whereupon only the tail correction shown shaded in Figure 10.4.2b need be applied to recover the true Lennard–Jones result. Yet another refinement of (10.4.2) which ensures that both the potential and its derivatives are continuous at R is

$$\Phi_{MD}(r) = \Phi_{LJ}(r) - \Phi_{LJ}(R) + (R - r)\left(\frac{\partial \Phi_{LJ}(r)}{\partial r}\right)_{r=R}. \qquad (10.4.3)$$

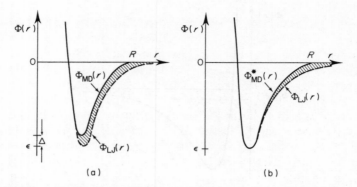

Figure 10.4.2 (a) Discrepancy between the shifted (10.4.2) and true Lennard–Jones interaction. (b) Discrepancy between the shifted and scaled interaction, and the true Lennard–Jones interaction

In the absence of these corrections we may anticipate a consistently 'stunted' estimate of the configurational properties in a molecular dynamics simulation, and a somewhat over-delocalized density profile, the latter being particularly sensitive to the attractive branch of the interaction. Indeed, little attention appears to have been given to the effects of truncation on the profile itself: the lack of attractive force between the molecules in both Monte Carlo and molecular dynamics simulations can only result in a more relaxed transition profile. Other than a qualitative broadening, the consequences for the surface thickness of the transition zone are difficult to anticipate. Miyazaki et al.[27] in their calculations of the surface excess free energy found that they obtained substantially more relaxed profiles for LJ(6:12) potentials truncated at $R = 2.5\sigma$ in comparison with $R = 5.0\sigma$, with similar consequences for the bulk density, being lower in the former case—a source of serious error in the calculation of surface excess functions.

Again, the total energy of the system is generally found to oscillate around a fixed value, whilst the use of a truncated but unshifted pair potential results in a slow monotonic drift.

10.5 Machine simulations of the liquid–vapour interface: structure

There would appear to be no particular advantage in reviewing each of the considerable number of simulations of the liquid surface which have been reported in the literature over the past few years:[8,13–15] all the early simulations may be criticized on one ground or another, although collectively they illustrated many of the specific difficulties which attend the simulation of the liquid surface, even if they did not directly assist in the detailed specification of the interphasal structure. We should also bear in mind that our simulation schemes are representative only of one of the subsystems illustrated in

Figure 10.2.1a, and bear *immediate* comparison with neither theory nor experiment, both of which purport to describe macroscopic assemblies.

Illustrative of the prevailing confusion are the two Monte Carlo simulations of Lee, Barker, and Pound and Abraham, Schreiber, and Barker, the former providing 'convincing evidence'[14] of the existence of pronounced surface layering (~7 layers), whilst the latter authors conclude that there exists 'no layer structure in the liquid region neighbouring the free surface'.[13] A similar reversal of opinion characterizes the two publications of Chapela *et al.*,[2] whilst Liu[15] interprets his rather smooth, if noisy, profile as showing 'a rather striking layered structure'. The consensus of opinion appears to be that any oscillatory features of the transition profile are essentially transient, and are the consequence of inadequate phase sampling—possibly compounded by wall and boundary constraints, or other constraints upon the simulation.

A molecular dynamics simulation of a Lennard–Jones liquid argon film ($\varepsilon_k = 119.4\,^{\circ}K$, $\sigma = 3.405\,\text{Å}$) in equilibrium with its vapour has been reported by Rao and Levesque[6] for systems of 1024 atoms (reduced mean temperature $T^* = 0.744 \equiv 84\,^{\circ}K$; $L = 6.78\sigma$, $h = 56.24\sigma$) shown in Figure 10.5.1. The particles interact through a shifted Lennard–Jones potential of the form (10.4.2), truncated at $R = 2.5\sigma$. The shift ensures continuity of the interaction as discussed in Section 10.4.

A small cubical 'droplet' of a couple of hundred atoms is first equilibrated and after 1600 time steps of 10^{-14} sec. is replicated in the z-direction to form the double-sided film shown in Figure 10.5.1. The usual periodic boundary conditions apply in the x and y directions. The system is allowed to evolve further until the free surfaces have reached equilibrium with the

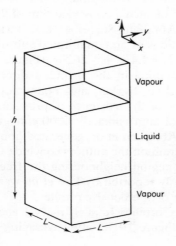

Figure 10.5.1 A Lennard–Jones film in equilibrium with its vapour within a basic cell

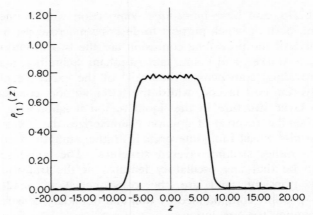

Figure 10.5.2 The density profile across the liquid argon film shown in Figure 10.5.1 after 39 000 molecular dynamics steps

vapour. The cross-sectional density profiles after 39 000 $(3.9 \times 10^{-10}$ sec.) time steps are shown in Figure 10.5.2 for the 1024 particle system $(T^* = 0.744)$. We point out that by virtue of the shift in the pair potential the system may be regarded rather as if it were at $\bar{T}^* \sim 0.756$. Both profiles are symmetrized about $z = 0$ 'to obtain better statistics'.[6] We have commented on this procedure in Section 10.3, which presumed considerable confidence in the fixed location of the centre of the film at $z = 0$ in the course of the simulation; otherwise such a practice can result in a smearing of surface structure, if it is present. Rao and Levesque conclude that there is no significant structure, the r.m.s. fluctuations in the final symmetrized density profile being only 2%—below that of Abraham et al.,[13] and well below that reported by Liu[15] ($\sim$7%) which Rao and Levesque suggest must be a spurious effect attributable to his lower cell boundary.

A Monte Carlo simulation of such a two-sided film of 256 argon atoms at 84 °K has been reported by Abraham, Schreiber, and Barker.[13] The film was assembled from four replications of an equilibrated sub-unit of 64 particles, the cell parameters being $L = 4.4\sigma$, $h = 26.5\sigma$ (Figure 10.5.1). Periodic boundary conditions were applied in the x and y directions. A previous Monte Carlo simulation by Lee, Barker, and Pound for the same system showed 'convincing evidence' of the existence of pronounced surface layering;[14] this was on the basis of approximately 6000 configurations per atom. The more extended run of Abraham et al. generated approximately 24 200 configurations per atom whereupon the authors conclude that there exists no layer structure in the liquid region neighbouring the free surface, and that the surface layering previously reported by Lee et al. was an artifact of the extremely slow convergence of the density profiles.

The single-particle density profiles $\{\rho_{(1)}(x), \rho_{(1)}(y)\}$ and the symmetrized profile $\rho_{(1)}(z)$ are shown in Figure 10.5.3 after averaging over 0.8, 2.6, 4.2,

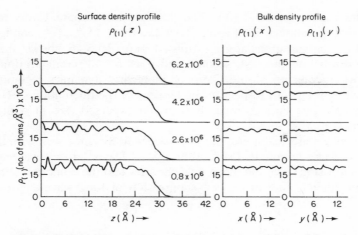

Figure 10.5.3 The single-particle densities $\rho_{(1)}(x)$, $\rho_{(1)}(y)$, and $\rho_{(1)}(z)$ after averaging over 0.8, 2.6, 4.2, and 6.2 million configurations for a two-sided film of 256 argon atoms at 84 °K

and 6.2 million configurations. Although the persistence of apparent layer-like structures is remarkable, Abraham *et al.* conclude that the structure is showing a monotonic decay, and that after ~ 10 million configurations the curve would have been smooth. As we observed earlier, the net drift in the centre of mass of the film $(\sim 1.3\sigma)$ in the course of the simulation yields a meaningless profile: the authors do not state how the cumulative profile was formed—relative to the cell or relative to the centre of mass. No mention of the real possibility of a slow numerical drift of the simulation in the x, y, and z directions is made which would, of course, also yield smooth profiles after 10 million configurations (see Section 10.4), although trial trajectories which cross the median plane are rejected in an attempt to spatially stabilize the simulation.[42]

Similar conclusions are drawn by Chapela *et al.*[2] on the basis of extended runs, having previously noted the development of density oscillations. Both Monte Carlo and molecular dynamic simulations are reported for Lennard–Jones systems of 255 ($L = 5\sigma$), 1020 ($L = 10\sigma$), and 4080 atoms ($L = 20\sigma$). In each case $h = 25\sigma$. The upper and lower boundaries ($z = 25\sigma$ and $z = 0$) are composed of a molecularly homogeneous substance whose density is a bulk vapour ρ_V for $z > 25\sigma$, or a bulk liquid ρ_L for $z < 0$. The interaction was truncated at $R = 2.5\sigma$. No attempt was made to correct for the various features involved in the truncation of Monte Carlo and molecular dynamics interactions discussed in Section 10.4.

In both the Monte Carlo and molecular dynamics runs it was found that although pronounced oscillations developed over the range $z > 5\sigma$, they do not stay fixed in phase but shift in a random manner with respect to the base of the cell, or, more particularly, with respect to the surface. Thus over a sufficiently long run the oscillations vanish and lead to a constant density

272

over the region $5\sigma < z < 10\sigma$. The 255 particle Monte Carlo profile at $T^* = 0.701$ is shown in Figure 10.5.4a after 15.2×10^6 configurations ($\sim 6 \times 10^4$ configurations/atom), whilst Figure 10.5.4b shows the profile broken down into the first, second, and third 5×10^6 configurations. Obviously, whether using Monte Carlo or molecular dynamics techniques, the run must be long with respect to the lifetime of the fluctuation ($\sim 10^{-10}$ sec.) or a sample size large compared with the expected size of a fluctuation (~ 3–4σ).[2]

Neglecting statistical noise, the profiles generally agree well with the mean field and perturbation theories, although as Davis[19] has observed, this is hardly surprising since the mean field theories and the computer simulations exert an implicit and an explicit suppression of long wavelength capillary

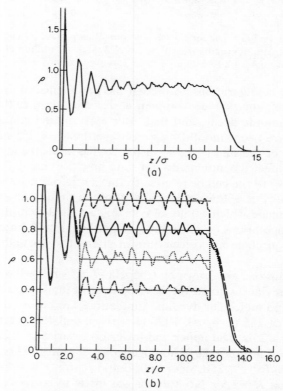

Figure 10.5.4 (a) The Monte Carlo density profile for a Lennard–Jones fluid at $T^* = 0.701$ for a 255 particle system after 15.2×10^6 configurations. (b) The profile after the first, second, and third 5×10^6 configurations. Whilst the profile shows considerable structure over these 'intervals', it appears that the *phase* of the structure varies randomly with respect to the surface

wave fluctuations, respectively (see Section 10.2). However, a simple odd function which arises naturally in the van der Waals theory of surface tension,[20] and which has been used for variational estimates of the surface excess free energy of subcritical fluids, is the tanh density profile

$$\rho_{(1)}(z) = \tfrac{1}{2}(\rho_L + \rho_V) - \tfrac{1}{2}(\rho_L - \rho_V)\tanh\left[2(z - z_0)/d\right] \qquad (10.5.1)$$

where d is a measure of interfacial surface thickness defined as

$$d = -(\rho_L - \rho_V)\left[\frac{d\rho_{(1)}(z)}{dz}\right]^{-1}_{z = z_0}$$

and where $\rho_{(1)}(z_0) = \tfrac{1}{2}(\rho_L + \rho_V)$.

Chapela et al.[2] fit (10.5.1) to their simulated profiles over the range $5\sigma \leqslant z \leqslant 20\sigma$ and find that the surface thickness diverges rapidly as the critical point is approached (Figure 10.5.5). Of particular interest is the apparent increase in thickness as the surface area L^2 is increased, in agreement with Davis' suggestion that this is a direct manifestation of the decreasing suppression of long wavelength fluctuations as L increases; the percentage increase of d with increasing L being $(L = 5\sigma) \xleftrightarrow{19\%} (L = 10\sigma)$ $\xleftrightarrow{4.3\%} (L = 20\sigma)$. It appears that the surface thickness is sensitive to the

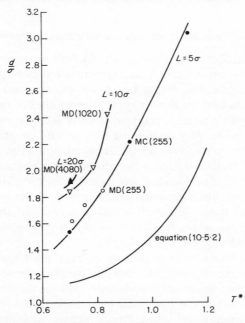

Figure 10.5.5 Divergence of the surface thickness as the critical point is approached for various values of the cell parameter L. Equation (10.5.2) is also shown for comparison

surface area, and reaches its thermodynamic limit in the vicinity of $L = 20\sigma^+$. Generally the interfacial thickness seems to be of the order of 2σ, in qualitative agreement with the estimate of Egelstaff and Widom,[21] also shown in Figure 10.5.4, based on the product of the surface tension γ and bulk isothermal compressibility χ_T (Section 1.4)

$$\gamma\chi_T \sim 0.07l \qquad (10.5.2)$$

where l is the correlation length, expected to be of similar range as the interfacial thickness.

A further comparison of the thickness of the surface may be made with the root of the second moment of the profile $(\mu_2)^{1/2}$ which for argon at 90 °K on the basis of equation (10.5.1) is $(\pi/4\sqrt{3})d$. Taking $d = 2.2\sigma$, $T^* = 0.75$ for their 4048 atom system, Chapela et al. obtain $(\mu_2)^{1/2} = 3.4$ Å, which is within the limits of 3.2–5.0 Å estimated by Lovett et al.[22] (Section 2.11). These are, however, no more than qualitative trends and do not contribute to the resolution of the central question regarding the structure of the profile. The response of the system to an increase in the interfacial area L^2 is, nevertheless, of great importance and clearly illustrates the caution which must be exercised in discerning the facts from the artifacts.

A comparative summary of these results is given in Table 10.3.1a,b.

10.6 Evidence of surface structure: a review

Direct experimental investigation of the surface structure at a molecular level does not yet seem feasible.[43] Such information as we do have is largely indirect and circumstantial, being based for example on the temperature derivative of the surface tension or on scattering studies from solid surfaces just below the melting point. Certainly there is little experimental evidence to support the proposition of a structured liquid surface, at least for the class of Lennard–Jones fluids. We have discussed the effect of low-entropy relatively ordered surface states on the temperature derivative of the surface tension (Section 1.4), and we shall not reiterate it here. But we do observe that there has never been reported an inversion of the $\gamma(T)$ characteristic for Lennard–Jones fluids, although we cannot rule out the possibility that there is nevertheless some reduction in the negative slope of the $\gamma(T)$ curve just beyond the triple point.

The majority of surface simulations have been for Lennard–Jones fluids, and initial indications of possible surface structure have given way to a general but nevertheless tentative consensus that the density profile is in fact represented as a monotonic transition from bulk liquid to bulk vapour, as more extensive sampling of phase space seems to indicate. Perhaps we should rather be asking whether the simulations demonstrate unequivocally the non-existence of surface structure; on the basis of the simulations so far reported, are we entitled to conclude that there is no surface structure? If there were structure at the surface would our extended runs have shown it?

Certainly the profiles are strikingly oscillatory over substantial periods of time ($3\text{--}5 \times 10^{-11}$ sec. or 5×10^6 configurations for a 1020 atom system[2] see Figure 10.5.3b). Such transient structure appears to develop in all simulations, no matter how protracted. The phase of the structure appears to fluctuate randomly,[2] resulting in a smearing of the oscillations to produce a smooth, if noisy, profile. Webb[23] interpreted the existing light scattering and reflection data for interfaces as implying 'that when the interface undergoes its equilibrium vibrations it does so as a membrane that has first been endowed with an intrinsic profile $\rho_{(1)}(z)$.' Whilst such a clear and non-arbitrary decoupling of the long and short wavelength components of the interfacial structure cannot be made, it nonetheless forms the basis of a number of recent capillary wave treatments,[3-5] and leads us to ask what it is we think we are simulating: the intrinsic, or the total profile? Clearly, the large area simulations $L \times L = 400\sigma^2$ are more nearly *total* profiles having a greater surface thickness, and are almost inevitably structureless, particularly if referenced from the cell base or a distant median plane. If the intrinsic profile had been structured would its presence have emerged despite the long wavelength undulations? Perhaps not. Indeed, if the long wavelength capillary amplitudes are $\geqslant 0.5\sigma$, then the structure is inevitably lost—a possibility discussed in Section 10.3. On the other hand, the small area simulations $L \times L = 25\sigma^2$ in suppressing the long wavelength fluctuations are more nearly *intrinsic* profiles. Despite the criticism that the constraint of small area and cyclic boundary conditions enforces an artificial planarity of the surface, it may be that only under these circumstances will any structure at a molecular level become apparent.

The effects of truncation of the pair interaction have already been mentioned (Section 10.4). In both the Monte Carlo and the molecular dynamics cases the lack of attractive force arising from truncation can only result in a more relaxed profile, as Miyazaki *et al.*[27] have found on the basis of Monte Carlo simulations using a LJ(6:12) potential truncated at $R = 2.5\sigma$ and 5.0σ. Unfortunately, their chains were not long enough to determine the smooth profile, but their observation of a substantial relaxation remains. The consequences are far-reaching, affecting the bulk fluid density, for example, in a way which does not arise in a constrained (bulk) simulation, and we have to remember that our interest centres on *excess* functions, which presumes a bulk simulation which is beyond suspicion. Certainly tail corrections for configurational properties can be made, but only on the basis of the suspect profile. Rao and Levesque applied an energy shift to their interaction to ensure continuity of the interaction at $R = 2.5\sigma$ for their molecular dynamic simulation, but the effects of truncation outlined above remain, with the additional feature of an effective temperature shift by a factor of 1.0166, implying further structural relaxation.

Let it be said once again that the object is not to suggest that the Lennard–Jones fluids do possess a structured interface, but whether on the basis of these simulations we can be quite certain they do not, and whether for a system which does have a structured interfacial zone we might not miss it altogether.

10.7 Machine simulations of the liquid–vapour interface: the surface tension

The statistical average which gives the surface tension is, on the basis of the stress tensor, or mechanical definition,[16,24]

$$\gamma_{sim} = \frac{1}{\mathscr{A}} \left\langle \sum_{i \neq j} \sum \frac{x_{ij}^2 - z_{ij}^2}{r_{ij}} \Phi'(r_{ij}) \right\rangle \tag{10.7.1}$$

where $\mathscr{A}$ is the area of the simulation, $\langle \cdots \rangle$ denotes a canonical average, and Φ' is the radial derivative of the pair potential. The above expression has to be corrected for the truncation in the pair potential which, for a Lennard–Jones interaction truncated at R, is[2]

$$\gamma_{tail} = 2\pi\varepsilon \int_{-\infty}^{\infty} \int_{-1}^{1} \int_{R}^{\infty} \rho_{(1)}(z_1)\rho_{(1)}(z_2)(1 - 3s^2)r^{-4} \, dr \, ds \, dz_1 \tag{10.7.2}$$

where $s = (z_1 - z_2)/r$.

By symmetry, of course, the sum in (10.7.1) vanishes in isotropic fluids—the entire contribution to γ coming from the surface region. In the simulations of Chapela et al.[2] contributions arising from $z < 5\sigma$ were excluded on the grounds that the region $0 \leq z \leq 5\sigma$ is structurally unrepresentative of the interfacial region (see Figure 10.5.4). The effect of this is to provide a second surface at $z = 5\sigma$, the surface tension of which is, of course, given exactly by the Kirkwood–Buff formula[26]

$$\gamma^{KBF} = \frac{\pi\rho_L^2}{8} \int_0^{\infty} r^4 \frac{d\Phi(r)}{dr} g_{(2)}^L(r) \, dr \tag{10.7.3}$$

where ρ_L and $g_{(2)}^L(r)$ are the bulk liquid density and radial distribution function, respectively. In fact, (10.7.3) has already been calculated in a Monte Carlo simulation by Freeman and McDonald[25] using a pair potential truncated at R, in which case a tail correction

$$\gamma_{tail}^{KBF} = \tfrac{3}{2}\pi\varepsilon\rho_L^2 R^{-2} \tag{10.7.4}$$

needs to be applied. The true surface tension is then given by[2]

$$\gamma = \gamma_{sim} + \gamma_{tail} - (\gamma^{KBF} - \gamma_{tail}^{KBF}). \tag{10.7.5}$$

For γ_{tail} (10.7.2), Chapela et al.[2] assume a hyperbolic tangent form (10.5.1) for $\rho_{(1)}(z)$; performing the integration over z_1 they obtain

$$\gamma_{tail} = 12\pi\varepsilon(\rho_L^2 - \rho_V^2) \int_0^{1} \int_{R}^{\infty} \tanh(2rs/\sigma)(3s^3 - s) \, dr \, ds. \tag{10.7.6}$$

The γ_{tail} and KBF terms almost cancel exactly in practice for Lennard–Jones systems. Their results for the reduced surface tension $\gamma^* = \gamma\sigma^3/\varepsilon$ are summarized in Figure 10.7.1 from which we see there is a general agreement with Toxvaerd's perturbation calculations based on an assumed tanh profile, although it is appropriate to remark that the good agreement may simply be a result of the smallness of the interfacial area chosen for simulation.[19] Unfortunately Chapela et al. quote no error bounds for their reduced surface tensions. It is known that estimates based on the mechanical stress-tensor definition (10.7.1) fluctuate over a wide range,[27] and the statistical uncertainty in the result, even with runs of reasonable length, is rather high—probably of the order of ±10%. Also shown in Figure 10.7.1 is the experimental $\gamma(T)$ curve for liquid argon plotted in reduced coordinates. As we anticipated on the basis of equation (10.2.1), suppression of long wavelength fluctuations in the course of simulation leads to an overestimate of the surface tension as do the mean field theories, a conclusion implicit in Davis'[5] relationship of these latter treatments to capillary wave representations of liquid interfaces. Moreover, the tail corrections increase the overestimate still further. In this regard, the molecular dynamics result of Rao and Levesque[6] is interesting in that it appears to *under*estimate the experimental result; however, they do not apply the long range tail correction which would increase γ by almost 50%. We observed in Section 10.4 that a simple energy shift to ensure continuity in the pair potential for the purposes of molecular dynamics simulation has roughly the effect of scaling the temperature upwards. The result of Rao and Levesque, then, should be compared

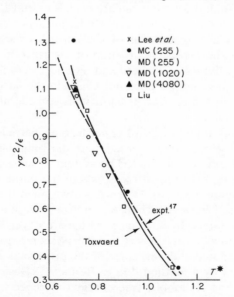

Figure 10.7.1 Comparison of reduced simulated surface tension of a Lennard–Jones fluid with experiment[17]

with the experimental value at $T^* = 0.717$ rather than 0.704 as reported; this observation applies equally to the molecular dynamics results of Chapela *et al.* shown in Figure 10.7.1.

Weeks[4] has determined the correction which needs to be applied to a simulated estimate of the surface tension in going over to a system of macroscopic size, that is, the effect of reintroducing the long wavelength contributions, as follows:

$$\frac{1}{kT}(\gamma - \gamma_{sim}) = -\frac{2 \ln (L/\sigma)}{L^2} + \frac{1}{2L^2} \ln \left(\frac{\beta \gamma \sigma^2}{4\pi(\rho_L - \rho_V)^2 \sigma^6}\right) + \frac{C(G^2)}{2L^2} \tag{10.7.7}$$

where

$$C(G^2) = \frac{1}{(2\pi)^2} \int_{-\pi}^{\pi} dq_x \int_{-\pi}^{\pi} dq_y \ln [4 - \phi(q) + G^2]$$

and where

$$\phi(q) = 2[\cos (q_x) + \cos (q_y)]$$
$$G^2 = 2mgL^2(\rho_L - \rho_V)/\gamma.$$

The value of G^2 near the triple point is extremely small, of the order of 10^{-11} using values appropriate to argon at earth normal gravity, but diverges as $T \to T_c$. Near the triple point the main correction to γ_{sim} comes from the first term in (10.7.7), and for the simulation of Rao and Levesque amounts to a lowering of about 3% of their reported value.

A direct Monte Carlo evaluation of the surface excess Helmholtz free energy required for the reversible creation of a surface in a bulk liquid has been reported by Miyazaki, Barker, and Pound.[27] Unlike earlier Monte Carlo calculations, this method does not evaluate the surface stress, and is capable of estimating the surface tension with considerably less statistical uncertainty.

Starting with a bulk liquid of 216 Lennard–Jones particles whose interaction is truncated at $R = 2.5\sigma$, the usual periodic boundary conditions are applied and the system equilibrated. The system is subsequently separated into two slabs creating two free surfaces, where the parameter Δ represents their separation (Figure 10.7.2b). When $\Delta = 0$ we have the initial reference state (Figure 10.7.2a); however when Δ is nonzero we have two *interacting* slabs of material contained by hard faces which are nevertheless transparent to intermolecular forces. When $\Delta \geq R$ the slabs become non-interacting and the free energy associated with this stage of the process as Δ varies from 0 to R may be calculated by a method due to Bennett.[28] Basically, Bennett has been able to relate the Helmholtz free energy difference in the two systems (a) and (b) in Figure 10.7.2 to the average acceptance ratio of an ordinary Monte Carlo move in combined and separated configurations. The cumulative excess Helmholtz free energy change involved in the separation process

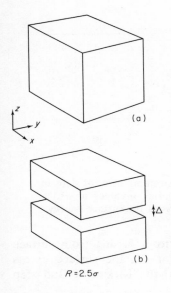

Figure 10.7.2 (a) Initial reference bulk liquid cube of 216 Lennard–Jones particles. (b) Configuration of separated slabs, separation $0 < \Delta < 3\sigma$

as Δ varies from 0 to 3σ is shown in Figure 10.7.3, from which it is seen that no further change occurs for $\Delta > 2.5\sigma$, the truncated range of the interaction.

It remains to determine the free energy difference between the slab constrained by the hard walls and that held together by its own cohesion. This is done by moving the walls out from the slab in symmetrical steps, allowing the system to relax, and determining the pressure at the wall, which is the free energy derivative with respect to the distance between the walls. This quantity is integrated out to a distance such that the density at the wall

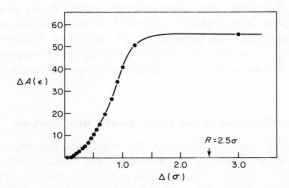

Figure 10.7.3 Increase in surface excess Helmholtz free energy with separation Δ. No further change occurs for $\Delta > 2.5\sigma$, the range of the truncated molecular interaction

280

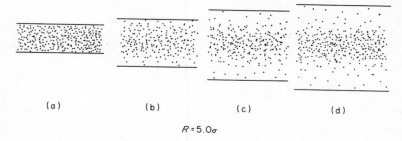

$R = 5.0\sigma$

Figure 10.7.4 (a)–(d) Relaxation of rigid constraining faces of slabs in Figure 10.7.2. This relaxation of the boundaries involves a -13% contribution to the surface tension, and $+36\%$ for the surface energy

is negligible (Figure 10.7.4). The contribution to γ arising from surface relaxation is large, being -13%, and $+36\%$ for U_s, the surface excess energy, these quantities being closely related to the Kirkwood–Buff step model discrepancies.

An important difference in the estimates of U_s arises from the choice of the interaction cut-off distance R. Retaining $R = 2.5\sigma$ throughout, Miyazaki *et al.* obtain $U_s = 54.2$ erg/cm^2 for liquid argon at the triple point, whilst increasing R to 5.0σ before performing the relaxation shown in Figure 10.7.4 yields $U_s = 38.9$ erg/cm^2 showing that the excess energy is extremely sensitive to the shape of the density profile, being somewhat less relaxed under the more attractive conditions applying in the case $R = 5\sigma$. (The effect of truncation on the profile was briefly discussed in Section 10.4.) The difference in the two results cannot be attributed to the long range tail: in both cases full tail corrections were applied.

The final estimates were $\gamma = 18.3 \pm 0.3$ dyn/cm (13.4 dyn/cm) and $U_s = 38.9 \pm 0.8$ erg/cm^2 (34.8 erg/cm^2), the experimental values[29] being given in brackets. The rather poor agreement is attributed to inadequacies of the Lennard–Jones pair potential. The authors do not, however, consider the possibility that suppression of long wavelength contributions is responsible for the overestimate: Weeks' correction is likely to be considerable in the present case.

10.8 Surface modification of the radial distribution function

Much of the difficulty in the description of the liquid–vapour interfacial structure centres on the specification of the anisotropic radial distribution function $g_{(2)}(z_1, \mathbf{r}_{12})$, as we have seen from the preceding chapters. Direct experimental investigation appears incapable of yielding little more than the gross structural features of the transition zone. Machine simulation, then, would appear to have a useful contribution to make, particularly in testing the various model closures against the simulated distributions. Very recently

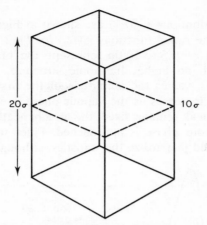

Figure 10.8.1 The fundamental cell

Thompson *et al.*[31] have analysed the interfacial anisotropic radial distribution on the basis of the harmonic expansion

$$g_{(2)}(z_1, \mathbf{r}_{12}) = \sum_{l=0}^{\infty} g_l(z_1, r_{12}) P_l(z_{12}/r_{12}) \tag{10.8.1}$$

where P_l are the unassociated Legendre functions. In the bulk fluid, of course $P_0 = 1$ and $g_0 \equiv g_{(2)}(r)$. At all levels throughout the transition zone we may expect the $l = 0$ contribution to dominate, although the higher harmonics must of course contribute to yield the angular anisotropy.

A double-sided film centred at 10σ in a rectangular prism of height 20σ and cross sectional area $39.69\sigma^2$ containing 432 atoms (Figure 10.8.1) at about 85 °K was followed over 60 000 time steps. The mean distributions were formed from both interfaces, yielding the single-particle profile shown in Figure 10.8.2. The interface was subdivided into slices of thickness 0.5σ;

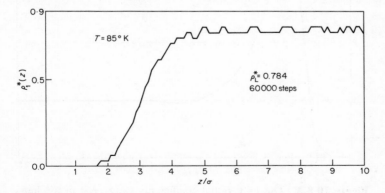

Figure 10.8.2 The single-particle density profile for a Lennard–Jones film of 432 atoms at 85 °K after 60 000 time steps

the zero order distributions $g_{(0)}(r_{12})$ are shown in Figure 10.8.3. In Figure 10.8.4 are shown the radial functions $g_l(r_{12})$ for $l = 2, 3, 4$.

The $g_{(0)}(z_1, r_{12})$ at $z = 9.5$ to 10.0σ represents the 'bulk' radial distribution in its entirety, with no higher harmonic structure. As we rise through successively higher z values the amplitude of the first peak in $g_{(0)}(z_1, \mathbf{r}_{12})$ decreases and then increases as the vapour phase is entered, and although the location of the peak remains fixed, the area beneath the peak appears to increase as the vapour phase is approached. Finer resolution of the slice than $\Delta z = 0.5\sigma$ would jeopardize the statistics, although it must be said that

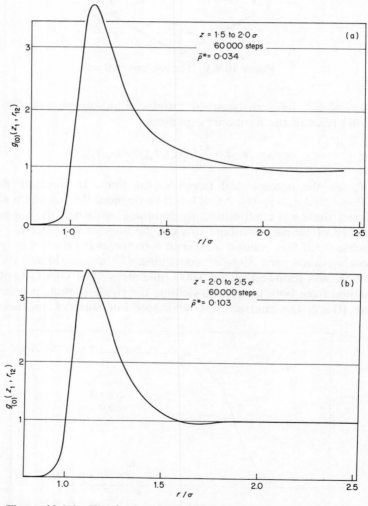

Figure 10.8.3 The $l = 0$ radial coefficients $g_{(0)}(z_1, r_{12})$ in the harmonic expansion (10.8.1) at various depths z_1 within the simulated film

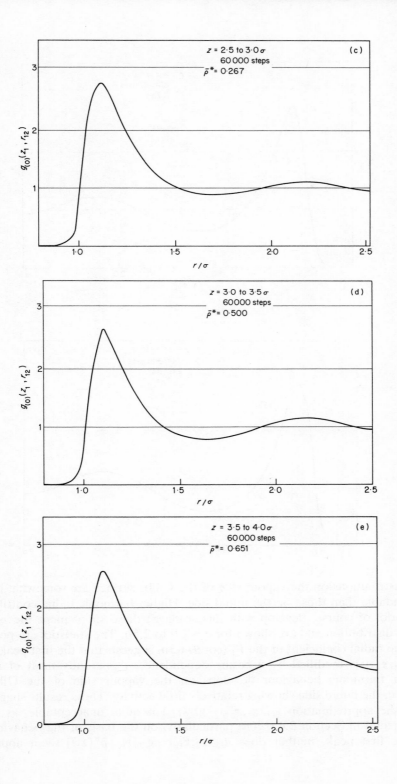

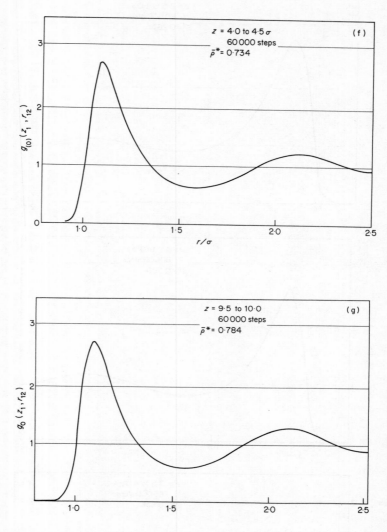

the distributions on the vapour side of the Gibbs surface are somewhat less dependable than those on the liquid side. Higher harmonic radial contributions do, of course, develop with increasing z; these supplement the zero order distribution and are shown for $z = 2.0$ to 2.5σ. The statistics are poor, but the radial coefficient of the $P_1 (\cos \theta)$ term suggests that the first peak of $g_{(0)}(z_1, \mathbf{r}_{12})$ is modified by a cosine dependence. Apparently most of the higher harmonic behaviour develops on the vapour side of the Gibbs surface, the liquid side showing relatively little activity. These results suggest that the approximation $g_{(2)}(z_1, \mathbf{r}_{12}) \sim g_{(2)}^{L}(r_{12})$ is quite unacceptable, as we anticipated in Section 2.4. More particularly, on the basis of the behaviour of the first peak, neither does $g_{(2)}(z_1, \mathbf{r}_{12}) \sim g_{(2)}[r_{12} | \bar{\rho}^*(z_1)]$ seem appropriate.

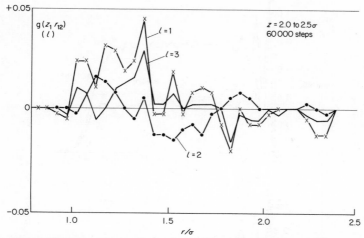

Figure 10.8.4 The $l = 1$, 2, 3 radial coefficients $g_{(l)}(z_1, r_{12})$ in the harmonic expansion (10.8.1) within the layer $2.0\sigma \leqslant z \leqslant 2.5\sigma$ of the simulated film

10.9   Simulation of a binary mixture surface

The only reported simulation of the free surface of an equimolar binary mixture appears to be that of Chapela *et al.*[2] obtained by Monte Carlo simulation (15×10^6 configurations) at $T_a^* = kT/\varepsilon_a = 0.918$, $T_b^* = 0.701$, and by molecular dynamics simulation (8.68×10^4 time steps) at $T_a^* = 0.934 \pm 0.002$, $T_b^* = 0.713 \pm 0.002$, both for 225 Lennard–Jones particles for which $\varepsilon_{aa}/\varepsilon_{bb} = 0.763$ (the ratio of the critical temperatures of argon and krypton). For ε_{ab} the usual geometric mean $(\varepsilon_{aa}\varepsilon_{bb})^{1/2}$ was taken. The diameters σ_{aa}, σ_{bb} and hence σ_{ab}, were taken to be the same, as were the masses.

The component and the total profiles are shown in Figure 10.9.1 for the Monte Carlo and molecular dynamics simulations. There is no evidence of surface structure; the more volatile component is clearly adsorbed from the mixture which extends about one atomic diameter into the gas phase beyond the other species, in qualitative agreement with the calculations of Plesner, Platz, and Christiansen[30] (Section 4.4). We note that the depth of the liquid is insufficient for the bulk equimolar densities to be attained, that is, for $\rho_a = \rho_b$, which prevents an unambiguous estimate of the surface adsorption, although the authors do calculate $(d\gamma/dx_b)$, the variation of reduced surface tension as a function of mole fraction, using Gibbs' adsorption isotherm which does, of course, presume a knowledge of the relative adsorption (Section 4.2).

10.10   Simulation of the surface of molecular fluids

The possibility of the development of surface oriented states was considered in Chapters 3 and 9, for slightly anisotropic molecular systems such as the

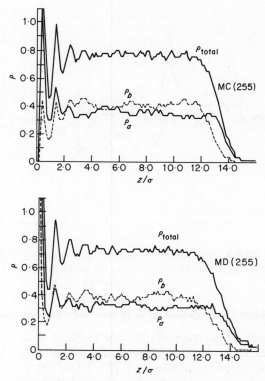

Figure 10.9.1 Density profile $\rho_{(1)}(z)$ for an equimolar binary total mixture at equivalent temperatures (a) MCC (255) $T_a^* = 0.918$, $T_b^* = 0.701$, (b) MD (255) $T_a^* = 0.933$, $T_b^* = 0.713$. The component profiles are shown by broken lines

homonuclear diatomics and for the strongly anisotropic liquid crystals, respectively. The particular case of liquid water as a strongly anisotropic polar system was discussed in some detail in Chapter 7. The general conclusion was that a degree of preferred molecular orientation at the liquid surface was likely to develop as a response to an interfacial torque field associated with the existence of a density gradient. *A priori* considerations led us to conclude that in the absence of pronounced electrostatic multipolar interactions we should expect the molecular axes to lie preferentially parallel to the liquid surface, the degree of orientational order relaxing with increasing temperature. These conclusions would, of course, be modified should strong multipolar interactions arise.

Molecular dynamic computer simulation of two molecular fluids, modelled on nitrogen and chlorine have been reported recently by Thompson[32] at three temperatures from the triple point upwards. Particular attention was paid to the orientation density profiles and the surface tensions of the fluids.

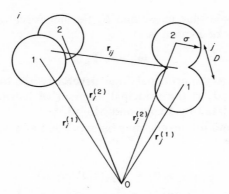

Figure 10.10.1 The site–site interaction geometry

A site–site potential was used[33] (Figure 10.10.1) where each molecule was regarded as a pair of interaction centres located on the atomic nuclei. The site–site interaction was of shifted Lennard–Jones form (see Section 10.4), whereupon the intermolecular potential may be written

$$\Phi(\mathbf{r}_{12}; \mathbf{e}_1, \mathbf{e}_2) = \sum_{i=1}^{2} \sum_{j=1}^{2} \Phi_{MD}(|\mathbf{r}_i^{(1)} - \mathbf{r}_j^{(2)}|) \qquad (10.10.1)$$

where $\mathbf{r}_i^{(k)}$ is the position vector of site i on molecule k, $\mathbf{r}_{12}$ is the centre-to-centre separation, and $\mathbf{e}_k = \{\theta_k \phi_k\}$ are the polar angles specifying the orientation of molecule k. The internuclear separations D (Figure 10.10.1) were taken to be 0.3292σ and 0.608σ for N_2 and Cl_2, respectively. For Φ_{MD} the shifted Lennard–Jones interaction (10.4.3) was used; we refer the reader to the literature for the various parameters.[33]

A double-sided film of 216 molecules was centred in a square rectangular prism of height 19σ for the two lower temperatures and 22σ for the highest temperature. The cell widths were 5.6584σ and 6.216σ for N_2 and Cl_2, respectively, and the usual periodic boundary conditions were applied in the x, y directions.

The initialization of the run is of interest on account of the additional rotational degrees of freedom, and so we describe it in some detail here. The initial translational and angular velocities are chosen randomly such that the correct total kinetic energy, appropriately partitioned, is obtained, and that the centre of mass of the system has no resultant translational or angular velocity. At the lowest temperature the molecular centres of mass are assigned coordinates on an fcc lattice of the appropriate liquid density, with empty vapour phases. The system is allowed to melt over an equilibration period $\sim 10^4$ steps, after which the statistics are allowed to accrue.

The equations of motion are of interest. As usual, the centre of mass satisfies

$$\ddot{\mathbf{r}}_i = m^{-1}\mathbf{F}_i \qquad (10.10.2)$$

where m is the molecular mass, and $\mathbf{F}_i$ the net force on molecule i. The rotational equation of motion is

$$\dot{\boldsymbol{\omega}}_{i_p} = I_p^{-1} \boldsymbol{\Gamma}_{i_p} \tag{10.10.3}$$

where $\boldsymbol{\omega}_{i_p}$ and $\boldsymbol{\Gamma}_{i_p}$ are the principal angular velocity and torque, respectively, on molecule i, and I_p is the principal moment of inertia. Its solution has been discussed in detail by Evans and Murad.[34]

The density-orientation profile may be expanded as a sum of unassociated Legendre functions

$$\rho_{(1)}(z, \theta) = \sum_{l=0}^{\infty} \rho_l(z) P_l(\cos \theta) \tag{10.10.4}$$

whilst the centre of mass density is simply

$$\rho_{(1)}(z) = \int_0^{\pi} \rho_l(z, \theta) \sin \theta \, d\theta$$

$$= \sum_{l=0}^{\infty} \rho_l(z) \frac{2\delta_{l0}}{2l+1} = 2\rho_0(z) \tag{10.10.5}$$

where δ_{l0} is the Kronecker delta. For symmetrical linear molecules $\rho_l(z)$ vanishes for odd l. The coefficients $\rho_l(z)$ may be determined in the course of simulation from the formula

$$\rho_l(z) = \tfrac{1}{4}(2l+1) \sum_{s=1}^{N_s} \sum_i P_l(\cos \theta_i)/(\mathscr{A} \Delta z N_s) \tag{10.10.6}$$

where the sum over s is over a sequence of N_s time steps whilst the sum over i is over those molecules within the stratum $z - \Delta z < z_i < z + \Delta z$, and $\mathscr{A}$ is the surface area. A value of $\Delta z = 0.05\sigma$ was used and the coefficients $l = 0, 2, 4$ were determined.

The surface tension is calculated from

$$\gamma = (N_s \mathscr{A})^{-1} \sum_{s=1}^{N} \sum_{i<j} \frac{r_{ij}^2 - 3z_{ij}^2}{2r_{ij}^2} \sum_{k=1}^{4} \frac{1}{r_k} \frac{\partial \Phi_{LJ}(r_k)}{\partial r_k} \mathbf{r}_{ij} \cdot \mathbf{r}_k. \tag{10.10.7}$$

If the sum is taken over all molecular pairs then the result must be divided by two since there are two free surfaces.

The density profiles $\rho_{(1)}(z)$ for N_2 and Cl_2 are shown in Figure 10.10.2. The structure is not considered significant. The simulated coexisting bulk densities are found to be slightly less than the appropriate experimental values: this, however, may be attributed to the use of a shifted potential. The second density harmonic coefficient $\rho_2(z)$ for Cl_2 at 171.1 °K is shown in Figure 10.10.3 and is seen to be extremely noisy. Nevertheless, Thompson regards the profile as providing evidence for specific molecular orientations within the interfacial zone. To second order, the density-orientation profile

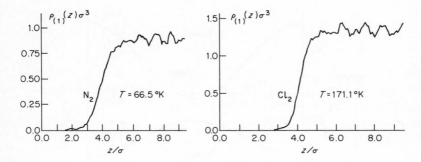

Figure 10.10.2 The simulated density profiles for (a) N_2 (66.5 °K) and (b) Cl_2 (171.1 °K). The structure is not considered significant

is, from equation (10.10.4),

$$\rho_{(1)}(z, \theta) = \rho_0(z) + \rho_2(z)P_2(\cos \theta)$$
$$= \tfrac{1}{2}\rho_{(1)}(z) + \rho_2(z)P_2(\cos \theta) \qquad (10.10.8)$$

the latter following from (10.10.5). The corresponding density orientation profiles at the points starred in Figure 10.10.3 are shown in Figure 10.10.4 for Cl_2 at 171.1 °K. A pronounced tendency for the chlorine molecules to adopt preferred orientations at various heights is apparent. In the liquid phase (0.85σ on the liquid side of the Gibbs surface) at the positive maximum in $\rho_2(z)$ the molecules have a tendency to orient themselves with their axes *perpendicular* to the surface, whilst in the vapour phase (0.05σ on the vapour side of the Gibbs surface) the molecular axes tend to orient

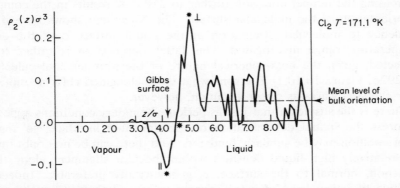

Figure 10.10.3 The second density harmonic coefficient $\rho_2(z_1)$ (equation 10.10.4) for liquid Cl_2 at 171.1 °K. The pronounced z dependence of the coefficient reflects the tendency for the chlorine molecules to adopt preferred orientations in the vicinity of the Gibbs surface: parallel ($\|$) and perpendicular ($\perp$) molecular orientations are indicated accordingly. We note that even in the 'bulk' molecular fluid there is a prevailing level or orientation which may be attributable to the constraint imposed by periodic boundary conditions

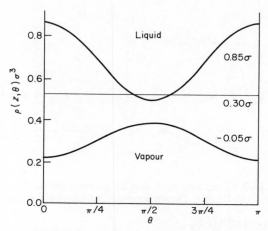

Figure 10.10.4 The density orientation profiles at the points marked * in Figure 10.10.3: $z = 0.85\sigma$, 0.30σ, and -0.05σ relative to the Gibbs dividing surface

themselves *parallel* to the surface: a strange situation indeed—precisely the converse of Parsons' conclusion (Section 9.2), though nevertheless in qualitative agreement with the perturbation treatment of Haile *et al.*[35] using a spherical reference potential with anisotropic overlap forces (Section 3.7).[37] Increasing the temperature to 195.8 °K has the effect of reducing the amplitude of the $\rho_2(z)$ profile, as might be expected. The qualitative features remain the same, although somewhat shifted towards the liquid phase. Increasing the temperature still further to 218.8 °K results in the complete disorientation of the molecular surface. The N_2 system shows little if any tendency to molecular orientation at the liquid surface over the entire temperature range investigated. This latter observation is rather to be expected, given the near-spherical nature of the nitrogen molecule ($D = 0.3292\sigma$, Figure 10.10.1) The somewhat more elongated chlorine molecule ($D = 0.608\sigma$) warrants closer attention, however.

There is the suspicion that just as periodic boundary conditions appear to suppress the spatial delocalization of the simulated interface, so angular disorientation may be similarly hindered. And there can be no doubt that at the relatively high liquid densities within the film alignment along the z direction, normal to the surface, is geometrically preferable. Indeed, a positive (if noisy) level of $\rho_2(z)$ appears to prevail throughout the entire liquid film, indicated by the horizontal broken line in Figure 10.10.3: it should of course be zero. On the vapour side of the Gibbs surface the molecules are less constrained geometrically, and the orientation anticipated by Parsons[36] develops. As in the case of the interfacial thickness, it would be interesting to see to what extent the degree of orientational order depends upon the cross-sectional area of the simulation—the present area is, after

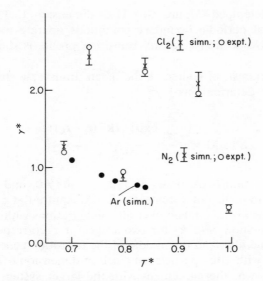

Figure 10.10.5 The reduced surface tensions of Cl_2, N_2, and Ar on the basis of machine simulation and experiment over a range of reduced temperatures

all, only $\sim 6\sigma \times 6\sigma$. Nevertheless, reasonable estimates of the reduced surface tension are obtained, to within rather large error bounds (Figure 10.10.5), although this is an insensitive criterion of the qualitative correctness of the profile as we have seen on several previous occasions. Multipolar effects have yet to be incorporated, however, and these may be expected to modify substantially the conclusions reached so far.

10.11 Interfacial transverse correlations

The development of very long range transverse correlations parallel to the surface within the interfacial zone has been suggested by a number of workers[41,42] (see Section 2.10), the consensus of opinion being that these long wavelength components bear interpretation as surface capillary waves. Whilst a qualitative resolution of the interfacial dynamics into capillary and thermal modes may be made, rather as for diffusive and vibratory modes in the bulk fluid, their interaction is subtle and certainly not a simple release of capillary modes upon an intrinsic (unmodulated) interface.

A direct molecular dynamics investigation of the vertical and horizontal correlations at the surfaces of a liquid argon film has been reported by Kalos *et al.*[7] on the basis of further analysis of the simulation of Rao and Levesque[6] in which a double-sided film of 1728 Lennard–Jones particles at a mean reduced temperature of $\bar{T}^* = 0.704$ (84 °K) is simulated within a rectangular parallelepiped (Figure 10.5.1) of dimension $13.15\sigma \times 13.15\sigma \times 39.45\sigma$. The usual periodic boundary conditions operate in the x and y directions; the cross-sectional density transition profile is shown in Figure 10.5.2.

Of particular interest, of course, is the mean 'transverse structure factor' which Kalos *et al.* determine as

$$\bar{S}^{\parallel}(z, \mathbf{k}) = \frac{\left\langle \sum_{j,k=1}^{N(\Delta z)} \exp\left[-i\mathbf{k} \cdot (\mathbf{r}_j - \mathbf{r}_k)\right] \right\rangle}{N(\Delta z)} \qquad (10.11.1)$$

where $\mathbf{k} = (2\pi n/L, \ 2\pi m/L, 0)$. $m, n = 0, 1, 2, \ldots$; $mn \neq 0$, and j, k run over the $N(\Delta z)$ atoms in a slice of volume $L \times L \times \Delta z$ centred at z. The angular brackets denote the average taken over all configurations within the slice. In Figure 10.11.1 we show $\bar{S}^{\parallel}(z, k)$ for two values of z corresponding to the median plane of the film, and the middle of the transition zone. In the latter case a calculation with 1024 particles in a cell of dimension $6.78\sigma \times 6.78\sigma \times 56.24\sigma$ is also shown: the agreement with the larger system is good, and suggests that the results are generally reliable. The divergence in the low-k region of $\bar{S}^{\parallel}(z, k)$ within the transition zone is clearly apparent, indicating the development of long wavelength cooperative modes up to the dimension of the box. The transverse function cannot, of course, be followed below $k = 2\pi/13.15$, corresponding to the fundamental cell dimension. The low-k divergence appears to be of k^{-2} form, which is more likely than the substantially stronger δ-function singularity at $\mathbf{k} = 0$ predicted on the basis of (2.11.4a), corresponding to perfect interfacial coherence. This latter expression was based on a discontinuous (Heaviside) liquid–vapour density transition: a weaker divergence as $\mathbf{k} \to 0$ is to be expected from a more delocalized (though nevertheless atomically sharp) transition in qualitative agreement with the k^{-2} behaviour observed.

The interface thus appears as a relatively sharp density transition, fluctuating in location under the action of long wavelength $(k \to 0)$ capillary waves (see Section 2.11). It is important to appreciate that the single-particle densities $\rho_V < \rho_{(1)}(z) < \rho_L$ are realized in terms of the statistical fluctuation in the location of a sharp density transition involving, virtually, only the coexisting densities ρ_L, ρ_V, and not distribution and associated functions based on intermediate local, and indeed metastable, densities. In the former case conditions of infinite compressibility will arise, whilst in the latter complex compressibility behaviour, involving negative compressibilities, associated with the van der Waals loop of the equation of state will develop. Evidently the total density profile cannot be formed by simple *ankylosis* (a term borrowed by Poincaré from pathology to denote the 'freezing out' of a degree of freedom) of surface capillary modes upon an intrinsic interface: such a clear resolution of intrinsic and extrinsic modes can, as we have suggested above, be only qualitatively correct. Nevertheless,

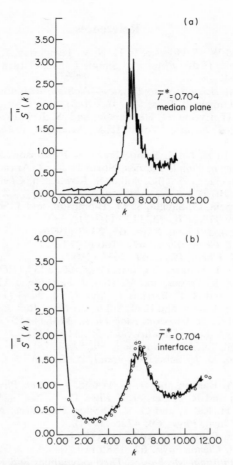

Figure 10.11.1 The mean transverse structure factor $\overline{S}^{\parallel}(z, k)$ taken parallel to the surface, determined by simulation of a double-sided Lennard–Jones film (a) at the median plane, (b) at the interface. The full lines and circles represent a 1728 ($\overline{T}^* = 0.704$) and 1024 ($\overline{T}^* = 0.744$) particle system, respectively

a 'snapshot' of an instantaneous surface configuration (Figure 2.11.3) resolved by the local energy criterion

$$\text{particle } i \text{ in vapour if } E = \sum_{j \neq i} \Phi(\mathbf{r}_i - \mathbf{r}_j) > 0$$
$$\text{particle } i \text{ in liquid if } E \qquad\qquad < 0$$

provides partial qualitative support for such an interpretation as an interfacial mosaic of liquid and vapour domains.

294

References

1. B. J. Alder and W. G. Hoover. In H. N. V. Temperley, J. S. Rowlinson, and G. S. Rushbrooke (Eds), *Physics of Simple Liquids*, North-Holland, Amsterdam (1968), Ch. 4.

   C. A. Croxton, *Liquid State Physics—A Statistical Mechanical Introduction*, Cambridge University Press (1974), Ch. 5.

   F. H. Ree. In H. Eyring, D. Henderson, and W. Jost (Eds), *Physical Chemistry, An Advanced Treatise*, Vol. VIIIA, Academic Press, London/New York (1971), Ch. 3.

   W. W. Wood, In H. N. V. Temperley, J. S. Rowlinson, and G. S. Rushbrooke (Eds), *Physics of Simple Liquids*, North-Holland, Amsterdam (1968), Ch. 5.
2. (a) G. A. Chapela, G. Saville, and J. S. Rowlinson, *Chem. Soc. Faraday Disc. No. 59*, 22 (1975).

   (b) G. A. Chapela, G. Saville, S. M. Thompson, and J. S. Rowlinson, *J. Chem. Soc. Faraday Trans. II*, **73**, 1133 (1977).
3. M. S. Wertheim, *J. Chem. Phys.*, **65,** 2377 (1976).
4. J. D. Weeks, *J. Chem. Phys.*, **67**, 3106 (1977).
5. H. T. Davis, *J. Chem. Phys.*, **67**, 3636 (1977).
6. M. Rao and D. Levesque, *J. Chem. Phys.*, **65**, 3233 (1976).
7. M. H. Kalos, J. K. Percus, and M. Rao, *J. Stat. Phys.*, **17**, 111 (1971).
8. C. A. Croxton and R. P. Ferrier, *J. Phys. C.*, **4**, 2433 (1971).
9. F. P. Buff, R. A. Lovett, and F. H. Stillinger, *Phys. Rev. Lett.*, **15,** 621 (1965).
10. T. Osborn and C. A. Croxton, *Mol. Phys.*, **34,** 841 (1977).
11. N. G. Parsonage, *Disc. Faraday Soc.*, **59,** 51 (1975).
12. J. S. Rowlinson, *Disc. Faraday Soc.*, **59,** 52 (1975).
13. F. F. Abraham, D. E. Schreiber, and J. A. Barker, *J. Chem. Phys.*, **62,** 1958 (1975).
14. J. K. Lee, J. A. Barker, and G. M. Pound, *J. Chem. Phys.*, **60,** 1976 (1974).
15. C. A. Croxton and R. P. Ferrier, *J. Phys. C.*, **4,** 2447 (1971).

    K. S. Liu, M. H. Kalos, and G. V. Chester, *Phys. Rev.* **A10,** 303 (1974).

    K. S. Liu, *J. Chem. Phys.*, **60,** 4226 (1974).

    A. C. L. Opitz, *Phys. Lett.*, **A47,** 439 (1974).

    S. Toxvaerd, *J. Chem. Phys.*, **62,** 1589 (1975).
16. O. K. Rice, *Statistical Mechanics, Thermodynamics and Kinetics*, Freeman, San Francisco (1967).
17. D. Stansfield, *Proc. Phys. Soc.*, **72,** 854 (1967).
18. B. J. Alder and T. E. Wainwright, *J. Chem. Phys.*, **31,** 459 (1959).

    A. Rahman, *Phys. Rev.*, **136,** A405 (1964).

    L. Verlet, *Phys. Rev.*, **159,** 98 (1967).
19. H. T. Davis, *J. Chem. Phys.*, **67,** 3636 (1977).
20. J. W. Cahn and J. E. Hilliard, *J. Chem. Phys.*, **28,** 258 (1958).

    S. Toxvaerd. In J. Singer (Ed.), *Statistical Mechanics*, Specialist Periodical Reports, Chem. Soc., (1975), Vol. 2. Ch. 4, p. 256.

    C. A. Leng, J. S. Rowlinson, and S. M. Thompson, *Proc. Roy. Soc.*, A., **352,** 1 (1976).
21. P. A. Egelstaff and B. Widom, *J. Chem. Phys.*, **53,** 2667 (1970).
22. F. P. Buff and R. A. Lovett, In H. L. Frisch and Z. W. Salsburg (Eds), *Simple Dense Fluids*, Academic Press, New York (1968), Ch. 2.
23. See B. Widom's article on 'Surface Tension of Fluids' in C. Domb and M. S. Green (Eds), *Phase Transitions and Critical Phenomena*, Vol. 2, Academic Press, New York (1972), Ch. 3.
24. F. P. Buff, *Z. Elektrochem.*, **56,** 311 (1952).
25. K. S. C. Freeman and I. R. McDonald, *Mol. Phys.*, **26,** 529 (1973).

295

26. J. G. Kirkwood and F. P. Buff, *J. Chem. Phys.*, **17**, 338 (1949).
    R. H. Fowler, *Proc. Roy. Soc.*, **A159**, 229 (1937).
27. J. Miyazaki, J. A. Barker, and G. M. Pound, *J. Chem. Phys.*, **64**, 3364 (1976).
28. C. H. Bennett, preprint cited in Reference 27.
29. F. B. Sprow and J. M. Prausnitz, *Trans. Faraday Soc.*, **62**, 1097 (1966).
30. I. W. Plesner, O. Platz, and S. E. Christiansen, *J. Chem. Phys.*, **48**, 5364 (1968).
31. S. M. Thompson, J. S. Rowlinson, and G. Saville, private communication (1978).
32. S. M. Thompson, *Faraday Disc. Chem. Soc.*, **66** (1978).
33. J. R. Sweet and W. A. Steele, *J. Chem. Phys.*, **47**, 3029 (1967).
34. D. J. Evans and S. Murad, *Mol. Phys.*, **34**, 327 (1977).
35. J. M. Haile, K. E. Gubbins, and C. G. Gray, *J. Chem. Phys.*, **64**, 1852 (1976).
36. J. D. Parsons, *J. de Phys.*, **37**, 1187 (1976).
37. J. M. Haile, *Ph.D. Thesis*, Gainesville, University, Florida (1976).
38. D. J. Evans, *Mol. Phys.*, **34**, 317 (1977).
    D. J. Evans and S. Murad, *Mol. Phys.*, **34**, 327 (1977).
39. P. S. Y. Cheung and J. G. Powles, *Mol. Phys.*, **30**, 921 (1975).
40. C. W. Gear, *Computational Methods in Ordinary Differential Equations*, Prentice Hall, Englewood Cliffs, New Jersey (1971).
41. M. S. Wertheim, *J. Chem. Phys.*, **65**, 2377 (1976)
    J. B. Weeks, *J. Chem. Phys.*, **67**, 3106 (1977).
42. J-P. Hansen and L. Verlet, *Phys. Rev.*, **184**, 151 (1969).
43. See, however, the recent X-ray grazing incidence experiments of B. C. Lu and S. A. Rice (*J. Chem. Phys.*, **68**, 5558 (1978); Section 5.6).

Appendix 10.1 Expressions for the force and torque for axially symmetric molecules: molecular dynamics simulation

With very few exceptions the so-called Euler angles have been used to specify the molecular orientations. However, as we shall see, the equations of motion expressed in these variables have some severe disadvantages. Such a molecule obeys the translational and rotational equations of motion

$$(\mathbf{F}_1)_{\mathbf{e}_1,\mathbf{e}_i} = m\left(\frac{\partial^2 \mathbf{r}_1}{\partial t^2}\right)_{\mathbf{e}_1,\mathbf{e}_i} = -\frac{\partial}{\partial \mathbf{r}_1} \sum_{j\neq 1} \Phi(\mathbf{r}_{1j};\mathbf{e}_1,\mathbf{e}_j) \qquad (A10.1.1)$$

$$(\mathbf{\Gamma}_1)_{\mathbf{r}_{1j}} = I\left(\frac{\partial^2 \boldsymbol{\theta}_1}{\partial t^2}\right)_{\mathbf{r}_{1j}} = -\frac{\partial}{\partial \boldsymbol{\Omega}_1} \sum_{j=1} \Phi(\mathbf{r}_{1j};\mathbf{e}_1,\mathbf{e}_j) \qquad (A10.1.2)$$

where $(\mathbf{F}_1)_{\mathbf{e}_1,\mathbf{e}_i}$ and $(\mathbf{\Gamma}_1)_{\mathbf{r}_{1j}}$ represent the total force and torque acting on particle 1 due to all the neighbouring molecules. I is the moment of inertia of the molecule, and $\partial/\partial\boldsymbol{\Omega}_1$ represents the angular gradient of the total intermolecular pair potential. (In the case of spherically symmetric particles only the translational response to the total force need be considered.)

Choice of the reference frame proves crucial as indicated above. If, for example, we choose an intermolecular reference frame with its z axis aligned along $\mathbf{r}_{12}$ (Figure A10.1.1) then the orientation dependence of $\mathbf{r}_{12}$ vanishes. Moreover, dependence of the potential upon azimuthal angles φ_1 and φ_2 reduces to the simple difference $\phi_1 - \varphi_2 = \varphi_{12}$. This frame also has the important advantage that the polar angles θ_i may be found from the dot product between a unit vector $\hat{\mathbf{r}}_{ij}$ along $\mathbf{r}_{ij}$ and a unit vector $\hat{\mathbf{h}}_i$ along the

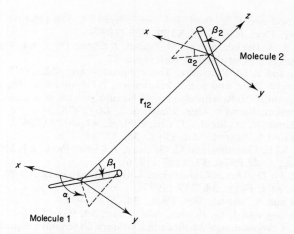

Figure A10.1.1 Geometry for the interaction of
two rod-like molecules

molecular axis

$$\cos \theta_i = \hat{\mathbf{h}}_i \cdot \hat{\mathbf{r}}_{1j}. \tag{A10.1.3}$$

From the law of cosines of spherical trigonometry we have

$$\cos \theta_{ij} = \frac{\cos \gamma - \cos \theta_i \cos \theta_j}{\sin \theta_i \sin \theta_j} \tag{A10.1.4}$$

where γ is the angle between the molecular axes i, j:

$$\cos \gamma = \hat{\mathbf{h}}_i \cdot \hat{\mathbf{h}}_j. \tag{A10.1.5}$$

It is clear, however, that there are computational difficulties in (A10.1.4) in
the vicinity of $\theta_i = 0$. However, equation (A10.1.5) implies that φ_{ij} may be
eliminated in favour of γ, in which case the cosines of θ_i, θ_j, and γ may be
used as the independent variables specifying the molecular orientation,
rather than the angles themselves. So, we need to express the pair potentials
in the form

$$\Phi_{ij} = \Phi(r_{ij}, \cos \theta_i, \cos \theta_j, \cos \gamma)$$
$$\equiv \Phi(r_{ij}, \hat{\mathbf{h}}_i \cdot \mathbf{r}_{ij}, \hat{\mathbf{h}}_j \cdot \mathbf{r}_{ij}, \hat{\mathbf{h}}_i \cdot \hat{\mathbf{h}}_j) \tag{A10.1.6}$$

and we list a number of these in Table (A10.1.1)

The general expressions for the force acting on an axially symmetric
molecule have been given by a number of workers,[38, 39] here we quote the
results of Haile[35]

$$\mathbf{F}_1 = \sum_{j \neq 1} \left\{ -\hat{\mathbf{r}}_{1j} \frac{\partial \Phi_{1j}}{\partial r_{1j}} + \frac{1}{r_{1j}} [\hat{\mathbf{r}}_{1j}(\hat{\mathbf{h}}_1 \cdot \hat{\mathbf{r}}_{1j}) - \hat{\mathbf{h}}_1] \frac{\partial \Phi_{1j}}{\partial \cos \theta_1} \right.$$
$$\left. + \frac{1}{r_{1j}} [\hat{\mathbf{r}}_{1j}(\hat{\mathbf{h}} \cdot \hat{\mathbf{r}}_{1j}) - \hat{\mathbf{h}}_j] \frac{\partial \Phi_{1j}}{\partial \cos \theta_j} \right\}. \tag{A10.1.7}$$

Table A10.1.1 Expressions for anisotropic potential models in the intermolecular frame using γ rather than Φ_{ij}[35]

$$\Phi_{DD}(ij) = \frac{\mu^2}{r_{ij}^3}[c(\gamma) - 3c_i c_j]$$

$$\Phi_{QQ}(ij) = \frac{3Q^2}{4r_{ij}^5}[1 - 5c_i^2 - 5c_j^2 - 15c_i^2 c_j^2 + 2\{c(\gamma) - 5c_i c_j\}^2]$$

$$\Phi_{DQ}(ij) = \frac{3}{2}\frac{\mu Q}{r_{ij}^4}[(c_i - c_j)(1 + 5c_i c_j - 2c(\gamma))]$$

$$\Psi_{over}(ij) = 4\delta\varepsilon\left(\frac{\sigma}{r_{ij}}\right)^6[3c_i^2 + 3c_j^2 - 2]$$

$$\Phi_{dis}(ij) = -2\kappa\varepsilon\left(\frac{\sigma}{r_{ij}}\right)^6[3c_i^2 + 3c_j^2 - 2]$$

$$-\frac{54}{35}\kappa^2\varepsilon\left(\frac{\sigma}{r_{ij}}\right)^6[1 - 5c_i^2 - 5c_j^2 - 15c_i^2 c_j^2 + 2\{c(\gamma) - 5c_i c_j\}^2]$$

D = dipole, Q = quadrupole, over = overlap, dis = dispersion, $c_i = \cos\theta_i$, $c(\gamma) = \cos\gamma$, μ and Q = dipole and quadrupole moment, δ = overlap parameter, κ = anisotropic polarizability.

Clearly, for an isolated pair of molecules $F_1 = -F_2$, as it should.

The general expression for the torque turns out to be

$$\mathbf{\Gamma}_1 = -\hat{\mathbf{h}}_1 \times \sum_{j \neq 1}\left\{\hat{\mathbf{r}}_{1j}\frac{\partial\Phi_{1j}}{\partial\cos\theta_1} + \hat{\mathbf{h}}_j\frac{\partial\Phi_{1j}}{\partial\cos\gamma}\right\}. \tag{A10.1.8}$$

Again, for an isolated pair of molecules equations (A10.1.7) and (A10.1.8) satisfy the condition for conservation of angular momentum:

$$\mathbf{\Gamma}_1 + \mathbf{\Gamma}_2 = -\mathbf{r}_{12} \times \mathbf{F}_1. \tag{A10.1.9}$$

Cheung and Powles[39] have described a method for the solution of the force and torque equations which have the advantage of replacing the second order equations of motion (A10.1.1, A10.1.2) by their first order counterparts in terms of linear and angular velocities. Thus, having determined, say, the torque from (A10.1.8) the angular velocity $\dot{\mathbf{\Omega}}$ may be determined from

$$\mathbf{\Gamma}_1 = I\ddot{\mathbf{\Omega}}. \tag{A10.1.10}$$

There only remains the determination of the molecular orientation $\hat{\mathbf{h}}_1$. This may be readily determined since it is known that the angular velocity, the angular acceleration, and the orientation of the molecular axis are mutually perpendicular. Thus, the unit vector $\hat{\mathbf{h}}_i$ is given by

$$\hat{\mathbf{h}}_i = \frac{(\dot{\mathbf{\Omega}}_i \times \ddot{\mathbf{\Omega}}_i)}{|\dot{\mathbf{\Omega}}_i \times \ddot{\mathbf{\Omega}}_i|}. \tag{A10.1.11}$$

The method used for the solution of the second order translational equation of motion and the first order rotational equation is the predictor–corrector algorithm of Gear.[40] In this algorithm the position $\mathbf{r}_i$ and orientation $\hat{\mathbf{h}}_i$ of each molecule and its first five time derivatives at time $t + \Delta t$ are *predicted* from their values at time t by means of a Taylor expansion:

$$\mathbf{r}_i^P(t+\Delta t) = \mathbf{r}_i(t) + \mathbf{r}_i^{(1)}(t)\Delta t + \frac{\mathbf{r}_i^{(2)}(t)(\Delta t)^2}{2!} + \cdots + \frac{\mathbf{r}_i^{(5)}(t)(\Delta t)^5}{5!}$$

$$\mathbf{r}_i^{P(1)}(t+\Delta t) = \mathbf{r}_i^{(1)}(t) + \mathbf{r}_i^{(2)}(t)\Delta t + \cdots + \frac{\mathbf{r}_i^{(5)}(t)(\Delta t)^4}{4!} \qquad \text{(A10.1.12)}$$

$$\vdots$$

$$\mathbf{r}_i^{P(5)}(t+\Delta t) = \mathbf{r}_i^{P(5)}(t).$$

The superscripts $P(n)$ refer to the order of the predicted time differential. The angular velocity and its first four time derivatives are similarly predicted:

$$\dot{\boldsymbol{\Omega}}_i^P(t+\Delta t) = \dot{\boldsymbol{\Omega}}_i(t) + \dot{\boldsymbol{\Omega}}_i^{(1)}(t)\Delta t + \cdots + \frac{\dot{\boldsymbol{\Omega}}_i^{(4)}(\Delta t)^4}{4!}$$

$$\vdots$$

$$\dot{\boldsymbol{\Omega}}_i^{P(5)}(t+\Delta t) = \dot{\boldsymbol{\Omega}}_i^{P(4)}(t) \qquad \text{(A10.1.13)}$$

and the orientation:

$$\hat{\mathbf{h}}_i^P(t+\Delta t) = \hat{\mathbf{h}}_i(t) + \hat{\mathbf{h}}_i^{(1)}(t)\Delta t + \cdots + \frac{\hat{\mathbf{h}}_i^{(5)}(\Delta t)^5}{5!}$$

$$\vdots$$

$$\hat{\mathbf{h}}_i^{P(5)}(t+\Delta t) = \hat{\mathbf{h}}_i^{(5)}(t). \qquad \text{(A10.1.14)}$$

The force and the torque on the molecule are then evaluated at the predicted positions $\mathbf{r}_i^P$ and $\mathbf{h}_i^P$ from (A10.1.7) and (A10.1.8). The results are then corrected by an amount proportional to the difference between the predicted and evaluated values:

$$\mathbf{r}_i^{(n)}(t+\Delta t) = \mathbf{r}_i^{P(n)}(t+\Delta t) + \alpha_n \Delta \mathbf{r}_i \left[\frac{(\Delta t)^n}{n!}\right]^{-1} \qquad \text{(A10.1.15)}$$

$$\dot{\boldsymbol{\Omega}}_i^{(n)}(t+\Delta t) = \boldsymbol{\Omega}_i^{P(n)}(t+\Delta t) + \beta_n \Delta \dot{\boldsymbol{\Omega}}_i \left[\frac{(\Delta t)^n}{n!}\right]^{-1} \qquad \text{(A10.1.16)}$$

where

$$\Delta \mathbf{r}_i = [\mathbf{r}_i^{(2)}(t+\Delta t) - \mathbf{r}_i^{P(2)}(t+\Delta t)]\frac{(\Delta t)^2}{2!} \qquad \text{(A10.1.17)}$$

$$\Delta \dot{\boldsymbol{\Omega}}_i = [\dot{\boldsymbol{\Omega}}_i^{(1)}(t+\Delta t) - \dot{\boldsymbol{\Omega}}_i^{P(1)}(t+\Delta t)]\Delta t. \qquad \text{(A10.1.18)}$$

The coefficients α_n, β_n are chosen to maintain stability of the solution, and depend upon the order of the differential equation and the degree of Taylor expansion in the predictor step. For the translational motion Gear[40] advocates the corrector coefficients

$$\alpha_0, \ldots, \alpha_5 = 3/16, 251/360, 1, 11/18, 1/6, 1/60$$

whilst for rotation

$$\beta_0, \ldots, \beta_4 = 251/720, 1, 11/12, 1/3, 1/24.$$

Appendix 10.2 Monte Carlo simulation of molecular fluids

Monte Carlo simulations may be performed in any of a number of ensembles, but most frequently the NVT or canonical ensemble is considered. NVT values and initial positions and orientations (in the case of molecular systems) are assigned to the particles, and the form of molecular interaction $\Phi(\mathbf{r}_{12}, \mathbf{e}_1, \mathbf{e}_2)$ chosen. It proves computationally more expedient if the direction cosines of the molecular axes, rather than the Euler angles themselves, are used. The simulation proceeds as follows:

(i) The system energy of the initial configuration is calculated in the pairwise approximation

$$\Phi_N^{(1)} = \sum_{i<j} \sum \Phi^{(1)}(\mathbf{r}_{ij}, \mathbf{e}_i, \mathbf{e}_j). \tag{A10.2.1}$$

The superscript indicates the current configuration.

(ii) One particle having coordinates $(\mathbf{r}_i^{(1)}, \mathbf{e}_i^{(1)})$ is selected either serially or at random, and moved to a new configuration

$$\mathbf{r}_i^{(2)} = \mathbf{r}_i^{(1)} + \boldsymbol{\xi}_k \Delta r \tag{A10.2.2}$$

$$\mathbf{e}_i^{(2)} = \mathbf{e}_i^{(1)} + \boldsymbol{\xi}_{k+1} \Delta e \tag{A10.2.3}$$

where $\boldsymbol{\xi}_k$ is a vector of random components uniformly distributed on $(-1, 1)$. Δr, Δe are the translational and rotational step lengths.

(iii) The new configurational energy is then calculated

$$\Phi_N^{(2)} = \sum_{i<j} \sum \Phi^{(2)}(\mathbf{r}_{ij}, \mathbf{e}_i, \mathbf{e}_j). \tag{A10.2.4}$$

(iv) The probabilities of the two configurations are calculated

$$\begin{aligned} P^{(1)} &= \exp\left(-\Phi_N^{(1)}/kT\right) \\ P^{(2)} &= \exp\left(-\Phi_N^{(2)}/kT\right) \end{aligned} \tag{A10.2.5}$$

and are accepted if $P^{(2)} > P^{(1)}$. If $P^{(2)} < P^{(1)}$ the new configuration is accepted with probability proportional to the Boltzmann factor. This is achieved by generating a random number ξ on $(0, 1)$ and if $P^{(2)}/P^{(1)} > \xi$ accept the new configuration, if $P^{(2)}/P^{(1)} < \xi$ reject it and return to the old configuration which is now counted as the new configuration.

This process is repeated over the entire length of the calculation.

CHAPTER 11

Recent Developments

11.1 Correlation functions in an inhomogeneous fluid

In Sections 2.7 and 2.10 attention was directed at the specification of the pair and direct correlation functions for an inhomogeneous fluid where density, and for that matter, composition, vary with position. Various local density models were proposed each of which was characterized, amongst other things, by the absence of coupling between particle correlation and density gradient, at least within an xy plane parallel to the interface. A recent investigation by Davis and Scriven[1] exploits the density functional techniques of Yang et al.[2] to establish a rigorous expression of the correlation in a density gradient representation.

Cahn–Hilliard descriptions of planar interfaces in terms of density gradient expansions of the interfacial correlations, truncating at second order, are already familiar. These, however, are appropriate only for near-critical systems. For a subcritical fluid we may form a functional Taylor expansion for the direct correlation $c_{(2)}(\mathbf{r}_1, \mathbf{r}_2)$ about the local density $\rho(\mathbf{r}_1)$ as follows:

$$c_{(2)}(\mathbf{r}_1, \mathbf{r}_2) = c_{(2)}^0(\mathbf{r}_1, \mathbf{r}_2)\bigg|_{\rho(\mathbf{r}_1)} + \sum_{k=1}^{\infty} \frac{1}{k!} \int \cdots \int \frac{\delta^k c_{(2)}(\mathbf{r}_1, \mathbf{r}_2)}{\delta\rho(\mathbf{r}_3) \cdots \delta\rho(\mathbf{r}_{k+2})}\bigg|_{\rho(\mathbf{r}_1)}$$

$$\times \prod_{i=3}^{k+2} (\rho(\mathbf{r}_i) - \rho(\mathbf{r}_1))\, d\mathbf{i} \qquad (11.1.1)$$

And since[3]

$$c_{(s)}(\mathbf{r}_1, \ldots, \mathbf{r}_s) = \frac{\delta^{s-2} c_{(2)}(\mathbf{r}_1, \mathbf{r}_2)}{\delta\rho(\mathbf{r}_3) \cdots \delta\rho(\mathbf{r}_s)}, \qquad s > 2$$

equation (11.1.1) becomes

$$c_{(2)}^0(\mathbf{r}_1, \mathbf{r}_2 \mid \rho(\mathbf{r}_1)) + \sum_{k=1}^{\infty} \frac{1}{k!} \int \cdots \int c_{(k+2)}^0(\mathbf{r}_1, \ldots, \mathbf{r}_{k+2} \mid \rho(\mathbf{r}_1)) \prod_{i=3}^{k+2} (\rho(\mathbf{r}_i) - \rho(\mathbf{r}_1))\, d\mathbf{i}$$

where $c_{(k+2)}^0$ represents the $(k+2)$-particle direct correlation in a homogeneous fluid of density $\rho(\mathbf{r}_1)$. Now, Taylor expanding about $\mathbf{r}_1$,

$$\rho(\mathbf{r}_i) - \rho(\mathbf{r}_1) = \mathbf{r}_{i1} \cdot \nabla\rho + \frac{1}{2!} \mathbf{r}_{i1}\mathbf{r}_{i1} : \nabla\nabla\rho + \cdots$$

where $\mathbf{r}_{1i} \equiv \mathbf{r}_i - \mathbf{r}_1$, $\rho = \rho(\mathbf{r}_1)$, $\nabla\rho \equiv \nabla_{\mathbf{r}_1}\rho(\mathbf{r}_1)$ etc. We may form the following gradient approximation from (11.1.1)

$$c(\mathbf{r}_1, \mathbf{r}_2) = c_{(2)}^0(\mathbf{r}_1, \mathbf{r}_2 \mid \rho) + \mathbf{A}(\mathbf{r}_1, \mathbf{r}_2 \mid \rho) \cdot \nabla\rho + \mathbf{B}(\mathbf{r}_1, \mathbf{r}_2 \mid \rho) : \nabla\nabla\rho$$
$$+ \mathbf{D}(\mathbf{r}_1, \mathbf{r}_2 \mid \rho) : \nabla\rho\nabla\rho + O(\nabla^3) \tag{11.1.2}$$

where

$$\mathbf{A} = \int c_{(3)}^0(\mathbf{r}_1, \mathbf{r}_2, \mathbf{r}_3 \mid \rho)\mathbf{r}_{31} \, d3$$

$$\mathbf{B} = \frac{1}{2} \int c_{(3)}^0(\mathbf{r}_1, \mathbf{r}_2, \mathbf{r}_3 \mid \rho)\mathbf{r}_{31}\mathbf{r}_{31} \, d3$$

$$\mathbf{D} = \frac{1}{2} \int c_{(4)}^0(\mathbf{r}_1, \mathbf{r}_2, \mathbf{r}_3, \mathbf{r}_4 \mid \rho)\mathbf{r}_{31}\mathbf{r}_{41} \, d3 \, d4$$

Similarly, the gradient expansion of $g_{(2)}(\mathbf{r}_1, \mathbf{r}_2)$ may be obtained by inserting (11.1.2) in the expanded version of the Ornstein–Zernike equation (2.10.3) yielding

$$g_{(2)}(\mathbf{r}_1, \mathbf{r}_2) = g_{(2)}^{(0)}(\mathbf{r}_1, \mathbf{r}_2 \mid \rho) + \mathbf{a}(\mathbf{r}_1, \mathbf{r}_2 \mid \rho) \cdot \nabla\rho$$
$$+ \mathbf{b}(\mathbf{r}_1, \mathbf{r}_2 \mid \rho) : \nabla\nabla\rho + \mathbf{d}(\mathbf{r}_1, \mathbf{r}_2 \mid \rho) : \nabla\rho\nabla\rho + O(\nabla^3) \tag{11.1.3}$$

where $\mathbf{a}, \mathbf{b}$, and $\mathbf{d}$ may be determined by inserting the gradient expansions for $c_{(2)}$, $g_{(2)}$, and $\rho(\mathbf{r}_3)$ in the Ornstein–Zernike equation (2.10.5) and equating coefficients of common gradients.

More particularly, it is shown that three of the most widely used closures, discussed at length in Section 2.7, *viz.*

$$g_{(2)}(\mathbf{r}_1, \mathbf{r}_2) = g_{(2)}\left[r_{12} \mid \rho\left(\frac{\mathbf{r}_1 + \mathbf{r}_2}{2}\right) \right]$$

$$g_{(2)}(\mathbf{r}_1, \mathbf{r}_2) = g_{(2)}\left[r_{12} \mid \frac{\rho(\mathbf{r}_1) + \rho(\mathbf{r}_2)}{2} \right]$$

$$g_{(2)}(\mathbf{r}_1, \mathbf{r}_2) = \tfrac{1}{2}\{g_{(2)}[r_{12} \mid \rho(\mathbf{r}_1)] + g_{(2)}[r_{12} \mid \rho(\mathbf{r}_2)]\}$$

all yield gradient expansions of the form

$$g_{(2)}(\mathbf{r}_1, \mathbf{r}_2) = g_{(2)}^0[r_{12} \mid \rho(\mathbf{r}_1)] + q^a \mid \mathbf{P}_{12} \cdot \nabla\rho \mid$$
$$+ s^a \mathbf{P}_{12} : \nabla\rho\nabla\rho + t^a \mathbf{P}_{12} : \nabla\nabla_\rho \tag{11.1.4}$$

where $\mathbf{P}_{12}$ is a projection operator onto the unit $\hat{\mathbf{r}}_{12}$ vector joining the two centres (i.e. $\mid \mathbf{P}_{12} \cdot \mathbf{v} \mid = \hat{\mathbf{r}}_{12} \cdot \mathbf{v}$ and $\mathbf{P}_{12} : \mathbf{vu} = (\hat{\mathbf{r}}_{12} \cdot \mathbf{v})(\hat{\mathbf{r}}_{12} \cdot \mathbf{u})$) and q^a, s^a, and t^a are simply scalar functions depending upon interparticle separation and local density at $\mathbf{r}_1$, which is to be contrasted with the exact gradient expansion (11.1.3), written in analogous form,

$$g_{(2)}(\mathbf{r}_1, \mathbf{r}_2) = g_{(2)}^0[r_{12} \mid \rho(\mathbf{r}_1)] + q \mid \mathbf{P}_{12} \cdot \nabla\rho \mid$$
$$+ [s_\parallel \mathbf{P}_{12} + s_\perp(\mathbf{1} - \mathbf{P}_{12})] : \nabla\rho\nabla\rho$$
$$+ [t_\parallel \mathbf{P}_{12} + t_\perp(\mathbf{1} - \mathbf{P}_{12})] : \nabla\nabla\rho \tag{11.1.5}$$

where, as before, q, $s_{\parallel}$, $s_{\perp}$, $t_{\parallel}$, and $t_{\perp}$ are scalar functions of interparticle separations and local density $\rho(\mathbf{r}_1)$. $\mathbf{1}$ is a unit isotropic tensor and $(\mathbf{1}-\mathbf{P}_{12})$ projects onto planes *perpendicular* to $\hat{\mathbf{r}}_{12}$. As Davis and Scriven observe, there are important qualitative and quantitative differences between (11.1.4) and (11.1.5) since $q \neq q^a$, $s_{\parallel} \neq s^a$, and $t_{\parallel} \neq t^a$. Moreover, generally $s_{\perp}$ and $t_{\perp}$ are non-zero.

We see that according to (11.1.4) correlation within a given density stratum is essentially that of an isotropic fluid at the local density $g_{(2)}^0[r_{12} \mid \rho(\mathbf{r}_1)]$ since there are no components of the projection of the density gradient onto the $\hat{\mathbf{r}}_{12}$ vector. In contrast, (11.1.5) shows us that in fact there is coupling to the density gradient for *all* orientations of the $\hat{\mathbf{r}}_{12}$ vector, which means that we are *not* entitled to regard the interfacial zone simply as a stack of thin homogeneous phases. Indeed, the authors go on to demonstrate that even in the low density limit serious discrepancies between the exact and approximate representations of $g_{(2)}(\mathbf{r}_1, \mathbf{r}_2)$ arise for a centrally symmetric pairwise interacting system. Of interest, however, is the fact that the Percus–Yevick approximation

$$g_{(2)}(\mathbf{r}_1, \mathbf{r}_2) = \{1 - \exp[\beta\Phi(r_{12})]\}^{-1} c_{(2)}(\mathbf{r}_1, \mathbf{r}_2)$$

in conjunction with (11.1.1) yields

$$\begin{aligned} g_{(2)}(\mathbf{r}_1, \mathbf{r}_2) = {}& g_{(2)}^0(r_{12} \mid \rho) + \{1 - \exp[\beta\Phi(r_{12})]\}^{-1} \\ & \times [\mathbf{A}(\mathbf{r}_{12} \mid \rho) \cdot \nabla\rho + \mathbf{B}(\mathbf{r}_{12} \mid \rho) : \nabla\nabla\rho \\ & + \mathbf{D}(\mathbf{r}_{12} \mid \rho) : \nabla\rho\nabla\rho + O(\nabla^3)] \end{aligned} \qquad (11.1.6)$$

which is exact in the low-density limit.

Although not explicitly discussed by Davis and Scriven, it does appear that the general form (11.1.3) is not inconsistent with the development of extremely long-range horizontal (surface capillary) correlations proposed by Kalos *et al.*[4] Clearly, long-range correlations, if they are to develop, are confined to the horizontal zone of density inhomogeneity by virtue of the gradient terms. Long wavelength behaviour of the Fourier transformed coefficients $\mathbf{a}(\mathbf{k})$, $\mathbf{b}(\mathbf{k})$ and $\mathbf{d}(\mathbf{k})$ may derive from their common denominator $\{1 - \rho(\mathbf{r}_1)c_{(2)}^0(\mathbf{k})\}^{-1}$, whose singular behaviour in the limit $k \to 0$ is known to characterize cooperative behaviour along the liquid–vapour tie line[9] (Section 10.11).

The nature of the intrinsic density profile in the context of a capillary wave description of the liquid–vapour interface has been discussed recently by Abraham[5] who argues that the 'bare' interfacial thickness in the Buff–Lovett–Stillinger[6] capillary model (Section 2.11) may be identified, within the framework of the Triezenberg–Zwanzig[7] theory, as the interfacial width associated with the 'van der Waal's-like' theories.

However, Abraham places a slightly different interpretation upon the Triezenberg–Zwanzig expressions for the equilibrium density profile $\rho_{(1)}^0(z)$, and applies the additional constraint that the superscript⁰ explicitly excludes any effects arising from capillary fluctuations. Thus the intrinsic density profile,

$$\int \frac{d\rho_{(1)}^0(z_2)}{dz_2} \int c^0(z_1, \mathbf{r}_{12}) \, d\mathbf{r}_2 \, dz_2 = 0 \qquad (11.1.7)$$

and the intrinsic surface tension

$$\gamma^0 = \frac{kT}{4} \int \frac{d\rho_{(1)}^0(z_1)}{dz_1} \frac{d\rho_{(1)}^{(0)}(z_2)}{dz_2} \int c^0(z_1, \mathbf{r}_{12}) r^2 \, d\mathbf{r} \, dz_1 \, dz_2 \qquad (11.1.8)$$

involve $c^0(z_1, \mathbf{r}_{12})$, the direct correlation functional of a non-uniform fluid *with the explicit exclusion of capillary wave contributions*. Clearly, such a functional may be approximated from a knowledge of the uniform phase functional at all densities, and consequently incorporates no features which may be associated with capillary fluctuations. And it is just such an approximation which underlies the van der Waals theories of the liquid–vapour interface.

More revealing is Abraham's comparison of the mean square density fluctuations of the component $\rho(k)$ in the bulk fluid of volume V

$$\langle |\rho_1(\mathbf{k})|^2 \rangle = \frac{k_B T}{V(a + bk^2)}, \qquad a \propto \chi_T^{-1} \qquad (11.1.9a)$$

and at the surface

$$\langle |\rho_1(\mathbf{k}, z)|^2 \rangle = \frac{k_B T}{\mathcal{A}_0 \gamma_0 k^2} \left(\frac{d\rho_{(1)}^0(z)}{dz} \right)^2 \qquad (11.1.9b)$$

$\mathcal{A}_0$ is the undistorted interfacial area, and b is a positive constant of proportionality which may depend upon density and temperature. The two expressions are quite distinct; the apparently divergent behaviour of (11.1.9b) at long wavelengths may be eliminated by a correct incorporation of gravitational effects.

Although single-particle distributions in the vicinity of a constraining wall have been excluded from discussion in this book, a recent comparison of longitudinal and transverse correlations for such a system with those of a free, unconstrained surface has been made by Rao et al.,[8] and appears consistent with prevailing capillary wave ideas of the interface.

It will be recalled from Section 2.11 that Kalos et al.[4] suggested that an intrinsic profile extending laterally over a coherence length may sweep back and forth with a distribution $f(Z)$ for its location (2.11.13a). The primary effect of a wall is to truncate the distribution $f(Z)$, whereupon we have

$$\rho_{(1)}^a(z) = \int_{-\infty}^a f(Z)\rho_{(1)}^0(z - Z) \, dZ \Big/ \int_{-\infty}^a f(Z) \, dZ \qquad (11.1.10)$$

where $\rho_{(1)}^0$ is the intrinsic profile, and $\rho_{(1)}^a(z)$ the density response arising from the insertion of an impenetrable wall at $z = a$. Clearly if the wing of the

304

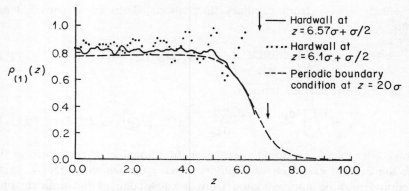

Figure 11.1.1 Resolution of interfacial structure arising from the insertion of a hard wall. The symmetrized density curves show a dramatic transition from monotonic to oscillatory form roughly as the contact density at the wall passes through the bulk fluid value, ρ_L.

distribution $f(Z)$ is sufficiently truncated so that the intrinsic profile is localized, structural features of the profile should be resolved: this is found to be the case in a Monte Carlo simulation (Figure 11.1.1).

To discuss the damping mechanism of the wall, in particular its suppression of transverse correlations associated with the capillary waves, would take us too far afield. Nevertheless, this Monte Carlo study of the density profile against a hard wall provides a severe test of the heuristic capillary wave theory of a free interface outlined in Section 10.11.

Finally, a Monte Carlo simulation of the free surface of liquid argon in equilibrium with its vapour at 110 °K has been reported by Rao and Berne[10] in which the two components $P_\perp(z)$ and $P_\parallel(z)$ of the pressure tensor across the interfacial zone were determined, together with the locations of the surface of tension and the Gibbs surface. The relative positions of these surfaces, together with a knowledge of the surface tension enabled an estimate of the curvature-dependence of the surface tension to be estimated.

The normal and tangential components of the pressure tensor are determined as

$$P_\perp(z) = \rho_{(1)}(z)kT - \frac{1}{2\mathscr{A}} \left\langle \sum_{i \neq j} \frac{|z_{ij}| \, \nabla\Phi(r_{ij})}{|r_{ij}|} \, \theta\!\left(\frac{z - z_i}{z_{ij}}\right) \theta\!\left(\frac{z_j - z}{z_{ij}}\right) \right\rangle \quad (11.1.10a)$$

$$P_\parallel(z) = \rho_{(1)}(z)kT - \frac{1}{4\mathscr{A}} \left\langle \sum_{i \neq j} \frac{[x_{ij}^2 + y_{ij}^2] \, \nabla\Phi(r_{ij})}{|r_{ij}|} \, \frac{\theta\!\left(\frac{z - z_i}{z_{ij}}\right) \theta\!\left(\frac{z_j - z}{z_{ij}}\right)}{|z_{ij}|} \right\rangle \quad (11.1.10b)$$

for a planar interface. $\theta(z)$ is the unit step function. The variation of the components across the interfacial zone are shown in Figure 11.1.2, together with their difference, the area beneath which provides an estimate of the surface tension. $P_\perp(z)$ should, of course, be constant and equal to the bulk system pressure: the scatter gives some idea of statistical noise ($\sim 7\%$).

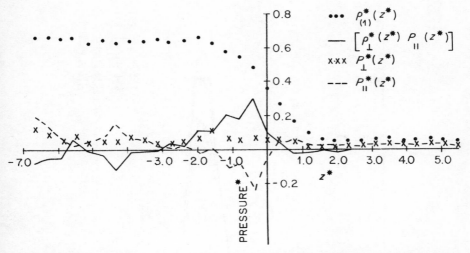

Figure 11.1.2 Variation of the reduced normal and tangential components of the pressure tensor $P_\perp^*(z^*)$ and $P_\parallel^*(z^*)$ across the interfacial transition zone. $P_\perp^*(z^*)$ should, of course, be constant throughout the system—the departure provides a qualitative estimate of the statistical noise. The difference $[P_\perp^*(z^*) - P_\parallel^*(z^*)]$ and the density profile $\rho_{(1)}(z^*)$ are also shown.

Attention is drawn to the general form of the pressure profiles which are seen to differ considerably from the theoretical curves shown in Figure 2.3.2. Of interest is the asymmetric disposition of the difference about the Gibbs surface, an important measure of which is given by the first moment

$$z_S = \frac{1}{\gamma} \int_{-L/2}^{L/2} [P_\perp(z) - P_\parallel(z)]z \, dz \qquad (11.1.11)$$

which defines the surface of tension, z_S. The quantity $(z_S - z_G) = \delta_\infty$ may then be readily determined, and establishes the curvature dependence of the surface tension according to

$$\gamma \sim \gamma_\infty \left(1 - \frac{2\delta_\infty}{r}\right) \qquad (11.1.12)$$

Since Rao and Berne[10] determine $\delta_\infty = +0.96 \pm 0.12$ it is evident that departure from the planar value of the surface tension γ_∞ with drop radius r will be relatively weak.

11.2 The interfacial properties of liquid water

Further to our discussion in Chapter 7 the dipole order parameter $\eta(z)$ across the interfacial zone of liquid water is shown in Figure 11.2.1 for the temperature range 4–200 °C. In all cases we see that the protons are preferentially oriented into the vapour phase, whilst the range of orientational order extends over −50Å (∼20σ), depending upon temperature and

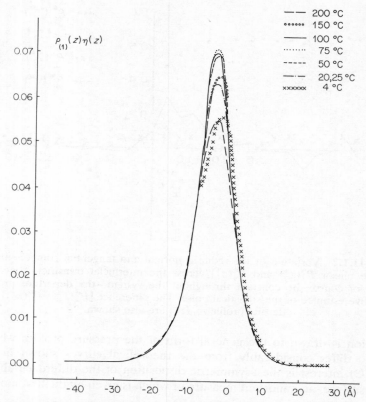

Figure 11.2.1 The dipole order parameter $\eta(z)$ across the interfacial zone of liquid water for the temperature range 4–200 °C. Positive values of the order parameter represent a preferential orientation of the water molecule with the protons towards the vapour phase. In each case the Gibbs surface is located at $z = 0$

density of the system. An estimate of the interfacial thickness L is provided by the surface tension–isothermal compressibility relationship $L \propto \gamma \chi_T$.[11] The thickness appears to show an anomalous minimum in the vicinity of 100 °C, at which L is arbitrarily set to 4σ (~11Å) (Figure 11.2.2). Whilst this aspect of the interfacial structure appears not to have been reported in the literature, it does seem to be directly attributable to the behaviour of the isothermal compressibility.

The components of the surface tension γ_{DD}, γ_{DQ}, γ_{QQ} and γ_{LJ} arising from the dipole–dipole, dipole–quadrupole, quadrupole–quadrupole, and Lennard–Jones interactions, respectively, are tabulated below (Table 11.2.1) for water at 100 °C.

The spatial contributions are shown in Figure 11.2.3 at 100 °C from which we see that the dipole–quadrupole component dominates, whilst the quadrupole–quadrupole component extends significantly deeper into the

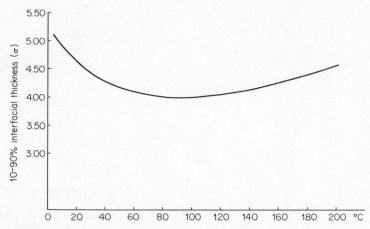

Figure 11.2.2 The interfacial thickness of the water surface as a function of temperature on the basis of the relationship $L \sim \gamma \chi_T$. The surface appears to show an anomalous minimum thickness at $\sim 100\,°C$, at which L is arbitrarily set to 4σ

bulk liquid phase. The surface potential associated with the interfacial polarization is also listed in Table 11.2.1. The observation has been made previously (Section 7.5) and elsewhere[12] that these estimates are significantly greater than other values reported in the literature, and undoubtedly arises from the neglect of free ion contributions present at an equilibrium concentration of $\sim 3 \times 10^{15}\,cm^{-3}$. Clearly the surface potential will be reduced substantially by the presence of free ions, and additional contributions to the interfacial torque field arising from charge–electropole interactions will participate in the free-energy minimization process. The incorporation of ionic contributions at a molecular level presents a formidable problem— substantially more complicated than at the liquid metal surface where the electronic component could at least be treated as a continuum.

In an essentially continuum approach, Fletcher[13] variationally determines the decay of orientational order into the bulk away from a discontinuously

Table 11.2.1. Components of surface tension and surface potential for liquid water at 100 °C

	Surface tension (dyn cm^{-1})
γ_{DD}	2.00
γ_{DQ}	7.84
γ_{QQ}	0.89
γ_{LJ}	22.03
Total	32.76
Experiment	58.91
	Surface potential (volts)
∇V	3.02

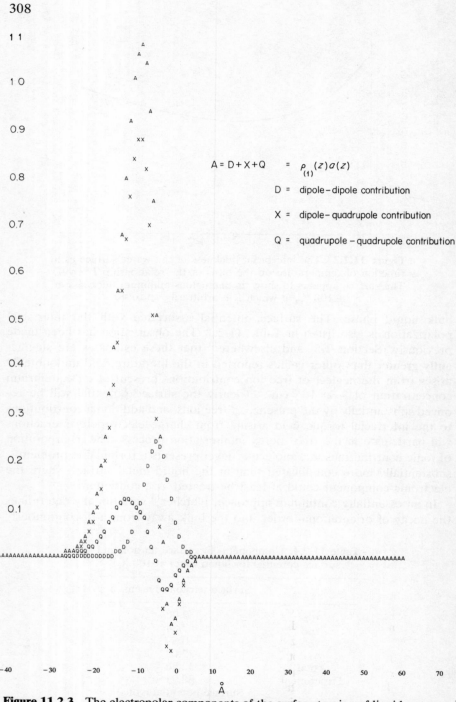

$$A = D + X + Q \quad = \quad \rho_{(1)}(z)\,a(z)$$

D = dipole–dipole contribution

X = dipole–quadrupole contribution

Q = quadrupole–quadrupole contribution

Figure 11.2.3 The electropolar components of the surface tension of liquid water at 100 °C

sharp interface of oriented fraction α_0 related to Croxton's order parameter η roughly as $\eta = 2(\alpha_0 - 0.5)$. He takes into account free energy contributions arising from the interfacial entropy defect associated with an exponential relaxation of orientational order, plus the response of the free ionic distributions to the associated development of an interfacial dipolar field. The dipolar and quadrupolar coupling to the field and its gradient complete the free energy contributions and the sum is minimized with respect to α_0 at a particular temperature. Fletcher finds $\alpha_0 = +0.74$ at 0 °C (roughly equivalent to $\eta \sim 0.48$), with an associated surface potential of $\Delta V \sim -0.1$V. The positive sign of α_0 indicates that the protons point outward, in agreement with Croxton's microscopic analysis but contrary to Stillinger and Ben-Naim's conclusion.[18] These latter authors, however, assume a *positive* quadrupole moment for the water molecule which undoubtedly accounts for their result. Fletcher's value for the oriented fraction of surface molecules is high and would be substantially reduced for a delocalized interface; however, the *range* of orientational correlation extends about 20σ into the bulk liquid in both treatments. Of course, Croxton's analysis neglects the additional contributions to the free energy arising from electropolar interaction with the ionic atmosphere whose associated electric field is

$$E(z) = \frac{4\pi q}{\varepsilon} \int_0^z [\rho^+(z') - \rho^-(z')] \, \mathrm{d}z' \qquad (11.2.1)$$

In the absence of a precise knowledge of the ionic distribution the specific response of the interfacial order parameter $\eta(z)$ and the surface potential is difficult to predict. However, the interaction of the quadrupole moment with the normal gradient of the ionic field (11.2.1) is seen to sum to zero across the transition on the grounds of charge neutrality.

The energetic contribution arising from the dipole–ionic interaction is obviously minimized under conditions of interfacial disorder when this component vanishes entirely, from which we conclude that the presence of free ions leads to a general reduction of the spontaneous interfacial polarization of the water molecules. Clearly, the surface potential will be correspondingly reduced and the surface excess free energy per unit area (surface tension) increased, in general agreement with experimental observation.

The specification of the dielectric constant ε is a central feature of the earlier analyses,[13,18] and the implicit assumption of Croxton in setting $\varepsilon = 1$ throughout warrants some comment. From the outset it has to be recognized that the introduction of a dielectric constant $\varepsilon > 1$ accounts for the effects of molecular and ionic polarizability and is essentially a cooperative phenomenon depending, amongst other things, upon local density. Clearly, then, we anticipate the introduction of such a device would involve the specification of a *local* dielectric constant $\varepsilon_L \geqslant \varepsilon(z) \geqslant \varepsilon_V \sim 1.0$. However, such an approach introduces complicated interfacial effects whereby a polarized molecule or ion interacts with both its own and its neighbour's images. The development of a dipole layer at the water surface together with the existence of free ions of both signs provides a compelling reason for an alternative description.

Such an approach is provided by the polarizable electropole model wherein cooperative effects modify the multipolar moments generally assumed for an *isolated* molecule. Unfortunately, machine simulation estimates reported by Barnes[23] so far extend only to dipolar enhancement; quadrupolar and higher electropolar polarizabilities remain unknown, but are thought to be small. Nevertheless, the introduction of a local, density-dependent dipole moment ranging from $M_L \sim 2.65$ Debye in the bulk liquid to $M_V \sim 1.84$ Debye in the vapour enables partial account to be taken of dielectric effects at the surface. We recall (p. 215) that the requirement that the boundary between two dielectric regions ε_L, $\varepsilon_V(=1.0)$ be an equipotential implies the existence of an image dipole $(\varepsilon_L - 1)/(\varepsilon_L + 1)$ times the original. In terms of the polarizable electropole model, to which the electrostatic image is formally equivalent, cooperative effects diminish as the surface is approached and $M_L \to M_V$. In this essentially molecular approach we only invoke the dielectric permittivity of free space between molecules $(\varepsilon = 1.0)$ and the image dipole $M_L(\varepsilon - 1)/(\varepsilon + 1)$ vanishes. Similar reasoning applies for the interaction of free ions with the free surface (see Section 11.4).

11.3 Landau–de Gennes theory of the nematic liquid crystal surface

As we discussed in Section 9.5, the bulk order parameter η may be expressed phenomenologically by the Landau–de Gennes expression for the free energy, in a slightly modified nomenclature:

$$A(\eta) = A_0 + \tfrac{1}{3}\bar{A}\eta^2 - \tfrac{2}{27}\bar{B}\eta^3 + \tfrac{1}{9}\bar{C}\eta^4 \qquad (11.3.1)$$

neglecting distortions in the η-field. This expression may be variationally minimized with respect to η to yield

$$\frac{\partial A}{\partial \eta} = \tfrac{2}{3}\bar{A}\eta - \tfrac{2}{9}\bar{B}\eta^2 + \tfrac{4}{9}\bar{C}\eta^3 = 0 \qquad (11.3.2)$$

from which we conclude that either $\eta = 0$ (isotropic phase) or $\eta = (\bar{B} \pm \sqrt{\bar{B}^2 - 24\bar{A}\bar{C}})/6\bar{A}$ (nematic phase), where $\bar{A} = a(T - T_{NI})$. $\bar{B}$ and $\bar{C}$ are the coefficients of expansion of the nematic and isotropic phases, respectively. At $T = T_{NI}(\bar{A} = 0)$ a first-order nematic–isotropic phase transition occurs, and the condition $\partial^2 A/\partial \eta^2 = 0$ yields a discontinuous jump $\Delta\eta = \bar{B}/3\bar{C}$ in the order parameter.

Mada and Kobayashi[14,15] have recently discussed the modification of the order parameter in the vicinity of a treated surface and propose a phenomenological relation for the mean surface order parameter $\bar{\eta}$:

$$A(\bar{\eta}) = A_0 - \tfrac{1}{3}\Delta\beta\bar{\eta} + \tfrac{1}{3}\bar{A}\bar{\eta}^2 - \tfrac{2}{27}\bar{B}\bar{\eta}^3 + \tfrac{1}{9}\bar{C}\bar{\eta}^4 \qquad (11.3.3)$$

$\Delta\beta$ represents an anisotropy of anchoring strength, and represents the principal difference between the bulk and surface energies.

The equilibrium state may again be determined by minimizing (11.3.3):

$$\frac{\partial A}{\partial \bar{\eta}} = -\tfrac{1}{3}\Delta\beta + \tfrac{2}{3}\bar{A}\bar{\eta} - \tfrac{2}{9}\bar{B}\bar{\eta}^2 + \tfrac{4}{9}\bar{C}\bar{\eta}^3 = 0; \qquad (11.3.4)$$

clearly now no isotropic state ($\bar{\eta} = 0$) develops at the surface. General solutions of (11.3.4) are

$$\bar{\eta} = \frac{\bar{B}}{6\bar{C}} + \alpha_1^{1/3} + \alpha_2^{1/3} \quad \text{and} \quad \frac{\bar{B}}{6\bar{C}} - \tfrac{1}{2}(\alpha_1^{1/3} + \alpha_2^{1/3}) \pm \frac{\sqrt{-3}}{2}(\alpha_1^{1/3} - \alpha_2^{1/3}) \quad (11.3.5)$$

where we note the second solution is complex. α_1 and α_2 are given by

$$\left. \begin{array}{c} \alpha_1 \\ \alpha_2 \end{array} \right\} = \frac{1}{6^3 \bar{C}^3}(\bar{B}^3 + 81\bar{C}^2\Delta\beta - 27\bar{A}\bar{B}\bar{C} \pm 9\bar{C}\sqrt{D})$$

where D is the discriminant of (11.3.4);

$$D = 72\bar{A}^3\bar{C} - 3\bar{A}^2\bar{B}^2 + 81\bar{C}^2\Delta\beta^2 - 54\bar{A}\bar{B}\bar{C}\Delta\beta + 2\bar{B}^3\Delta\beta$$

The equilibrium states must, of course, be real solutions, and if $D = 0$, $\alpha_1 = \alpha_2$ giving

$$\bar{\eta} = \frac{\bar{B}}{6\bar{C}} + 2\gamma^{1/3} \quad \text{or} \quad \frac{\bar{B}}{6\bar{C}} - \gamma^{1/3}$$

where

$$\gamma = \frac{\bar{B}}{6^3\bar{C}^3}(\bar{B}^3 + 81\bar{C}^2\Delta\beta - 27\bar{A}\bar{B}\bar{C})$$

whilst if $D > 0$

$$\bar{\eta} = \frac{\bar{B}}{6\bar{C}} + \alpha_1^{1/3} + \alpha_2^{1/3}.$$

Since the discriminant varies with $\Delta\beta$ and temperature, Mada and Koybayashi are able to determine the temperature dependence of the order parameter for various values of $\Delta\beta$ (Figure 11.3.1). For $\Delta\beta > 0.03\bar{B}^3/\bar{C}^2$, $\bar{\eta}$ varies continuously across the transition point, whilst for $0 \leqslant \Delta\beta < 0.03\bar{B}^3/\bar{C}^2$ the transition is first order, and the discontinuity is of magnitude $(\bar{B}^3 + 81\bar{C}^2\Delta\beta - 27\bar{A}\bar{B}\bar{C})^{1/3}/2\bar{C}$, occurring at a temperature $T_C = T_{NI} + (\bar{B}^3 + 81\bar{C}^2\Delta\beta)/27\bar{A}\bar{B}\bar{C}$, increasing linearly with anchoring, $\Delta\beta$.

The bulk and mean surface order parameters were measured by Mada and Kobayashi for 4-n-heptyl-4'-cyanobiphenyl (7CB) by an interferometric method;[16] the results are shown in Figure 11.3.2. The optical cell containing the liquid crystal was 24 μm thick, and the determination was made at $\lambda = 533$ nm. The existence of an enhanced surface order parameter with

312

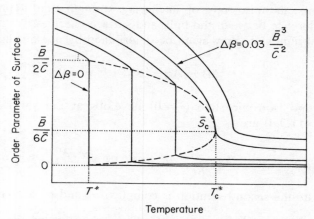

Figure 11.3.1 Temperature dependence of the order parameter for various values of the surface anchoring field, $\Delta\beta$. The order parameter exhibits a first-order phase transition for subcritical values of the anchoring field $(\Delta\beta \leqslant 0.03 B^3/C^2)$. $\Delta\beta = 0$ corresponds to a bulk liquid nematic system

respect to the bulk at all temperatures, and a residuum of interfacial order at $T > T_{NI}$ are clearly demonstrated, and provides strong support for the earlier propositions of Croxton and Chandrasekhar[17] (Section 9.4.) who advocated the existence of a local order parameter $\eta(z)$ enhanced at the surface by the development of interfacial torque fields—a role assumed by the parameter $\Delta\beta$ in this analysis. Of course, Mada and Kobayashi's development is not complicated by the density variations which arise at a free surface, and so their treatment is incomplete in that sense. Indeed, they restrict themselves to the discussion of a *mean* surface order parameter rather than a local

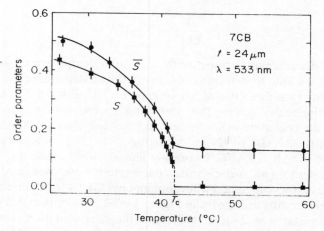

Figure 11.3.2 The bulk and mean surface order parameters for 4-n-heptyl-4′-cyanobiphenyl (7CB)

point function $\eta(z)$. Nevertheless, encouraging qualitative agreement between theory and experiment has been obtained.

A very recent development by Croxton[34] expresses $\Delta\beta$ in terms of the local density and the square of its gradient (Section 9.2). It is found that qualitative descriptions of the pre- and post-transitional contributions to the orientational component of the surface tension are obtained which are in good agreement with experiment. In all cases it is found that the surface is characterized by a *critical divergence of its interfacial thickness* in the vicinity of the nematic-isotropic transition. This effect appears not to have been previously reported and would seem essential in any adequate description of the liquid crystal surface.

11.4 Interfacial properties of molten salts and ionic solutions

Calculation of the bulk and interfacial properties of ionic solutions and molten salts is complicated by the existence of long-range Coulombic interactions and the requirement of local charge neutrality which characterizes the system. Debye–Hückel screening appropriate to a completely ionized gas has to be substantially modified to incorporate the essentially geometrical features of the ionic cores in any statistical specification of local charge neutrality, and such a condition provides an additional constraint upon the bulk and interfacial distributions.

Goodisman and Pastor[19] consider a system of two oppositely charged species i, j interacting through a long-range Coulombic and short-range hard-sphere potential:

$$\Phi(r) = \frac{q_i q_j}{r_{ij}} + \Phi_{HS}(r)$$

where

$$\Phi(r) = +\infty \qquad r \leq \sigma$$
$$= 0 \qquad r > \sigma \qquad (11.4.1)$$

Both species are assumed to have the same ionic diameter σ, and the bulk distributions $g_{(2)}^{++}(r) = g_{(2)}^{--}(r)$ and $g_{(2)}^{+-}(r) = g_{(2)}^{-+}(r)$ are determined in the mean spherical approximation. An initial attempt to predict the interfacial properties of NaCl at 1128 °K on the basis of the Kirkwood–Buff–Stillinger model which, of course, retains the bulk distributions to a surface of density discontinuity, led to disastrous results (Table 11.4.1).

Table 11.4.1. Calculated surface properties of NaCl ($T = 1128$ °K)

	γ (dyn cm^{-1})	U_s (erg cm^2)
Experiment[20]	111.3	216.2
KBF model[21]	−79.58	907.52
$f(z_1)(g_{(2)}^{+-}(r_{12}) - g_{(2)}^{++}(r_{12}))^{21}$	42.37	111.87
$f(u)(g_{(2)}^{+-}(r_{12}) - g_{(2)}^{++}(r_{12}))^{21}$	99.36	288.08

Rough qualitative agreement with experiment is generally achieved by the KBF model for simple systems in the vicinity of the triple point, and it does appear that the problem may be largely attributed to a violation of the local charge neutrality condition in the vicinity of the interface. Local charge neutrality in the bulk is ensured by the condition

$$\frac{2}{\rho} \int [\rho_{(2)}^{++}(z_1, \mathbf{r}_{12}) - \rho_{(2)}^{+-}(z_1, \mathbf{r}_{12})]\, d2 = 1 \tag{11.4.2}$$

in an obvious notation. Clearly, if the bulk isotropic distributions are retained in place of the correct interfacial functions $\rho_{(2)}^{++}(z_1, \mathbf{r}_{12})$, $\rho_{(2)}^{+-}(z_1, \mathbf{r}_{12})$ at the surface and the KBF step discontinuity is retained at $z = 0$, the region of integration in (11.4.2) is truncated by the interfacial plane and charge neutrality is violated. Indeed, the right-hand-side of (11.4.2) becomes

$$-4\pi\rho\left[\frac{1}{2}\int_0^\infty (g_{(2)}^{+-}(r_{12}) - g_{(2)}^{++}(r_{12}))r^2\, dr + \frac{1}{4}\int_{|z_1|}^\infty (g_{(2)}^{+-}(r_{12}))(-1 + |z_1|\, r^{-1})r^2\, dr\right]$$

$$\tag{11.4.3}$$

which is a function of $|z_1|$; the electroneutrality condition (11.4.2) is recovered when $|z_1|$ is large.

If now we replace $(g_{(2)}^{+-}(r_{12}) - g_{(2)}^{++}(r_{12}))$ in (11.4.3) by $f(z_1)(g_{(2)}^{+-}(r_{12}) - g_{(2)}^{++}(r_{12}))$, where $f(z_1)$ is the *inverse* of (11.4.3), then of course, we recover the electroneutrality condition for all $z_1 < 0$. Goodisman and Pastor[21] find a substantial improvement in the predicted values of the interfacial functions (Table 11.4.1) although they note that whilst correcting for electroneutrality, this latter modification is not symmetrical with respect to particle exchange, and they further propose

$$[f(z_1)f(z_2)]^{1/2}(g_{(2)}^{+-}(r_{12}) - g_{(2)}^{++}(r_{12}) \quad \text{and}$$

$$f(u)(g_{(2)}^{+-}(r_{12}) - g_{(2)}^{++}(r_{12})) \quad \text{where} \quad u = \tfrac{1}{2}(z_1 + z_2) \tag{11.4.4}$$

both of which satisfy the symmetry requirement, although the latter appears better at ensuring overall electroneutrality, particularly in the immediate vicinity of the surface. In each of these expressions the functional form of $[f(z)]^{-1}$ has been given by equation (11.4.3) which is in itself only approximate. Goodisman[21] subsequently sets up and variationally solves equations specifying $f(z)$ and the electroneutrality, and notes only marginal improvements in the calculated functions.

It is appropriate to observe that the usual inequalities appropriate to the KBF model $\gamma_{KBF} > \gamma_{(expt)}$, $U_{s\,KBF} < U_{s(expt)}$ appear not to hold (Chapter 2, p. 29), suggesting that the simple adoption of a more realistic profile may provide no more than a partial improvement, and that the model remains fundamentally incomplete.

The model itself amounts to a specification of an interfacial two-particle distribution function which satisfies certain symmetry and electroneutrality conditions, which Pastor and Goodisman[21] then invert to yield the density profile through the Born–Green–Yvon equation:

$$\frac{d\rho^i_{(1)}(z_1)}{dz_1} = \frac{1}{kT} \sum_j \nabla\Phi_{ij}(r_{12})\frac{z_{12}}{r_{12}} \rho^{ij}_{(2)}(z_1, \mathbf{r}_{12})\, d\mathbf{2} \qquad (11.4.5)$$

where i, j refer to the cation and anion species. Quite inconsistent with the initial KBF assumptions, these authors predict an *oscillatory* profile regardless as to whether the electroneutrality condition is incorporated, and is undoubtedly an artifact of the approximations.

Throughout these calculations the dielectric constant, which introduces the effects of ion–ion polarizability, is set at unity. Whilst this affords considerable simplification in the analysis, the implications should be carefully considered. Adoption of a dielectric constant $\varepsilon > 1$ simply divides the electrostatic interaction by ε, and values of 2 or 3 have been suggested which appear to provide a slight improvement in the unmodified KBF results. However, a number of important image corrections have to be incorporated if the usual boundary conditions are to be satisfied at the interfacial boundary between the two dielectric regions. Buff and Stillinger[22] have shown that the interfacial ions interact with both their own and their neighbours' images. The self-image interaction amounts to the usual one-particle potential

$$\Phi(z) = \frac{\varepsilon_L - 1}{\varepsilon_L + 1}\frac{q^2}{4\varepsilon_L z}, \qquad z < 0 \qquad (11.4.6)$$

Clearly, $\Phi(z)$ becomes infinite as an ion approaches the surface and accounts for the decrease in ionic density in an ionic solution as $z \to 0$. This in itself suggests that the dielectric constant itself tends to unity as the surface is approached, requiring a self-consistent recalculation of the image potential. Of course, the introduction of a realistic density profile further complicates the dielectric specification and electrostatic image representation of the surface; the motivation for retaining $\varepsilon = 1.0$ throughout the analysis is understandable. However, it may well prove more realistic, and indeed simpler, to recognize that ionic and solvent polarization will develop and modify as the surface is approached in much the same way as for the dipole moment of the water molecule (Section 11.2). Under these circumstances ε may legitimately be set equal to unity, appropriate for the dielectric permittivity of free space, throughout the entire system.

11.5 Decoupling of translational and rotational degrees of freedom for strongly anisotropic molecules

In our earlier consideration of anisotropic molecular systems, mean field analyses and their perturbative expansions were, *faute de mieux*, adopted in an attempt to decouple the molecular interactions and their associated spatial correlations for the purposes of integration. In these systems the interaction is generally of the form $\Phi(\mathbf{r}_{12}; \mathbf{e}_1, \mathbf{e}_2)$, depending explicitly upon the separation of the centres $\mathbf{r}_{12}$ and their orientations $\mathbf{e}_1, \mathbf{e}_2$. The problem is particularly severe in the case of liquid crystals for which no convenient reference interaction exists about which a perturbative expansion may be made.

Parsons[24] has shown that for a system of anisotropic molecules interacting with a pair potential of the form $\Phi(r/\sigma)$, and a pair correlation $g_{(2)}$ which scales as $g_{(2)}(r/\sigma)$, where r is the centre-of-mass distance and σ is an angle-dependent collision diameter, then the translational and orientational degrees of freedom decouple to all orders of density.

Parsons considers an interaction of the form $\Phi(\mathbf{r}_{12}; \mathbf{e}_1, \mathbf{e}_2) = \Phi(r/\sigma)$ where the collision diameter $\sigma = \sigma(\hat{\mathbf{r}}; \mathbf{e}_1, \mathbf{e}_2)$ and $\hat{\mathbf{r}}$ is a unit vector along $\mathbf{r}_{12}$. For fixed orientation of the two molecules $(\mathbf{e}_1, \mathbf{e}_2)$ and orientation of the $\hat{\mathbf{r}}$ vector it is clear that a unique collision diameter is specified along the vector joining the two centres of mass. Parsons proposes that the pair interaction is, say, of Lennard–Jones form scaled by the appropriate collision diameter $\sigma(\hat{\mathbf{r}}; \mathbf{e}_1, \mathbf{e}_2)$. The decoupling approximation amounts to setting $g_{(2)}(\hat{\mathbf{r}}_{12}; \mathbf{e}_1, \mathbf{e}_2)$ which is exact at low density since then $g_{(2)} \sim \exp(-\Phi/kT)$, but cannot of course be exact at higher density.

Holding $(\mathbf{e}_1, \mathbf{e}_2)$ *fixed*, we see that an integral over all locations of the centre of mass of the second particle about the first may be decoupled as follows:

$$\int \frac{\partial \Phi}{\partial r}(r) g_{(2)}(r/\sigma) r^n \, d\mathbf{r} = \int_0^\infty y^{n+2} \frac{\partial \Phi(y)}{\partial y} g(y) \, dy \int \sigma^{n+2}(\hat{\mathbf{r}}; \mathbf{e}_1, \mathbf{e}_2) \, d\hat{\mathbf{r}} \quad (11.5.1)$$

where $y = r/\sigma$. Subsequently integration over $(\mathbf{e}_1, \mathbf{e}_2)$ may be performed.

Of course it remains to specify $\sigma(\hat{\mathbf{r}}; \mathbf{e}_1, \mathbf{e}_2)$; for spheroidal molecules Berne and Pechukas[25] have suggested:

$$\sigma^2(\hat{\mathbf{r}}; \mathbf{e}_1, \mathbf{e}_2) = \sigma_0^2 \left(1 - \frac{\chi(\hat{\mathbf{r}} \cdot \mathbf{e}_1)^2 + (\hat{\mathbf{r}} \cdot \mathbf{e}_2)^2 - 2(\hat{\mathbf{r}} \cdot \mathbf{e}_1)(\hat{\mathbf{r}} \cdot \mathbf{e}_2)(\mathbf{e}_1 \cdot \mathbf{e}_2)}{1 - \chi^2(\mathbf{e}_1 \cdot \mathbf{e}_2)^2} \right)^{-1} \quad (11.5.2)$$

where $\mathbf{e}_1$ and $\mathbf{e}_2$ are unit vectors along the symmetry axes of the two interacting spheroids. For a prolate spheroid of long and short axes $2b$ and $2a$, respectively

$$\chi = (k^2 - 1)/(k^2 + 1), \qquad k = b/a \geqslant 1$$

and $\sigma_0 = 2a$.

We should point out that for a bulk isotropic fluid only the *relative* orientation of the two particles matters. In a region of density anisotropy, however, such as at the liquid surface, interfacial torque fields will establish preferred orientations for the molecular axes and an explicit expression will be necessary for the execution of subsequent integrals over the interfacial distributions $f(z, \mathbf{e})$.

11.6 The Cahn–Hilliard/van der Waals interface

For a one-dimensional density gradient the local free-energy density may be written as a series of derivatives of all orders, the range and convergence of which remains uncertain:[2,26]

$$a(z_1) = a^\dagger[\rho_{(1)}(z_1)] + A\left(\frac{\partial\rho_{(1)}(z_1)}{\partial z_1}\right)^2 + \cdots; \qquad A > 0 \qquad (11.6.1)$$

Truncation of this series at the second term yields the familiar 'squared-gradient' representation of van der Waals, subsequently reconsidered by Cahn and Hilliard,[27] and discussed extensively in Section 2.4. More recent considerations by Abraham,[28] Telo da Gama and Evans,[29] and Croxton[30] have provided a statistical mechanical expression of (11.6.1) in terms of microscopic quantities (Section 2.4) from which it is apparent that the coefficient A becomes functionally independent of the local density only as $T \to T_c$, when we recover van der Waals' initial assumption. Croxton[30] concludes that the range of applicability of the van der Waals representation extends substantially below the region of incipient criticality, characterized by small density gradients, as (11.6.1) might imply. Indeed, there appears a growing use of (11.6.1) in subcritical liquid regions which *a priori* considerations would suggest to be quite inappropriate. Clearly, some assessment of the approximation throughout the liquid phase would be of value.

A comparison of the van der Waals and mean field approximations through the temperature range $0 \leqslant T \leqslant T_c$ has recently been made by Harrington and Rowlinson[31] on the basis of the penetrable sphere model. The mean field result is known to be correct at zero temperature,[31] and coincides with results based on the van der Waals' assumption (11.6.1) in the critical limit. Consequently, the mean field treatment provides a subcritical locus against which an assessment of the van der Waals' approximation may be made. Unfortunately, for reasons of tractability, analysis is necessarily restricted to the somewhat pathological system of penetrable line segments confined to the z-axes. For this case the mean field density profile, which holds at all temperatures, is known to be of tanh form:

$$\rho_{(1)}(z_1) = \tfrac{1}{2}(\rho_L + \rho_V) + \Delta \tanh\left(\tfrac{1}{4}\Delta(z - z_G)/z_G\right) \qquad (11.6.2)$$

$\Delta = \tfrac{1}{2}(\rho_L - \rho_V)$ and z_G locates the Gibbs surface. The mean field surface tension for such a system of penetrable line segments is

$$\gamma_{MF} = \frac{2kT}{\varepsilon} \int_0^\Delta (1 - \Delta'^2 \operatorname{cosech}^2 \Delta') \, d\Delta' \qquad (11.6.3)$$

the temperature dependence of which is shown in Figure 11.6.1. It should be emphasized that the points $\gamma_{MF}(T=0)$ and $\gamma_{MF}(T=T_c)$ are exact for this system.

Now, since the surface tension is the surface excess free energy determined with respect to the Gibbs surface, we have

$$\gamma_{VDW} = \int_{-\infty}^{z_G} [a(z) - a(\rho_L)] \, dz - \int_{z_G}^\infty [a(z) - a(\rho_V)] \, dz$$

$$= \int_{-\infty}^\infty [a(z) - \bar{a}] \, dz \qquad (11.6.4)$$

where $\bar{a}$ is the mean free energy of a two-phase liquid–vapour system having an overall density $\rho_{(1)}(z)$. The adoption of a profile which minimizes γ_{VDW} subject to the number of particles remaining fixed enables us to write (11.6.4) in two equivalent forms:

$$\gamma_{VDW} = 2A \int_{-\infty}^{\infty} (\rho'_{(1)}(z))^2 \, dz \tag{11.6.5a}$$

$$= 2 \int_{-\infty}^{\infty} [a(\rho_{(1)}(z)) - \bar{a}(\rho_{(1)}(z))] \, dz \tag{11.6.5b}$$

Previous analysis shows[32]

$$a(\rho) = 1 - \rho - e^{-\rho} + \frac{kT}{\varepsilon}(\rho - \rho \ln \rho)$$

$$\bar{a}(\rho) = \tfrac{1}{2}(a_L + a_V) + (\rho - \bar{\rho})\left(\frac{kT}{\varepsilon} \ln \frac{\varepsilon}{kT} - 1\right) \tag{11.6.6}$$

where $\bar{\rho} = \tfrac{1}{2}(\rho_L + \rho_V)$. We draw attention to the implicit assumption of z-independence of A in (11.6.5a). If we adopt the mean field density profile (11.6.2), rather than the profile which minimizes the free energy, equivalence of (11.6.5a) and (11.6.5b) is destroyed. Nevertheless, we may write approximately

$$A = 2z_G^2 kT/3\varepsilon \tag{11.6.7}$$

whereupon we may form two van der Waals' estimates of the surface tension on the basis of (11.6.5a) and (11.6.5b):

$$\gamma_{VDWa} = \tfrac{4}{9} z_G \frac{kT}{\varepsilon} \Delta^3 \tag{11.6.8a}$$

$$\gamma_{VDWb} = \frac{kT}{\varepsilon} \int_{-\infty}^{\infty} f[\rho_{(1)}(z)] \, dz \tag{11.6.8b}$$

where

$$f[\rho_{(1)}(z)] = \rho_L + \rho_V + \rho_L\rho_V - \frac{\varepsilon e^{-\rho}}{kT} - \rho \ln\left(\frac{e\varepsilon}{kT\rho}\right) \tag{11.6.9}$$

A comparison of the three estimates γ_{MF}, γ_{VDWa} and γ_{VDWb} is shown in Figure 11.6.1. The use of a density profile which *does* minimize the free energy, restoring the consistency of equations (11.6.5a) and (11.6.5b) is also shown (γ_{VDW}) and is seen to bring the mean field and van der Waals estimates into closer coincidence. However, the surface tension still diverges at zero temperature.

We observed in Section 2.4 and remarked above that departure from criticality introduces a functional dependence of A upon the local density. Indeed, (11.6.5a) might be more correctly written

$$\gamma_{YFG} = 2 \int_{-\infty}^{\infty} A[\rho_{(1)}(z_1)](\rho'_{(1)}(z))^2 \, dz \tag{11.6.10}$$

This proposition of Yang, Fleming, and Gibbs[2] in conjunction with a result

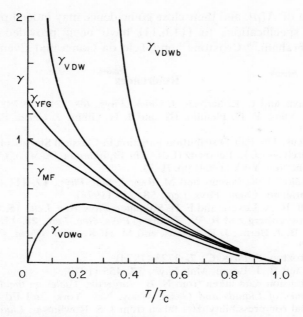

Figure 11.6.1 The reduced surface tension for a fluid of penetrable line segments as a function of T/T_c in various approximations. The mean field result, γ_{MF}, is known to be exact in the limits $T = 0$, T_c, and provides a reference locus against which the various van der Walls estimates may be assessed in the subcritical range

of Lebowitz and Percus[26]

$$A(\rho) = \frac{kT}{4\varepsilon d} \int c[r \mid \rho] r^2 \, d\mathbf{r} \qquad (11.6.11)$$

where d is the dimensionality of the system, yields for a system of interpenetrable rods[31]

$$\gamma_{YFG} = \frac{2kT}{3\varepsilon} \int_{\rho_V}^{\rho_L} e^{-\rho/2} [f(\rho)]^{1/2} \, d\rho \qquad (11.6.12)$$

This result is also shown in Figure 11.6.1. It is clear that the estimate of Yang *et al.* represents a substantial improvement upon the more primitive van der Waals treatments, and to a certain extent provides vindication of the squared-gradient representation of the local free energy density throughout the subcritical region. Of course, the model is highly pathological and the conclusions do not bear immediate extension to realistic systems, despite widespread use of the truncated approximation (11.6.1) at temperatures far below the critical when the approximation is in doubt.

Any agreement of the γ_{MF} and γ_{YFG} curves is contingent upon the specification of $A(\rho)$, and their close coincidence may indeed prove illusory: alternative specifications to (11.6.11) have been provided by van der Waals,[33] Abraham,[28] Croxton,[30] and Telo da Gama and Evans.[29]

References

1. H. T. Davis and L. E. Scriven, *J. Chem. Phys.*, **69**, 5215 (1978).
2. A. J. M. Yang, P. D. Fleming III, and J. H. Gibbs, *J. Chem. Phys.*, **64**, 3732 (1976).
3. J. K. Percus, The Pair Distribution Function in Classical Statistical Mechanics, in H. C. Frisch and J. L. Lebowitz (Eds.) *The Equilibrium Theory of Classical Fluids*, Benjamin, New York (1964) pp. II–33.
4. M. H. Kalos, J. K. Percus, and M. Rao, *J. Stat. Phys.*, **17**, 111 (1977).
5. F. F. Abraham, *Chem. Phys. Lett.*, **58**, 259 (1978).
6. F. P. Buff, R. A. Lovett, and F. H. Stillinger, *Phys. Rev. Lett.*, **15**, 621 (1965).
7. D. G. Triezenberg and R. W. Zwanzig, *Phys. Rev. Lett.*, **28**, 1183 (1972).
8. M. Rao, B. J. Berne, J. K. Percus, and M. H. Kalos, *J. Chem. Phys.*, **71**, 3802 (1979).
9. C. A. Croxton, *J. Phys. C.*, **7**, 3723 (1974).
10. M. Rao and B. J. Berne, *Mol. Phys.*, **37**, 455 (1979).
11. Surface tension data taken from N. B. Vargaftik, *Tables on the Thermophysical Properties of Liquids and Gases*, Wiley, New York, 2nd Ed. (1975) p. 53. Isothermal compressibility data taken from J. S. Rowlinson; *Liquids and Liquid Mixtures*, Butterworths, London, 2nd Ed. (1969) p. 55.
12. C. A. Croxton, *Phys. Lett. A.*, **74**, 325 (1979).
13. N. H. Fletcher, *Phil. Mag.*, **7**, 255 (1962).
    N. H. Fletcher, *Phil. Mag.*, **18**, 1287 (1968).
14. H. Mada and S. Kobayashi, *Proc. 2nd. Intl. Conf. Liquid Crystals Bangalore, December 1979*.
15. H. Mada and S. Kobayashi, *Appl. Phys. Lett.*, **35**, 4 (1979).
16. H. Mada and S. Kobayashi, *Mol. Cryst. Liq. Cryst.*, **51**, 43 (1979).
17. C. A. Croxton and S. Chandrasekhar, *Pramāna Suppl.*, No. **1**, 237 (1975).
18. F. H. Stillinger and A. Ben-Naim, *J. Chem. Phys.*, **47**, 4431 (1967).
19. J. Goodisman and R. W. Pastor, *J. Phys. Chem.*, **82**, 2078 (1978).
20. G. Janz, *Molten Salts Handbook*, Academic Press, New York, (1967). U_S calculated from $U_S = \gamma - T\, d\gamma/dT$.
21. R. W. Pastor and J. Goodisman, *J. Chem. Phys.*, **68**, 3654 (1978).
    J. Goodisman, *J. Chem. Phys.*, **69**, 5341 (1978).
22. F. P. Buff and F. H. Stillinger, *J. Chem. Phys.*, **25**, 312 (1956).
23. P. Barnes, Private communication, January, 1980.
24. J. D. Parsons, *Phys. Rev. A.*, **19**, 1225 (1979).
25. B. J. Berne and P. Pechukas, *J. Chem. Phys.*, **56**, 4213 (1972).
26. J. L. Lebowitz and J. K. Percus, *J. Math. Phys.*, **4**, 116 (1963).
27. J. W. Cahn and J. E. Hilliard, *J. Chem. Phys.*, **28**, 258 (1958).
28. F. F. Abraham, *J. Chem. Phys.*, **63**, 157 (1975).
29. M. M. Telo da Gama and R. Evans, *Mol. Phys.*, **38**, 687 (1979).
30. C. A. Croxton, *J. Phys. C*, **12**, 2239 (1979).
31. J. M. Harrington and J. S. Rowlinson, *Proc. Roy. Soc. (Lond.)*, **A367**, 15 (1979).
32. C. A. Leng, J. S. Rowlinson and S. M. Thompson, *Proc. Roy. Soc. (Lond.)*, **A352**, 1 (1976).
33. J. D. van der Waals, *Z. Phys. Chem.*, **13**, 657 (1894), (English translation *J. Stat. Phys.*, **20**, 197 (1979)).
34. C. A. Croxton, *J. Phys. C.* (in press, 1980).

Bibliography and Author Index

The figures shown in square brackets indicate pages of the text on which the reference is cited.

ABE, R., *Prog. Theor. Phys. (Kyoto)*, **19,** 57 (1958); *Prog. Theor. Phys. (Kyoto)*, **19,** 407 (1958). [189]

ABRAHAM, F. F., *J. Chem. Phys.*, **63,** 157 (1975). [39, 317, 320]

ABRAHAM, F. F., *Chem. Phys. Lett.*, **58,** 259 (1978). [302]

ABRAHAM, F. F., BARKER, J. A., and HENDERSON, D. *Mol. Phys.*, **31,** 1291 (1976). *See* HENDERSON, D.

ABRAHAM, F. F., SCHREIBER, D. E., and BARKER, J. A., *J. Chem. Phys.*, **62,** 1958 (1975). [263, 264, 268, 269, 270]

ABRAHAM, F. F., and SINGH, Y., *J. Chem. Phys.*, **67,** 537 *See* SINGH.

ABRAHAM, F. F., and SINGH, Y., *J. Chem. Phys.* **67,** 5960 (1977). *See* SINGH.

ALDER, B. J., and HOOVER, W. G., *Physics of Simple Liquids.* (Eds. Temperley, H. N. V., Rowlinson, J. S., and Rushbrooke G. S., North-Holland Publ. Co., Amsterdam, 1968, Ch. 4. [259]

ALDER, B. J., and WAINWRIGHT, T. E., *J. Chem. Phys.*, **31,** 459 (1959). [267]

ALLDREDGE, G. P., and KLEINMAN, L., *Phys. Rev.*, **B10,** 559 (1974). [145]

ALLDREDGE, G. P., and KLEINMAN, L., *Phys. Lett.*, **48A,** 337 (1974). [145]

ALLEN, B. C., *Liquid Metals Chemistry and Physics.* (Ed. Beer, S. Z., Dekker, New York, 1972, p. 161. [141, 154, 157, 159]

ALLEN, J. W., and RICE, S. A., *J. Chem. Phys.*, **67,** 5105 (1977). [146]

ALLEN, J. W., and RICE, S. A., *J. Chem. Phys.*, **68,** 5053 (1978). [149]

ALLISON, S. K., and COMPTON, A. K., *X-Rays in Theory and Experiment*, 2nd ed. D. Van Nostrand Co., Inc., Princeton N.J. *See* COMPTON.

AMIT, D. J., *Phys. Lett.*, **23,** 665 (1966); *J. Low Temp. Phys.*, **3,** 645 (1970). [183]

ANANTH, M. S., GUBBINS, K. E., and GRAY, C. G., *Mol. Phys.*, **28,** 1005 (1974). [96]

ANDERSON, H. C., CHANDLER, D., and WEEKS, J. D., *J. Chem. Phys.*, **56,** 3812 (1972). [65]

ANDERSON, H. C., WEEKS, J. D., and CHANDLER, D., *J. Chem. Phys.*, **54,** 5237 (1971). *See* WEEKS. [25]

ANDERSON, P. W., MOREL, P., BRUECKNER, K. A., and SODA, T., *Phys. Rev.*, **118,** 1442 (1960). *See* BRUECKNER.

ANDREEV, A. F., *Zh. Eksp. i Teor. Fiz.*, **50,** 1415 (1966) (*Sov. Phys. JETP*, **23,** 939 (1966)). [193]

APPELBAUM, J. A., and HAMANN, D. R., *Phys. Rev.*, **B6,** 2166 (1972). [145, 146, 158]

ASHCROFT, N. W., *Phys. Letters*, **23,** 48 (1966). [144]

ASHCROFT, N. W., and LANGRETH, D. C., *Phys. Rev.*, **155,** 682 (1967). [159]

ATKINS, K. R., *Can. J. Phys.*, **31,** 1165 (1953). [172, 173, 175, 176, 177, 178, 179, 183, 185, 188, 193, 195]

ATKINS, K. R., and NARAHARA, Y., *Phys. Rev.*, **138,** A437 (1965). [178, 179]

AVEYARD, R., and HAYDON, D. A., *An Introduction to the Principles of Surface Chemistry.* Cambridge Chemistry Texts, Cambridge University Press, 1973, Ch. 2. [116]

322

AVEYARD, R., and HAYDON, D. A., *An Introduction to the Principles of Surface Chemistry*. Cambridge Chemistry Texts, Cambridge University Press, 1973, §3.6. [199]

AZMAN, A., and BORSTNIK, B., *Mol. Phys.*, **29**, 1165 (1975). *See* BORSTNIK.

BAGCHI, A., *Phys. Rev.*, **A3**, 1133, (1971). [196]

BAKKER, G., *Handbook of Experimental Physics, Vol 6: Capillarity and Surface Tension*. Akadvertagsgesellschaft, Leipzig, 1928. [3]

BAND, W., *Introduction to Mathematical Physics*. van Nostrand, Princeton, 1959, pp. 191–203. [200]

BARDEEN, J., *Phys. Rev.*, **49**, 653 (1936). [142]

BARDEEN, J., BAYM, G., and PINES, D., *Phys. Rev.*, **156**, 207 (1967). [196]

BARDEEN, J., COOPER, L. N., SCHRIEFFER, J. R., *Phys. Rev.*, **108**, 1175 (1957). [196]

BARKER, J. A., ABRAHAM, F. F., and SCHREIBER, D. E., *J. Chem. Phys.*, **62**, 1958 (1975). *See* ABRAHAM.

BARKER, J. A., and HENDERSON, D., *J. Chem. Phys.*, **47**, 2856 (1967). [25, 34]

BARKER, J. A., HENDERSON, D., and ABRAHAM, F., *Mol. Phys.*, **31**, 1291 (1976) *See* HENDERSON, D.

BARKER, J. A., POUND, G. M., and LEE, J. K., *J. Chem. Phys.*, **60**, 1976 (1974). *See* LEE.

BARKER, J. A., POUND, G. M., and MIYAZAKI, J., *J. Chem. Phys.*, **64**, 3364 (1976). *See* MIYAZAKI.

BARKER, J. A., SMITH, W. R., and HENDERSON, D., *J. Chem. Phys.*, **55**, 4027 (1971). *See* SMITH.

BARNES, P., *Progress in Liquid Physics*. (Ed. Croxton, C. A.), John Wiley & Sons, Ltd., London, 1978, Ch. 9. [198, 199, 200, 207]

BARNES, P., Private communication, January, 1980. [310]

BAXTER, R. J., *J. Chem. Phys.*, **49**, 2770 (1968). [68]

BAYM, G., PINES, D., and BARDEEN, J., *Phys. Rev.*, **156**, 207 (1967). *See* BARDEEN.

BEARMAN, R. J., THROOP, G. J., *J. Chem. Phys.*, **44**, 1423 (1966). *See* THROOP.

BELLEMANS, A., *J. Polymer Sci.*, **C39**, 305 (1972). [221, 225]

BELLEMANS, A., EVERETT, D. H., DEFAY, R., and PRIGOGINE, I., *Surface Tension and Adsorption*. Longmans, 1966. *See* DEFAY.

BELTON, J. W., and EVANS, M. G., *Trans. Faraday Soc.*, **41**, 1 (1945). [124]

BEN-NAIM, A., *Progress in Liquid Physics*. (Ed. Croxton, C. A.), John Wiley & Sons, Ltd., London, 1978, Ch. 10. [198]

BEN-NAIM, A., and STILLINGER, F. H., *J. Chem. Phys.*, **47**, 4431. *See* STILLINGER.

BEN-NAIM, A., and STILLINGER, F. H., *Structure and Transport Processes in Water and Aqueous Solutions*. (Ed. Horne, R. A.), Wiley–Interscience, New York, 1972. [215]

BENNEMANN, K. H., and KETTERSON, J. P. (Eds.), *The Physics of Liquid and Solid Helium, Part I*. Wiley, New York, 1976. [172, 185]

BENNEMANN, K. H., MORAN-LOPEZ, J. L., and KERKER, G., *J. Phys. F.*, **5**, 1277 (1975). *See* MORAN-LOPEZ.

BENNETT, A. J., and DUKE, C. B., *Phys. Rev.*, **160**, 541 (1967). [142]

BENNETT, C. H., Preprint cited in Miyazaki, Baker, and Pound. [278]

BEREZNYAK, N. G., and ESEL'SON, B. N., *Dokl. Akad. Nauk SSSR*, **98**, 564 (1954). *See* ESEL'SON.

BERG, R. D., and WILETS, L., *Proc. Phys. Soc.*, **A68**, 229 (1955). [142]

BERNAL, J. D., *Proc. Roy. Soc.*, **A 280**, 299 (1964). [44]

BERNAL, J. D., and FOWLER, R. H., *J. Chem. Phys.*, **1**, 515 (1933). [199, 217]

BERNE, B. J., and PECHUKAS, P., *J. Chem. Phys.*, **56**, 4213 (1972). [218, 316]

BERNE, B. J., PERCUS, J. K., KALOS, M. H., and RAO, M., *J. Chem. Phys.*, **71**, 3802 (1979). *See* RAO.

BERNE, B. J., and RAO, M., *Mol. Phys.*, **37,** 455 (1979). *See* RAO.

BERRY, M. V., DURRANS, R. F., and EVANS, R. J. *Phys.*, **A5,** 166 (1972). [29]

BIRD, R. B., HIRSCHFELDER, J. O., and CURTISS, C. F., *Molecular Theory of Gases and Liquids.* New York, Wiley, 1954. *See* HIRSCHFELDER.

BLOCH, A., and RICE, S. A., *Phys. Rev.*, **185,** 933 (1969). [168]

BLUM, L., and STELL, G., *J. Stat. Phys.*, **15,** 439 (1976). [68]

BOHDANSKY, J., *J. Chem. Phys.*, **49,** 2982 (1968). [166]

BOKUT, B. V., and FISHER, I. Z., *Zh. Fiz. Khim*, **30,** 2747 (1956). *See* FISHER.

BOKUT, B. V., and FISHER, I. Z., *Zh. Fiz. Khim*, **31,** 200 (1957). *See* FISHER

BOLDAREV, S. T., and PESHKOV, V. P., *Physica*, **69,** 141 (1973). [178]

BOLDAREV, S. T., and ZINOV'EVA, N. K., *Zh. Eksp. i Teor. Fiz.*, **56,** 1088 (1969); *Sov Phys. JETP*, **29,** 585 (1969). *See* ZINOV'EVA.

BORN, M., and GREEN, H. S., *Proc. Roy. Soc.*, **A188,** 10 (1946). [42, 45]

BORSTNIK, B., and AZMAN, A., *Mol. Phys.*, **29,** 1165 (1975). [44, 50, 52, 61]

BOUCHIAT, M. A., and LANGEVIN, D., *Phys. Lett.*, **34A,** 331 (1971). [246, 253, 257]

BROUT, R., and NAUNBERG, M., *Phys. Rev.*, **112,** 1451 (1958). [187]

BROWN, R. C., and MARCH, N. H., *J. Phys. C.*, **6,** L363 (1973). [150, 154]

BROWN, R. C., and MARCH, N. H., *Physics Reports*, **24,** 77, (1976). [150, 166]

BRUECKNER, K. A., SODA, T., ANDERSON, P. W., and MOREL, P., *Phys. Rev.*, **118,** 1442 (1960). [196]

BUCHAN, G. D., *Phys. Lett.*, **59,** A35 (1976). [193]

BUCHAN, G. D., and CLARK, R. C., *J. Phys. C.*, **10,** 3081 (1977). [187, 193]

BUFF, F. P., *Phys. Rev.*, **82,** 773(T) (1951). [28]

BUFF, F. P., *Z. Elekrochem.*, **56,** 311 (1952). [276]

BUFF, F. P., and KIRKWOOD, J. G., *J. Chem. Phys.*, **17,** 338 (1949). *See* KIRKWOOD.

BUFF, F. P., and LOVETT, R. A., *Simple Dense Fluids.* (Eds. Frisch, H. L., and Salsburg, Z. W.), Academic Press, New York, 1968, Ch. 2. [72, 76, 274]

BUFF, F. P., and LOVETT, R. A., 1966 *Saline Water Conversion Report.* p. 26, U.S. Government Printing Office, Washington D.C. [76]

BUFF, F. P., LOVETT, R., DE HAVEN, P. W., and VIECELI, J. J., *J. Chem. Phys.*, **58,** 1880 (1973). *See* LOVETT.

BUFF, F. P., LOVETT, R. A., and STILLINGER, F. H., JR., *Phys. Rev. Lett.*, **14,** 491 (1965). [177]

BUFF, F. P., LOVETT, R. A., and STILLINGER, F. H., JR., *Phys. Rev. Lett.*, **15,** 621 (1965). [83, 260, 302]

BUFF, F. P., and STILLINGER, F. H., *J. Chem. Phys.*, **25,** 312 (1956). [315]

BUFF, F. P., and STILLINGER, F. H., *J. Chem. Phys.*, **39,** 1911 (1963). [33]

BUONGIORNO, V., and DAVIS, H. T., *Phys. Rev.*, **A12,** 2213 (1975). [40]

BURNET, G., and MAZE, C., *Surf. Sci.*, **27,** 411 (1971). *See* MAZE.

CAHN, J. W., and HILLIARD, J. E., *J. Chem. Phys.*, **28,** 258 (1958). [24, 25, 273, 317]

CAHN, J. W., and HILLIARD, J. E., *J. Chem. Phys.*, **31,** 688 (1959). [24]

CAMPBELL, C. E., *Progress in Liquid Physics.* (Ed. Croxton, C. A.), John Wiley, London, 1978, Ch. 6. [172, 185]

CHADWICK, G. A., and SOUTHIN, R. T., *Scripta. Metall.*, **3,** 541 (1969). *See* SOUTHIN.

CHALMERS, J. A., and PASQUILL, F., *Phil. Mag.*, **23,** 88 (1937). [217]

CHAN, S. L., *Can. J. Phys.*, **50,** 1139 (1972). [178]

CHANDLER, D., ANDERSON, H. C., and WEEKS, J. D., *J. Chem. Phys.*, **54,** 5237 (1971). *See* WEEKS.

CHANDLER, D., WEEKS, J. D., and ANDERSON, H. C., *J. Chem. Phys.*, **56,** 3812 (1972). *See* ANDERSON.

CHANDRASEKHAR, S., and CROXTON, C. A., *Proc. 1st Intl. Conf. Liquid Crystals, Bangalore* 1973, *Pramana Suppl. No. 1*, p. 237. *See* CROXTON.

CHANG, C. C., and COHEN, M., *Phys. Rev.*, **A8,** 1930 (1973). [184, 187, 193]

324

CHAPELA, G. A., SAVILLE, G., and ROWLINSON, J. S., *Chem. Soc. Faraday Disc.*, No. 59, 22 (1975). [134, 261, 262, 263, 264, 269, 271, 272, 275, 276, 285]

CHAPELA, G. A., SAVILLE, G., THOMPSON, S. M., and ROWLINSON, J. S., *J. Chem. Soc. Faraday Trans. II*, **73,** 1133 (1977). [83, 261, 262, 263, 264, 265, 269, 271, 272, 275, 276, 285]

CHAPYAK, E. J., *J. Chem. Phys.*, **57,** 4512 (1972). [66]

CHESTER, G. V., LIU, K. S., and KALOS, M. H., *Phys. Rev.*, **A10,** 303 (1974). *See* LIU.

CHESTER, G. V., LIU, K. S., and KALOS, M. H., *Phys. Rev.*, **12,** 1715 (1975). *See* LIU.

CHESTER, G. V., and REATTO, *Phys. Rev.*, **155,** 88 (1967). *See* REATTO.

CHEUNG, P. S. Y., and POWLES, J. G., *Mol. Phys.*, **30,** 921 (1975). [296, 297]

CHIHARA, J., *2nd Intl. Conf. on Liquid Metals*. Taylor & Frances, London, 1973, p. 137. [140]

CHRISTIANSEN, S. E., PLESNER, I. W., and PLATZ, O., *J. Chem. Phys.*, **48,** 5364 (1968). *See* PLESNER.

CLARKE, R. C., and BUCHAN, G. D., *J. Phys. C*, **10,** 3081 (1977). *See* BUCHAN.

CO, K. U., KOZAK, J. J., and LUKS, K. D., *J. Chem. Phys.*, **66,** 1002 (1977). [69]

COHEN, M., and CHANG, C. C., *Phys. Rev.*, **A8,** 1930 (1973). *See* CHANG.

COLE, G. H. A., *An Introduction to the Statistical Theory of Classical Simple Dense Fluids*. Pergamon, Oxford, 1967. [1, 3, 10, 18]

COLE, M. W., *Phys. Rev.*, **A1,** 1838 (1970). [176]

COLE, M. W., and PADMORE, T. C., *Phys. Rev.*, **A9,** 802 (1974). *See* PADMORE

COMPTON, A. H., *Bull. Nat. Res. Council*, **20,** 48 (1922). [168]

COMPTON, A. K., and ALLISON, S. K., *X-Rays in Theory and Experiment*, 2nd ed. D. Van Nostrand Co. Inc. Princeton, N.J. [168]

COOPER, L. N., SCHRIEFFER, J. R., and BARDEEN, J., *Phys. Rev.*, **108,** 1175 (1957). *See* BARDEEN.

CRAIG, R. A., *Phys. Rev.* **B6,** 1134 (1972). [145, 146, 158]

CRAIG, R. A., *Solid State Commun.*, **13,** 1517 (1973) [158]

CROXTON, C. A., *Ph.D. Thesis*. University of Cambridge, 1969. [167]

CROXTON, C. A., *Phys. Lett.*, **A41,** 413 (1972). [188]

CROXTON, C. A., *J. Phys. C*, **6,** 411 (1973). [188]

CROXTON, C. A., *Adv. Phys.*, **22,** 385 (1973). [32, 54, 103, 140, 160]

CROXTON, C. A., *Liquid State Physics—A Statistical Mechanical Introduction*. Cambridge University Press, London and New York, 1974. [1, 3, 10, 15, 42, 64, 70, 103, 140, 152, 160, 162, 189, 259]

CROXTON, C. A., *J. Phys. C*, **7,** 3723 (1974). [48, 83, 302]

CROXTON, C. A., *Introduction to Liquid State Physics*. John Wiley & Sons Ltd., London, 1975. [1, 3, 10, 15, 64, 70]

CROXTON, C. A., *Introductory Eigenphysics*. John Wiley & Sons Ltd., London, 1975, p. 140. [204]

CROXTON, C. A., *Physical Adsorption in Condensed Phases*; Faraday General Discussion, April, 1975. [66]

CROXTON, C. A., *Phys. Lett. A*, **59A,** 359 (1976). [231, 232]

CROXTON, C. A., *Phys. Lett.*, **60A,** 215 (1977) [127]

CROXTON, C. A., *Progress in Liquid Physics*. (Ed. Croxton, C. A.), John Wiley & Sons Ltd., London, 1978, Chap. 2. [23, 36, 43, 110, 127, 140, 160, 162, 231, 232]

CROXTON, C. A., *Mol. Phys.* (to be published 1980) [134]

CROXTON, C. A., *J. Phys. C*, **12,** 2239 (1979). [24, 25, 41, 165, 317, 320]

CROXTON, C. A., *Phys. Lett. A*, **74,** 325 (1979). [307]

CROXTON, C. A., *J. Phys. A*, **12,** 2475 (1979). [232, 236]

CROXTON, C. A., *J. Phys. A*, **12,** 2487 (1979). [232, 236]

CROXTON, C. A., *J. Phys. A*, **12,** 2497 (1979). [232, 236]

CROXTON, C. A., *Mol. Cryst. Liq. Cryst.* (in press 1980). [238]

CROXTON, C. A., *J. Phys, C.* (in press 1980). [313]

CROXTON, C. A., and CHANDRASEKHAR, S., *Proc. 1st Intl. Conf. Liquid Crystals, Bangalore 1975, Pramana Suppl. No.* 1, p. 237. [146, 151, 152, 312]

CROXTON, C. A., and FERRIER, R. P., *J. Phys. C*, **4**, 1909 (1971). [54].

CROXTON, C. A., and FERRIER, R. P., *J. Phys. C.*, **4**, 1921 (1971). [54, 55, 56]

CROXTON, C. A., and FERRIER, R. P., *J. Phys. C.*, **4**, 2433, (1971). [268]

CROXTON, C. A., and FERRIER, R. P., *J. Phys. C.*, **4**, 2447 (1971). [28, 264, 268, 269, 270]

CROXTON, C. A., and FERRIER, R. P., *Phil. Mag.*, **24**, 489 (1971). [54]

CROXTON, C. A., and FERRIER, R. P., *Phil Mag.*, **24**, 493 (1971). [54, 55, 56]

CROXTON, C. A., and FERRIER, R. P., *Phys. Lett. A.*, **35**, 330 (1971). [28, 55]

CROXTON, C. A., and OSBORN, T. R., *Phys. Lett.*, **55A**, 415 (1976). [106]

CROXTON, C. A., and OSBORN, T. R., *Mol. Phys.* (In press, 1980). *See* OSBORN.

CURRO, J. G., *J. Chem. Phys.*, **64**, 2496 (1976). [234]

CURTISS, C. F., BIRD, R. B., and HIRSCHFELDER, J. O., *Molecular Theory of Gases and Liquids.* New York, Wiley, 1954. *See* HIRSCHFELDER.

CYROT-LACKMANN, F., DESJONQUERS, M. C., and GASPARD, J. P., *J. Phys. C.*, **7**, 925 (1974). [154]

DAVIS, H. T., *J. Chem. Phys.*, **62**, 3412 (1975). [90, 93, 214, 234]

DAVIS, H. T., *J. Chem. Phys.*, **67**, 3636 (1977). [79, 260, 262, 272, 275, 277]

DAVIS, H. T., and BUONGIORNO, V., *Phys. Rev.*, **A12**, 2213 (1975). *See* BUONGIORNO.

DAVIS, H. T., and SALTER, S. J., *J. Chem. Phys.*, **63**, 3295 (1975). *See* SALTER, S. J.

DAVIS, H. T., and SCRIVEN, L. E., *J. Chem. Phys.*, **69**, 5215 (1978). [300]

DEFAY, R., PRIGOGINE, I., BELLEMANS, A., and EVERETT, D. H., *Surface Tension and Adsorption*, Longmans, 1966. [116, 122, 124, 225]

de GENNES, P. G., *Phys. Lett.*, **30A**, 454 (1969). [256]

DE HAVEN, P. W., VIECELI, J. J., BUFF, F. P., and LOVETT, R., *J. Chem. Phys.*, **58**, 1880 (1973). *See* LOVETT.

DESJONQUERS, M. C., GASPARD, J. P., and CYROT-LACKMANN, F., *J. Phys. C.*, **7**, 925 (1974). *See* CYROT-LACKMANN.

DRUDE, P., *Theory of Optics.* Longmans, Green & Co., New York, 1907, p. 292. [76]

DUKE, C. B., and BENNETT, A. J., *Phys. Rev.*, **160**, 541 (1967). *See* BENNETT.

DURRANS, R. F., EVANS, R., and BERRY, M. V., *J. Phys.*, **A5**, 166 (1972). *See* BERRY.

EBNER, C., and SAAM, W. F., *Phys. Rev.*, **B12**, 923 (1975). [76]

ECHINIQUE, P. M., and PENDRY, J. B., *J. Phys. C.*, **9**, 3183 (1976). [189]

EDWARDS, D. O., EKARDT, J. R., and GASPARINI, F. M., *Phys. Rev.*, **A9**, 2070 (1974). [178]

EDWARDS, D. O., FATOUROS, P., IHAS, G. G., MROZINSKI, P., SHEN, S. Y., GASPARINI, F. M., and TAM, C. P., *Phys. Rev. Lett.*, **34**, 1153 (1975). [189]

EDWARDS, D. O., SARWINSKI, R. E., TOUGH, J. T., and GUO, H. M., *Phys. Rev. Lett.*, **27**, 1259 (1971). *See* GUO.

EDWARDS, D. O., SHEN, S. Y., GASPARINI, F. M., and EKARDT, J. R., *J. Low Temp. Phys.*, **13**, 437 (1973). *See* GASPARINI.

EGELSTAFF, P. A., *An Introduction to the Liquid State.* Academic Press, London, 1967. [1, 3, 10, 15]

EGELSTAFF, P. A., and WIDOM, B., *J. Chem. Phys.*, **53**, 2667 (1970). [19, 274]

EINSTEIN, A., *Ann. Physik*, **4**, 513 (1901). [10]

EIRICH, F. R., SIMHA, R., and FRISCH, H. L., *J. Phys. Chem.*, **57**, 584 (1953). *See* SIMHA.

EISENBERG, D., and KAUZMANN, W., *The Structure and Properties of Water.* Oxford University Press, New York, 1969, p. 44. [199]

326

EISENSTEIN, A., and GINGRICH, N. S., *Phys. Rev.*, **62**, 261 (1942). [27]

EKARDT, J. R., *Ph.D. Dissertation*, Ohio State University (unpublished). [178, 179]

EKARDT, J. R., EDWARDS, D. O., SHEN, S. Y., and GASPARINI, F. M., *J. Low Temp. Phys.*, **13**, 437 (1973). See GASPARINI.

EKARDT, J. R., GASPARINI, F. M., and EDWARDS, D. O., *Phys. Rev.*, **9**, A2070 (1974). See EDWARDS.

ELEY, D. D., and EVANS, M. G., *Trans. Faraday Soc.*, **34**, 1093 (1938). [217]

EMERY, V. J., and SESSLER, A. M., *Phys. Rev.*, **119**, 43 (1960). [196]

ESEL'SON, B. N., and BEREZNYAK, N. G., *Dokl. Akad. Nauk SSSR*, **98**, 564 (1954). [193]

ESEL'SON, B. N., IVANTSOV, V. G., and SHVETS, A. D., *Zh. Eksop. Teor. Fiz.*, **44**, 483 (1963) (*Sov. Phys. JETP*, **17**, 330 (1963)). [193]

EVANS, D. J., *Mol. Phys.*, **34**, 317 (1977). [296]

EVANS, D. J., and MURAD, S., *Mol. Phys.*, **34**, 327 (1977). [288, 296]

EVANS, M. G., and BELTON, J. W., *Trans. Faraday Soc.*, **41**, 1 (1945). See BELTON.

EVANS, M. G., and ELEY, D. D., *Trans Faraday Soc.*, **34**, 1093 (1938). See ELEY.

EVANS, R., *J. Phys. C.*, **7**, 2808 (1974). [159]

EVANS, R., BERRY, M. V., and DURRANS, R. F., *J. Phys.*, **A5**, 166 (1972). See BERRY.

EVANS, R., and KUMARADIVAL, R., *J. Phys. C.*, **8**, 793, (1975). See KUMARADIVAL.

EVANS, R., and KUMARADIVAL, R., *J. Phys. C.*, **9**, 1891 (1976). [151, 155, 157]

EVANS, R., and TELO da GAMA, M. M., *Mol. Phys. See* TELO da GAMA

EVANS, R., and UPSTILL, C. E., *J. Phys. C. See* UPSTILL.

EVERETT, D. H., DEFAY, R., PRIGOGINE, I., and BELLEMANS, A., *Surface Tension & Adsorption*. Longmans, 1966. See DEFAY.

EWALD, P. P., and JURETSCHKE, H., In *Structure & Properties of Solid Surfaces*. (Eds. Gomer, R., and Smith, C. S.) University of Chicago Press, Chicago, 1953, p. 82. [142]

FABER, T. E., *An Introduction to the Theory of Liquid Metals*. Cambridge University Press, London, and New York, 1972. [9, 18, 161]

FATOUROS, P., IHAS, G. G., MROZINSKI, P., SHEN, S. Y., GASPARINI, F. M., TAM, C. P. and EDWARDS, D. O., *Phys. Rev. Lett.*, **34**, 1153 (1975). See EDWARDS.

FEIBELMAN, P. J., *Phys. Rev.*, **176**, 551 (1968). [146, 158]

FEIBELMAN, P. J., *Solid State Commun.*, **13**, 319 (1973). [146, 158]

FELDERHOF, B. U., *Phys. Rev.*, **A1**, 1185 (1970). [44]

FERRIER, R. P., and CROXTON, C. A., *J. Phys. C.*, **4**, 1909 (1971). See CROXTON.

FERRIER, R. P., and CROXTON, C. A., *J. Phys. C.*, **4**, 1921 (1971). See CROXTON.

FERRIER, R. P., and CROXTON, C. A., *J. Phys. C.*, **4**, 2433, (1971). See CROXTON.

FERRIER, R. P., and CROXTON, C. A., *J. Phys. C.*, **4**, 2667 (1971). See CROXTON.

FERRIER, R. P., and CROXTON, C. A., *Phil. Mag.*, **24**, 493, (1971). See CROXTON.

FERRIER, R. P., and CROXTON, C. A., *Phys. Lett.*, **A35**, 330 (1971). See CROXON.

FINNIS, M. W., *J. Phys. F.*, **4**, L37 (1975). [145, 151]

FINNIS, M. W., and HEINE, V., *J. Phys. F.*, **5**, 2227 (1974). [145]

FISCHER, J., *Mol. Phys.*, **33**, 75 (1977). [66]

FISHER, I. Z., *Statistical Theory of Liquids*. University of Chicago Press, Chicago, 1961. [1, 3, 10, 15]

FISHER, I. Z., *Statistical Theory of Liquids*. University of Chicago Press, Chicago, 1964, p. 170. [44]

FISHER, I. Z., and BOKUT, B. V., *Zh. Fiz. Khim*, **30**, 2747 (1956). [44, 66]

FISHER, I. Z., and BOKUT, B. V., *Zh. Fiz. Khim*, **31**, 200 (1957). [44]

FISK, S., and WIDOM, B., *J. Chem. Phys.*, **50**, 3219 (1969). [37]

FITTS, D. D., *A. Rev. Phys. Chem.*, **17**, 59 (1966). [90]

FITTS, D. D., *Physica*, **42**, 205 (1969). [29, 187]

FITTS, D. D., SMITH, W. R., and MADDEN, W. G., *Chem. Phys. Lett.*, **36**, 195, (1975). See SMITH, W. R.

FITTS, D. D., SMITH, W. R., and MADDEN, W. G., *Mol. Phys.*, **35,** 1017 (1978). *See* MADDEN, W. G.

FITTS, D. D., SMITH, W. R., NEZEBA, I., and MELNYK, T. W., *Faraday Dis. Chem. Soc.*, **66**/8 (1978). *See* SMITH, W. R.

FITTS, D. D., and WELSH, W. J., *Chem. Phys.*, **26,** 379 (1977). [17]

FLEMING, P. D., GIBBS, J. H., and YANG, A. J. M., *J. Chem. Phys.*, **64,** 3722 (1976). *See* YANG.

FLETCHER, N. H., *Phil. Mag.*, **7,** 255 (1962); *Phil. Mag.*, **18,** 1287 (1968). [307, 309]

FLOOD, E. A., *Canad. J. Chem.*, **33,** 979 (1955); *The Solid/Gas Interface, Vol 1.* Dekker, New York, 1966. [3]

FLORY, P. J., *J. Chem. Phys.*, **9,** 660 (1941). [124, 224]

FLYNN, C. P., *J. Appl. Phys.*, **35,** 1641 (1964). [150]

FORSMAN, W. C., and HUGHES, R. E., *J. Chem. Phys.*, **38,** 2118 (1963); *J. Chem. Phys.*, **38,** 2123 (1963). [225]

FOWLER, R. H., *Proc. Roy. Soc. London,* **A159,** 229 (1937). [8, 17, 27, 92, 276]

FOWLER, R. H., and BERNAL, J. D., *J. Chem. Phys.*, **1,** 515 1933. *See* BERNAL.

FRANKS, F. (Ed.), *Water—A Comprehensive Treatise.* Plenum Press, London and New York, 1973–75, Vols. 1–5. [198]

FREEMAN, K. S. C., and McDONALD, I. R., *Mol. Phys.*, **26,** 529 (1973). [276]

FRENKEL, J., *Phil. Mag.*, **33,** 297 (1917). [141]

FRENKEL, J., *Z. Phys.*, **51,** 232 (1928). [142]

FRENKEL, J., *Kinetic Theory of Liquids.* Dover, New York, 1942, p. 308. [166]

FRENKEL, J., *Kinetic Theory of Liquids.* Dover, New York, 1955, Ch. 6. [175, 198, 216]

FRISCH, H. L., EIRICH, F. R., and SIMHA, R., *J. Phys. Chem.*, **57,** 584 (1953). *See* SIMHA.

FRISCH, H. L., LEBOWITZ, J. L., and REISS, H., *J. Chem. Phys.*, **31,** 361 (1959). *See* REISS.

FRISCH, H. L., LEBOWITZ, J. L., and HELFAND, E., *J. Chem. Phys.*, **34,** 1037 (1961). *See* HELFAND.

FRUMKIN, A. N., JOFA, Z. A., and GERVICH, M. A., *Zh. Fiz. Khim*, **30,** 1455 (1956). [217]

FULTON, R. L., *J. Chem. Phys.*, **64,** 1857 (1976). [216]

GASPARD, J. P., CYROT-LACKMANN, F., and DESJONQUERS, M. C., *J. Phys. C.*, **7,** 925 (1974). *See* CYROT-LACKMANN.

GASPARINI, F. M., EDWARDS, D. O., and EKARDT, J. R., *Phys. Rev.*, **9,** A2070 (1974). *See* EDWARDS.

GASPARINI, F. M., EKARDT, J. R., EDWARDS, D. O., and SHEN, S. Y., *J. Low Temp. Phys.*, **13,** 437 (1973) [178, 179]

GASPARINI, F. M., TAM, C. P., EDWARDS, D. O., FATOUROS, P., IHAS, G. G., MROZINSKI, P., and SHEN, S. Y., *Phys. Rev. Lett.*, **34,** 1153, (1975). *See* EDWARDS.

GEAR, C. W., *Computational Methods in Ordinary Differential Equations.* Prentice Hall, Englewood Cliffs, New Jersey, 1971. [298]

GERNER, D., and MAYER, H., *Z. Phys.*, **210,** 391 (1968). [161]

GERVICH, A. N., JOFA, Z. A., and FRUMKIN, A. N., *Zh. Fiz. Khim.*, **30,** 1455 (1956). *See* FRUMKIN.

GIBBS, J. H., YANG, A. J. M., and FLEMING, P. D., *J. Chem. Phys.*, **64,** 3722, (1976). *See* YANG.

GIBBS, J. W., *Collected Works, Vol.* 1. Yale University Press, New Haven, (1928). [3]

GINGRICH, N. S., and EISENSTEIN, A., *Phys. Rev.*, **62,** 261 (1942). *See* EISENSTEIN.

GOOD, J. W., *J. Phys. Chem.*, **61,** 810 (1957). [198]

GOODISMAN, J., *J. Chem. Phys.*, **69,** 5341 (1978). [314]

GOODISMAN, J., *J. Chem. Phys.*, **68,** 3654 (1978). *See* PASTOR.

GOODISMAN, J., and PASTOR, R. W., *J. Phys. Chem.*, **82**, 2078 (1978). [313]

GOODMAN, R. M., and SAMORJAI, G. A., *J. Chem. Phys.*, **52**, 6331 (1970). [167]

GRAY, C. G., ANANTH, M. S., and GUBBINS, K. E., *Mol. Phys.*, **28**, 1005 (1974). *See* ANANTH.

GRAY, C. G., and GUBBINS, K. E., *Mol. Phys.*, **23**, 187 (1972). *See* GUBBINS.

GRAY, C. G., HAILE, J. M., and GUBBINS, K. E., *J. Chem. Phys.*, **64**, 1852 (1976). *See* HAILE.

GRAY, P., and RICE, S. A., *The Statistical Mechanics of Simple Liquids*. Interscience, New York, 1965. *See* RICE.

GREEN, H. S., *The Molecular Theory of Fluids*, §6.1. North-Holland Publ. Co., Amsterdam, 1952. [26]

GREEN, H. S., *Hand Phys.*, **10**, 79 (1960). [22, 34, 127]

GREEN, H. S., and BORN, M., *Proc. Roy. Soc.*, **A188**, 10 (1946). *See* BORN.

GUBBINS, K. E., and GRAY, C. G., *Mol. Phys.*, **23**, 187 (1972). [96]

GUBBINS, K. E., GRAY, C. G., and ANANTH, M. S., *Mol. Phys.*, **28**, 1005 (1974). *See* ANANTH.

GUBBINS, K. E., GRAY, C. G., and HAILE, J. M., *J. Chem. Phys.*, **64**, 1852 (1976). *See* HAILE.

GUBBINS, K. E., and REED, T. M., *Applied Statistical Mechanics*. McGraw-Hill Kogakusha, Tokyo, 1973. *See* REED.

GUGGENHEIM, E. A., *Trans. Farady Soc.*, **36**, 397, (1940); *Thermodynamics*. North Holland, Amsterdam, 1959. (The criticisms have been omitted from the 5th edition, 1967.) [3, 7]

GUGGENHEIM, E. A., *Mixtures*. Oxford, 1952, Ch. 9. [116, 121, 123]

GUGGENHEIM, E. A., *Mol. Phys.*, **9**, 43 (1965). [32]

GULLY, W. J., RICHARDSON, R. C., LEE, D. M., and OSHEROFF, D. D., *Phys. Rev. Lett.*, **29**, 920 (1972). *See* OSHEROFF.

GUO, H. M., EDWARDS, D. O., SARWINSKI, R. E., and TOUGH, J. T., *Phys. Rev. Lett.*, **27**, 1259 (1971). [187, 195]

GUTTMAN, C. L., McCRACKIN, F. L., and MAZUR, J., *Macromolecules*, **6**, 859, (1973). *See* McCRACKIN.

HAILE, J. M., *Ph.D. Thesis*, Gainesville, Univ. of Florida (1976). [90, 95, 98, 99, 103, 104, 108, 109, 110, 299]

HAILE, J. M., GUBBINS, K. E., and GRAY, C. G., *J. Chem. Phys.*, **64**, 1852 (1976). [290, 297]

HAMANN, D. R., and APPELBAUM, J. A., *Phys. Rev.*, **B6**, 2166 (1972). *See* APPELBAUM.

HANKINS, D., MOSKOWITZ, J. W., and STILLINGER, F. H., *J. Chem. Phys.*, **53**, 4544 *see also* Erratum, *J. Chem. Phys.*, **59**, 995 (1973). [199]

HANSEN, J.-P., and VERLET, L., *Phys. Rev.*, **184**, 151 (1969). [271, 292]

HARASIMA, A., *Proc. Int. Conf. Theor. Phys.*, Science Council of Japan (1954). [29]

HARASIMA, A., *Advances in Chemical Physics*, Vol. 1, **203**. (Ed. Prigogine, I.), New York, Interscience, 1958. [16, 29, 45]

HARRINGTON, J. M., and ROWLINSON, J. S., *Proc. Roy. Soc. (Lond.)*, **A367**, 15 (1979). [317, 319]

HARRISON, W. A., *Pseudopotentials in the Theory of Metals*. Benjamin, New York, 1966. [151]

HARRISON, W. A., *Phys. Rev.*, **181**, 1036 (1969). [154]

HAYDOCK, R., and KELLY, M. J., *Surf. Sci.*, **38**, 139 (1973). [154]

HAYDON, D. A., and AVEYARD, R., *An Introduction to the Principles of Surface Chemistry*. Cambridge Chemistry Texts, Cambridge University Press, 1973, Ch. 2. *See* AVEYARD.

HAYWARD, D. O., and TRAPNELL, G. M. W., *Chemisorption*. Butterworths, London, 2nd edition, 1964. [116]

HEINE, V., and FINNIS, M. W., *J. Phys. F.*, **5**, 2227 (1974). *See* FINNIS.

HEINRICHS, J., *Solid State Commun.*, **13**, 1599 (1973). [146, 158]

HELFAND, E., FRISCH, H. L., and LEBOWITZ, J. L., *J. Chem. Phys.*, **34**, 1037 (1961). [32, 63]

HENDERSON, D., ABRAHAM, F., and BARKER, J. A., *Mol. Phys.*, **31**, 1291 (1976). [67]

HENDERSON, D., and BARKER, J. A., *J. Chem. Phys.*, **47**, 2856 (1967). *See* BARKER.

HENDERSON, D., BARKER, J. A., and SMITH, W. R., *J. Chem. Phys.*, **55**, 4027 (1971). *See* SMITH, W. R.

HENDERSON, D., LEBOWITZ, J., and WAISMAN, E., *Mol. Phys.*, **32**, 1373 (1976). *See* WAISMAN.

HENDERSON, J. R., and LEKNER, J., *Mol. Phys.*, **34**, 333 (1977). *See* LEKNER.

HENDERSON, J. R., and LEKNER, J., *Mol. Phys.*, **36**, 781 (1978). *See* LEKNER.

HENNIKER, J. C., *Rev. Mod. Phys.*, **21**, 322 [199]

HENSHAW, D. G., and HURST, D., *Can. J. Phys.*, **33**, 797 (1955). [173, 175, 176, 189, 190, 192, 193, 194, 195]

HILL, T. L., *J. Chem. Phys.*, **19**, 261 (1951). [28]

HILL, T. L., *J. Chem. Phys.*, **19**, 1203 (1951). [28]

HILL, T. L., *J. Chem. Phys.*, **20**, 141 (1952). [31]

HILL, T. L., *J. Chem. Phys.*, **20**, 1510. (1952). [28]

HILL, T. L., *J. Chem. Phys.*, **30**, 1521 (1959). [2, 12, 16, 30, 33, 47]

HILLIARD, J. E., and CAHN, J. W., *J. Chem. Phys.*, **28**, 258 (1958). *See* CAHN. [24]

HILLIARD, J. E., and CAHN, J. W., *J. Chem. Phys.*, **31**, 688 (1959). *See* CAHN. [24]

HIRSCHFELDER, J. O., *Adv. Chem. Phys.*, **12**, (1967). [90]

HIRSCHFELDER, J. O., CURTISS, C. F., and BIRD, R. B., *Molecular Theory of Gases and Liquids*. New York, Wiley, 1954. [90]

HOEVE, C. A. J., *J. Polym. Sci.*, **C30**, 361 (1970); *J. Polym. Sci.*, **C34**, 1 (1971). [224, 225]

HOHENBERG, P. C., and KOHN, W., *Phys. Rev.*, **136**, B864 (1964). [183, 184]

HOOVER, W. G., and ALDER, B. J., *Physics of Simple Liquids*. (Eds. Temperley, H. N. V., Rowlinson, J. S., and Rushbrooke, G. S.), North-Holland Publ. Co., Amsterdam, 1968, Ch. 4. *See* ALDER.

HUANG, J. S., and WEBB, W. W., *J. Chem. Phys.*, **50**, 3677 (1969). [76]

HUANG, K., and WYLLIE, G., *Proc. Phys. Soc.*, **A62**, 180 (1949). [142]

HUGHES, R. E., and FORSMAN, W. C., *J. Chem. Phys.*, **38**, 2118 (1963); *J. Chem. Phys.*, **38**, 2123 (1963). *See* FORSMAN.

HUMPHREYS, C. W., and McBAIN, J. W., *J. Phys. Chem.*, **36**, *See* McBAIN.

HUNTINGDON, H. B., *Phys. Rev.*, **31**, 1035 (1951). [142]

HURST, D., and HENSHAW, D. G., *Can. J. Phys.*, **33**, 797 (1951). *See* HENSHAW.

HUTCHINSON, P., MARCH, N. H., and JOHNSON, M. D., *Proc. Roy. Soc.*, **A282**, 283, (1964). *See* JOHNSON.

IHAS, G. G., MROZINSKI, P., SHEN, S. Y., GASPARINI, F. M., TAM, C. P., EDWARDS, D. O., and FATOUROS, P. *Phys. Rev. Lett.*, **34**, 1153, (1975). *See* EDWARDS.

INGLESFIELD, J. E., and WIKBORG, E., *Sol. Stat. Commun.*, **16**, 335 (1975). *See* WIKBORG.

IRVING, J. H., and KIRKWOOD, J. G., *J. Chem. Phys.*, **18**, 817 (1950). [45]

IVANTSOV, V. G., SHVETS, A. D., and ESEL'SON, B. N., *Zh. Eksp. Teor. Fiz.*, **44**, 483 (1963) (*Sov. Phys. JETP*, **17**, 330 (1963)). *See* ESEL'SON.

JAHNIG, F., *Proc. 1st Intl. Conf. Liquid Crystals Bangalore*, 1975, *Pramana Suppl.* No. 1, p. 246. [253]

JANZ, G., *Molten Salts Handbook*. Academic Press, New York, 1967. [313]

JASTROW, R., *Phys. Rev.*, **98**, 1479 (1955). [185]

JOFA, Z. A., GERVICH, M. A., and FRUMKIN, A. N., *Zh. Fiz. Khim.*, **30**, 1455 (1956). *See* FRUMKIN.

330

JOHNSON, M. D., HUTCHINSON, P., and MARCH, N. H., Proc. Roy. Soc., **A283,** (1964). [151, 160]

JONSON, M., and SRINIVASAN, G., Phys. Lett., **43A,** 427 (1973). [146, 158]

JONSON, M., and SRINIVASAN, G., Physica Scripta, **10,** 262 (1974). [146, 158]

JOUANIN, C., Acad. Sci. Paris, **B268,** 1597 (1969). [29, 50]

JURETSCHKE, H., and EWALD, P. P., In Structure & Properties of Solid Surfaces. (Eds. Gomer, R., and Smith, C. S.), University of Chicago Press, Chicago, 1953, p. 82. See EWALD.

KADANOFF, L. P., GOTZE, W., HAMBLEN, D., HECHT, R., LEWIS, E. A. S., PALCIAUSKAS, V. V., RAYL, M., SWIFT, J., ASPRES, D., and KANE, J., Rev. Mod. Phys., **39,** 395 (1967). [41]

KALKSTEIN, D., and SOVEN, P., Surf. Sci., **26,** 85 (1971). [154]

KALOS, M. H., CHESTER, G. V., and LIU, K. S., Phys. Rev., **A10,** 303 (1974). See LIU.

KALOS, M. H., CHESTER, G. V., and LIU, K. S., Phys. Rev., **A12,** 1715 (1975). See LIU.

KALOS, M. H., PERCUS, J. K., and RAO, M., J. Stat. Phys., **17,** 111 (1977). [23, 25, 27, 57, 79, 80, 260, 291, 302, 303]

KAMIENSKI, B., Electrokhim. Acta., **3,** 208 (1960). [217]

KAUZMANN, W., and EISENBERG, D., The Structure and Properties of Water. Oxford University Press, New York, 1969, p. 44. See EISENBERG.

KELLY, M. J., Surf. Sci., **43,** 587 (1974). [154]

KELLY, M. J., and HAYDOCK, R., Surf. Sci., **38,** 139 (1973). See HAYDOCK.

KERKER, G., BENNEMANN, K. H., and MORAN-LOPEZ, J. L., J. Phys. F., **5,** 1277 (1975). See MORAN-LOPEZ.

KESTNER, N. R., and MARGENAU, H., Theory of Intermolecular Forces. Pergamon, New York, 1969. See MARGENAU.

KETTERSON, J. P., and BENNEMANN, K. H. (Eds.), The Physics of Liquid and Solid Helium, Part I. Wiley, New York, 1976. See BENNEMANN.

KIRKWOOD, J. G., J. Chem. Phys., **3,** 300 (1935). [12, 43, 45]

KIRKWOOD, J. G., and BOGGS, E. M., J. Chem. Phys., **10,** 394 (1942) [12]

KIRKWOOD, J. G., and BUFF, F. P., J. Chem. Phys., **17,** 338 (1949). [13, 17, 27, 28, 92, 276]

KIRKWOOD, J. G., and IRVING, J. H., J. Chem. Phys., **18,** 817 (1950). See IRVING.

KIRKWOOD, J. G., and MAZO, R. M., J. Chem. Phys., **41,** 204 (1955); J. Chem. Phys., **28,** 644 (1958). See MAZO.

KOBAYASHI, S., and MADA, H., Appl. Phys. Lett., **35,** 4 (1979); Mol. Cryst. Liq. Cryst., **51,** 43 (1979). See MADA.

KOHN, W., and HOHENBERG, P. C., Phys. Rev., **136,** B864 (1964). See HOHENBERG.

KONDO, S., and ONO, S., Hand. Phys., **10,** 179 (1960). See ONO.

KONDO, S., and ONO, S., Hand. Phys., **10,** 219 See ONO.

KOZAK, J. J., LUKS, K. D., and CO, K. U., J. Chem. Phys., **66,** 1002 (1977). See CO.

KLEINMAN, L., Phys. Rev., **160,** 585 (1967). [157]

KLEINMAN, L., and ALLDREDGE, G. P., Phys. Lett., **48A,** 337 (1974). See ALLDREDGE.

KLEINMAN, L., and ALLDREDGE, G. P., Phys. Rev., **B10,** 559 (1974). See ALLDREDGE.

KOHN, W., Solid State Commun., **13,** 323 (1973).

KOHN, W., Collective Excitations. Nobel Symposium. (Eds. Lundqvist, B. & S.), Academic Press, 1974. [142, 159]

KOHN, W., and LANG, N. D., Phys. Rev., **B1,** 4555 (1970). See LANG.

KONDO, S., and ONO, S., Hand. der Phys., **10,** 237 See ONO.

KRISHNASWAMY, S., and SHASHIDHAR, R., Mol. Cryst. Liq. Cryst., **35,** 253 (1976); Mol. Cryst. Liq. Cryst., **38,** 711 (1977). [253, 255]

KUMARADIVAL, R., and EVANS, R., *J. Phys. C.*, **8**, 793 (1975). [154, 157, 158, 159]

KUMARADIVAL, R., and EVANS, R., *J. Phys. C.*, **9**, 1891 (1976). *See* EVANS.

KWAIT, E. I., *Zh. Fiz. Khim*, **28**, (1415) (1954). *See* MISCENKO.

LANDAU, L. D., and LIFSHITS, E. M., *Statistical Physics.* Pergamon Press, Oxford, 1969, p. 457. [172, 173, 175, 176, 177, 178, 179, 183, 185, 188, 193, 195]

LANG, N. D., and KOHN, W., *Phys. Rev.*, **B1**, 4555 (1970). [142, 159]

LANGEVIN, D., and BOUCHIAT, M. A., *Phys. Lett.*, **34A**, 331 (1971). *See* BOUCHIAT.

LANGRETH, D. C., and ASHCROFT, N. W., *Phys. Rev.*, **155**, 682 (1967). *See* ASHCROFT.

LEBOWITZ, J. L., HELFAND, E., and FRISCH, H. L., *J. Chem. Phys.*, **34**, 1037 (1961). *See* HELFAND.

LEBOWITZ, J. L., and PERCUS, J. K., *J. Math. Phys.*, **4**, 116 (1963). [316, 319]

LEBOWITZ, J. L., REISS, H., and FRISCH, H. L., *J. Chem. Phys.*, **31**, 361 (1959). *See* REISS.

LEBOWITZ, J., WAISMAN, E., and HENDERSON, D., *Mol. Phys.*, **32**, 1373 (1976). *See* WAISMAN.

LEE, D. M., OSHEROFF, D. D., GULLY, W. J., and RICHARDSON, R. C., *Phys. Rev. Lett.*, **29**, 920, (1972). *See* OSHEROFF.

LEE, D. M., OSHEROFF, D. D., and RICHARDSON, R. C., *Phys. Rev. Lett.*, **28**, 885 (1972). *See* OSHEROFF.

LEE, J. K., BARKER, J. A., and POUND, G. M., *J. Chem. Phys.*, **60**, 1976 (1974). [83, 264, 266, 267, 268, 269, 270]

LEKNER, J., *Phil. Mag.*, **22**, 669 (1970). [193]

LEKNER, J., *Prog. Theor. Phys.*, **45**, 42 (1971). [35, 185]

LEKNER, J., and HENDERSON, J. R., *Mol. Phys.*, **34**, 333 (1977); *Physica*, **94A**, 545 (1978). [72, 74, 187]

LEKNER, J., and HENDERSON, J. R., *Mol. Phys.*, **36**, 781 (1978). [76]

LENG, C. A., ROWLINSON, J. S., and THOMPSON, S. M., *Proc. Roy. Soc.*, **A352**, 1 (1976). [273, 318]

LENG, C. A., ROWLINSON, J. S., and THOMPSON, S. M., *Proc. Roy. Soc.*, **A358**, 267 (1977). [75]

LENZ, B. R., and SCHERAGA, H. A., *J. Chem. Phys.*, **58**, 5296 (1973). [199]

LESLIE, F. M., *Proc. 2nd. Intl. Conf. Liquid Crystals, Bangalore*, 1973, *Pramana Suppl.* No. 1, p. 252. [253]

LEVESQUE, D., and RAO, M., *J. Chem. Phys.*, **65**, 3233 (1976). *See* RAO.

LEVINE, I. N., *Quantum Chemistry*, Vol. 1. Allyn and Bacon, Boston, 1970.

LIFSHITS, E. M., and LANDAU, L. D., *Statistical Physics.* Pergamon Press, Oxford, 1969, p. 457. *See* LANDAU.

LIU, K. S., *J. Chem. Phys.*, **60**, 4226 (1974). [268, 269, 270]

LIU, K. S., KALOS, M. H., and CHESTER, G. V., *Phys. Rev.*, **B12**, 1715 (1975). [187, 188]

LIU, K. S., KALOS, M. H., and CHESTER, G. V., *Phys. Rev.*, **A10**, 303 (1974). [67, 264, 268, 269, 270]

LLOPIS, J., *Modern Aspects of Electrochemistry.* (Eds. Bockris, J. O'M., and Conway, B. E.), Plenum, New York, 1971, Ch. 2. [216]

LONGUET-HIGGINS, H. C., and WIDOM, B., *Mol. Phys.*, **8**, 549 (1964). [32]

LOVETT, R. A., and BUFF, F. P., *Simple Dense Fluids.* (Eds. Frisch, H. L., and Salsburg, Z. W.), Academic Press, New York, 1968, Ch. 2. *See* BUFF.

LOVETT, R. A., and BUFF, F. P., 1966 *Saline water conversion Report.* p. 26, U.S. Government Printing Office, *See* BUFF.

LOVETT, R., DE HAVEN, P. W., VIECELI, J. J., and BUFF, F. P., *J. Chem. Phys.*, **58**, 1880, (1973). [72]

LOVETT, R. A., STILLINGER, JR., F. H., and BUFF, F. P., *Phys. Rev. Lett.*, **15**, 621 (1965). *See* BUFF.

332

LOVETT, R. A., STILLINGER, JR., F. H., and BUFF, F. P., *Phys. Rev. Lett.*, **14**, 491 (1965). *See* BUFF.

LU, B. C., and RICE, S. A., *J. Chem. Phys.*, **68**, 5558 (1978). [149, 168, 169, 274]

LUCAS, A. A., and SCHMIT, J., *Solid State Commun.*, **11**, 415 (1972). *See* SCHMIT.

LUKS, K. D., CO., K. U., KOZAK, J. J., *J. Chem. Phys.*, **66**, 1002 (1977). *See* CO.

MADA, H., and KOBAYASHI, S., *Appl. Phys. Lett.*, **35**, 4 (1979); *Mol. Cryst. Liq. Cryst.*, **51**, 43 (1979). [310, 311]

MADDEN, W. G., FITTS, D. D., and SMITH, W. R., *Chem. Phys. Lett.*, **36**, 195, (1975). *See* SMITH, W. R.

MADDEN, W. G., FITTS, D. D., and SMITH, W. R., *Mol. Phys.*, **35**, 1017 (1978). [93]

MAHAN, G. D., *Phys. Rev.*, **B12**, 5585 (1975). [146]

MAIER, W., and SAUPE, A., *Z. Naturf.*, **13a**, 564 (1958); *Z. Naturf.*, **14a**, 882 (1959); *Z. Naturf.*, **15a**, 287 (1960). [238]

MARC DE CHAZAL, L. E., SHOEMAKER, P. D., and PAUL, G. W., *J. Chem. Phys.*, **52**, 491 (1970). *See* SHOEMAKER.

MARCH, N. H., *Liquid Metals. Pergamon Press, Oxford*, 1968. [150]

MARCH, N. H., *Orbital Theories of Molecules and Solids.* (Ed. March, N. H.), Oxford, Clarendon Press, 1974. [142]

MARCH, N. H., and BROWN, R. C., *J. Phys. C.*, **6**, L363 (1973). *See* BROWN.

MARCH, N. H., and BROWN, R. C., *Physics Reports*, **24**, 77 (1976). *See* BROWN.

MARCH, N. H., JOHNSON, M. D., and HUTCHINSON, P., *Proc. Roy. Soc.*, **A282**, 283, (1964). *See* JOHNSON.

MARCH, N. H., and TOSI, M. P., *Ann. of Physics*, **81**, 414 (1973). *See* TOSI.

MARCH, N. H., and TOSI, M. P., *Nuovo Cimento*, **15B**, 308 (1973). *See* TOSI.

MARCH, N. H., and TOSI, M. P., *Atomic Dynamics in Liquids*, p. 260. London, Macmillan, 1976. [74]

MARCH, N. H., TOSI, M. P., and PARRINELLO, M., *Nuovo Cimento.*, **23B**, 135 (1974). *See* TOSI.

MARÉCHAL, J., and PRIGOGINE, I., *J. Colloid Sci.*, **7**, 122, (1952). *See* PRIGOGINE.

MAREGAWA and WATABE, M., *2nd Intl. Conf. on Liquid Metals.* Taylor & Francis, London, 1973, p. 133. *See* WATABE.

MARGENAU, H., and KESTNER, N. R., *Theory of Intermolecular Forces.* Pergamon, New York, 1969. [90]

MASON, E. A., and SPURLING, T. H., *J. Chem. Phys.*, **46**, 322 (1967). *See* SPURLING.

MAYER, H., and GERNER, D., *Z. Phys.*, **210**, 391 (1968). *See* GERNER.

MAYER, J. E., and PRESSING, J., *J. Chem. Phys.*, **59**, 2711 (1973). *See* PRESSING.

MAZE, C., and BURNET, G., *Surf. Sci.*, **27**, 411 (1971). [162]

MAZO, R. M., and KIRKWOOD, J. G., *Proc. Nat. Acad. Sci. U.S.* **41**, 204 (1955); *J. Chem. Phys.*, **28**, 644 (1958). [174]

MAZUR, J., GUTTMAN, C. L., and McCRACKIN, F. L., *Macromolecules*, **6**, 859 (1973). *See* McCRACKIN.

McBAIN, J. W., and HUMPHREYS, C. W., *J. Phys. Chem.*, **36**, 300, (1932). [130]

McBAIN, J. W., and SWAIN, R. C., *Proc. Roy. Soc.*, **A154**, 608 (1936). [130]

McCRACKIN, F. L., *J. Chem. Phys.*, **47**, 1980 (1965). [223]

McCRACKIN, F. L., MAZUR, J., and GUTTMAN, C. L., *Macromolecules*, **6**, 859 (1973). [219]

McDONALD, I. R., and FREEMAN, K. S. C., *Mol. Phys.* **26**, 529 (1973) *See* FREEMAN.

McMILLAN, W. L., *Phys. Rev.*, **A4**, 1238 (1971). [148]

MELNYK, T. W., FITTS, D. D., SMITH, W. R., and NEZEBA, I., *Faraday Disc. Chem. Soc.*, **66/8** (1978). *See* SMITH, W. R.

MICHAELI, I., and PLESNER, I. W., *J. Chem. Phys.*, **60**, 3016 (1974). *See* PLESNER.

MILLER, M. D., *Ph.D. Thesis*, Northwestern Univ. (1973). [178]

MISCENKO, K. P., and KWAIT, E. I., *Zh. Fiz. Khim.*, **28**, 1451 (1954). [217]

MIYAZAKI, J., BARKER, J. A., and POUND, G. M., *J. Chem. Phys.*, **64,** 3364 (1976). [268, 275, 277, 278]

MORÁN-LÓPEZ, J. L., KERKER, G., and BENNEMANN, K. H., *J. Phys. F.*, **5,** 1277 (1975). [154]

MOREL, P., BRUECKNER, K. A., SODA, T., and ANDERSON, P. W., *Phys. Rev.*, **118,** 1442 (1960). See BRUECKNER.

MOSKOWITZ, J. W., STILLINGER, F. H., and HANKINS, D., *J. Chem. Phys.*, **53,** 4544 (1970). See also Erratum, *J. Chem. Phys.*, **59,** 995 (1973). See HANKINS.

MROZINSKI, P., SHEN, S. Y., GASPARINI, F. M., TAM, C. P., EDWARDS, D. O., FATOUROS, P., and IHAS, G. G., *Phys. Rev. Lett.*, **34,** 1153 (1975). See EDWARDS.

MURAD, S., and EVANS, D. J., *Mol. Phys.*, **34,** 327, (1977). See EVANS.

NANENBERG, M., and BROUT, R., *Phys. Rev.*, **112,** 1415 (1958). See BROUT.

NARAHARA, Y., and ATKINS, K. R., *Phys. Rev.*, **138,** A437 (1965). See ATKINS.

NARANG, H., STELL, G., and RASAIAH, J. C., *Mol. Phys.*, **27,** 1393 (1974). See STELL.

NAZARIAN, G. M., *J. Chem. Phys.*, **56,** 1408 (1972). [51, 52]

NEZEBA, I., MELNYK, T. W., FITTS, D. D., and SMITH, W. R., *Faraday Dis., Chem. Soc.*, **66/8** (1978). See SMITH, W. R.

ONO, S., and KONDO, S., *Hand. Phys.*, **10,** (1960). [30, 42, 180]

ONO, S., and KONDO, S., *Hand. Phys.*, **10,** 219 (1960). [30]

ONSAGER, L., *J. Am. Chem. Soc.*, **58,** 1486 (1936). [107]

OPITZ, A. C. L., *Phys. Lett.*, **A47,** 439 (1974). [264, 268, 269, 270]

OSBORN, T. R., and CROXTON, C. A., *Phys. Lett.*, **55A,** 415, (1976). See CROXTON.

OSBORN, T., and CROXTON, C. A., *Mol. Phys.* (in press, 1980). [26, 44, 58, 60, 61, 265]

OSHEROFF, D. D., GULLY, W. J., RICHARDSON, R. C., and LEE, D. M., *Phys. Rev. Lett.*, **29,** 920 (1972). [196]

OSHEROFF, D. D., RICHARDSON, R. C., and LEE, D. M., *Phys. Rev. Lett.*, **28,** 885 (1972). [196]

PADMORE, T. C., and COLE, M. W., *Phys. Rev.*, **A9,** 802 (1974). [183]

PARRINELLO, M., MARCH, N. H., and TOSI, M. P., *Nuovo Cimento,* **23B,** 135 (1974). See TOSI.

PARSONAGE, N. G., *Disc. Faraday Soc.*, **59,** (1975). [263]

PARSONS, J. D., *Phys. Rev.*, **A19,** 1225 (1979). [98, 109, 209, 242, 315]

PARSONS, J. D., *Phys. Rev. Lett.*, **41,** 877 (1978). [256]

PARSONS, J. D., *J. de Phys.*, **37,** 1187 (1976). [246, 290]

PARSONS, R., *Modern Aspects of Electrochemistry.* (Eds. Bockris, J. O'M., and Conway, B. E.,). Academic Press, New York, 1954. Vol. 1, pp. 123–124. [198]

PASQUILL, F., and CHALMERS, J. A., *Phil. Mag.*, **23,** 88 (1937). See CHALMERS.

PASSOTH, G., *Z. Phys. Chem.*, **203,** 275 (1954). [217]

PASTOR, R. W., and GOODISMAN, J., *J. Phys. Chem.*, **82,** 2078 (1978). See GOODISMAN.

PASTOR, R. W., and GOODISMAN, J., *J. Chem. Phys.*, **68,** 3654 (1978). [313, 314]

PATTERSON, D., and SIOW, K. S., *J. Phys. Chem.*, **77,** 356 (1973). See SIOW.

PAUL, G. W., MARC DE CHAZAL, L. E., and SHOEMAKER, P. D., *J. Chem. Phys.*, **52,** 491 (1970). See SHOEMAKER.

PECHUKAS, P., and BERNE, B. J., *J. Chem. Phys.*, **56,** 4213 (1972). See BERNE.

PENDRY, J. B., and ECHINIQUE, P. M., *J. Phys. C.*, **9,** 3183 (1976). See ECHINIQUE.

PERCUS, J. K., *Lecture Notes on Non-Uniform Fluids* (Enseignement du 3 éme Cycle de Physique en Suisse Romande, Eté 1975); *J. Stat. Phys.*, **15,** 423 (1976). [68]

PERCUS J. K., *The Equilibrium Theory of Classical Fluids.* (Eds. Frisch, H. L., and Lebowitz, J. L.) Benjamin, New York, 1964, p. II–33. [73, 300]

PERCUS, J. K., and LEBOWITZ, J. L., *J. Math. Phys.*, **4,** 116. (1963). See LEBOWITZ.

334

PERCUS, J. K., RAO, M., and KALOS, M. H., *J. Stat. Phys.*, **17,** 111 (1977). *See* KALOS. [23, 25, 27]

PERCUS, J. K., RAO, M., KALOS, M. H., and BERNE, B. J., *J. Chem. Phys.*, **71,** 3802 (1979). *See* RAO.

PERRAM, J. W., and SMITH, E. R., *Chem. Phys. Lett.*, **35,** 138 (1975); *Chem. Phys. Lett.*, **39,** 328 (1976); *Phys. Lett. A.*, **59,** 11 (1976). [68]

PERRAM, J. W., and WHITE, L. R., *Faraday General Discussion, Physical Adsorption in Condensed Phases*, April, 1975. [63]

PESHKOV, V. P., and BOLDAREV, S. T., *Physica*, **69,** 141, (1973). *See* BOLDAREV.

PINES, D., BARDEEN, J., and BAYM, G., *Phys. Rev.*, **156,** 207 (1967), *See* BARDEEN.

PLATZ, O., CHRISTIANSEN, S. E., and PLESNER, I. W., *J. Chem. Phys.*, **48,** 5364 (1968). *See* PLESNER.

PLESNER, I. W., and MICHAELI, I., *J. Chem. Phys.*, **60,** 3016 (1974). [33]

PLESNER, I. W., and PLATZ, O., *J. Chem. Phys.*, **48,** 5361 (1968). [28, 32]

PLESNER, I. W., PLATZ, O., and CHRISTIANSEN, S. E., (1968). *J. Chem. Phys.* **48,** 5364 [33, 129, 131, 132, 285].

POPLE, J. A., *Proc. Roy. Soc.*, **A221,** 498 (1954). [95]

POUND, G. M., LEE, J. K., and BARKER, J. A., *J. Chem. Phys.*, **60,** 1976 (1974). See LEE.

POUND, G. M., MIYAZAKI, J., and BARKER, J. A., *J. Chem. Phys.*, **64,** 3364 (1976). *See* MIYAZAKI.

POWLES, J. G., and CHEUNG, P. S. Y., *Mol. Phys.*, **30,** 921 (1975). *See* CHEUNG.

PRAUSNITZ, J. M., and SPROW, F. B., *Trans. Faraday Soc.*, **62,** 1097 (1966). *See* SPROW.

PRAUZNITZ, J. M., and SPROW, F. B., *Trans. Faraday Soc.*, **62,** 1105 (1966). *See* SPROW.

PRESENT, R. D., *J. Chem. Phys.*, **61,** 4267 (1974). [19]

PRESSING, J., and MAYER, J. E., *J. Chem. Phys.*, **59,** 2711 (1973). [45, 53]

PRIEL, Z., and SILBERBERG, A., *Polym. Prepr. Amer. Chem. Soc. Div. Polym. Chem.*, **11,** 1405 (1970). [224, 225, 230]

PRIGOGINE, I., BELLEMANS, A., EVERETT, D. H., and DEFAY, R., *Surface Tension & Adsorption.* Longmans, 1966. *See* DEFAY.

PRIGOGINE, I., and MARÉCHAL, J., *J. Colloid Sci.*, **7,** 122 (1952). [225]

PUGACHEVICH, P. P., and ZADUMKIN, S. N., *Proc. Acad. Sci. USSR Physics & Chemistry Section*, **146,** 743 (1962). *See* ZADUMKIN.

RAHMAN, A., *Phys. Rev.*, **136,** A405 (1964). [267]

RAHMAN, A., and STILLINGER, F. H., *J. Chem. Phys.*, **60,** 1545, (1974). *See* STILLINGER.

RAO, M., and BERNE, B. J., *Mol. Phys.*, **37,** 455 (1979). [304, 305]

RAO, M., BERNE, B. J., PERCUS, J. K., and KALOS, M. H., *J. Chem. Phys.*, **71,** 3802 (1979). [303]

RAO, M., KALOS, M. H., and PERCUS, J. K., *J. Stat. Phys.*, **17,** 111 (1977). *See* KALOS. [23, 25, 27]

RAO, M., and LEVESQUE, D., *J. Chem. Phys.*, **65,** 3233 (1976). [262, 263, 264, 267, 269, 270, 277, 291]

RAPAPORT, D. C., *J. Phys.*, **A9,** 1521 (1976). [219, 223]

RASAIAH, J. C., NARANG, H., and STELL, G., *Mol. Phys.*, **27,** 1393 (1974). *See* STELL.

RAYLEIGH, LORD, *Proc. London Math. Soc.*, **17,** 4 (1885). [165]

REATTO, L., and CHESTER, G. V., *Phys. Rev.*, **155,** 88 (1967). [193]

REE, F. H., *Physical Chemistry, an Advanced Treatise.* (Eds. Eyring, H., Henderson, D., and Jost, W.), Academic Press, London/New York, 1971, Vol. VIIIA, Ch. 3. [259]

REED, T. M., and GUBBINS, K. E., *Applied Statistical Mechanics.* McGraw-Hill Kogakusha, Tokyo, 1973. [1, 3, 10, 15]

REGGE, T., *J. Low Temp. Phys.*, **9**, 123 (1972). [183]

REIF, F., *Statistical and Thermal Physics*. McGraw-Hill, New York, 1965, p. 314. [30]

REISS, H., FRISCH, H. L., and LEBOWITZ, J. L., *J. Chem. Phys.*, **31**, 369 (1959). [32, 64]

RICE, O. K., *Statistical Mechanics, Thermodynamics and Kinetics*. Freeman, San Francisco, 1967. [276]

RICE, S. A., and ALLEN, J. W., *J. Chem. Phys.*, **67**, 5105 (1977). See ALLEN.

RICE, S. A., and ALLEN, J. W., *J. Chem. Phys.*, **68**, 5053, (1978). See ALLEN

RICE, S. A., and BLOCH, A., *Phys. Rev.*, **185**, 933 (1969). See BLOCH.

RICE, S. A., and GRAY, P., *The Statistical Mechanics of Simple Liquids*. Interscience, New York, 1965. [1, 3, 10, 15]

RICE, S. A., and LU, B. C., *J. Chem. Phys.*, **68**, 5558 (1978). See LU.

RICHARDSON, R. C., LEE, D. M., and OSHEROFF, D. D., *Phys. Rev. Lett.*, **28**, 885 (1972). See OSHEROFF.

RICHARDSON, R. C., LEE, D. M., OSHEROFF, D. D., and GULLY, W. J., *Phys. Rev. Lett.*, **29**, 920, (1972). See OSHEROFF.

ROWLINSON, J. S., *Liquids and Liquid Mixtures*, 2nd Ed. Butterworth, London, 1969. [1, 3, 10, 15, 122, 306]

ROWLINSON, J. S., *Disc. Faraday Soc.*, **59**, 52 (1975). [263]

ROWLINSON, J. S., CHAPELA, G. A., and SAVILLE, G., *Chem. Soc. Faraday Disc.*, No. 59, 22 (1975). See CHAPELA.

ROWLINSON, J. S., CHAPELA, G. A., SAVILLE, G., and THOMPSON, S. M., *J. Chem. Soc., Faraday Trans. II*, **73**, 1133 (1977). See CHAPELA.

ROWLINSON, J. S., and HARRINGTON, J. M., *Proc. Roy. Soc. (Lond)*, **A352**, 15 (1979). See HARRINGTON.

ROWLINSON, J. S., SAVILLE, G., and THOMPSON, S. M., (private communication, 1978). See THOMPSON.

ROWLINSON, J. S., THOMPSON, S. M., and LENG, C. A., *Proc. Roy. Soc.*, **A352**, 1 (1976). See LENG.

ROWLINSON, J. S., THOMPSON, S. M., and LENG, C. A., *Proc. Roy. Soc.*, **A358**, 267 (1977). See LENG.

RUBIN, R. J., *J. Chem. Phys.*, **43**, 2392 (1965). [219, 223]

SAAM, W. F., *Phys. Rev.*, **A4**, 1278 (1971). [196]

SAAM, W. F., *Phys. Rev.*, **A8**, 1048 (1973). [178, 179]

SAAM, W. F., and EBNER, C., *Phys. Rev.*, **B12**, 923, (1975). See EBNER.

SALTER, S. J., and DAVIS, H. T., *J. Chem. Phys.*, **63**, 3295 (1975). [29, 83]

SAMORJAI, G. A., and GOODMAN, R. M., *J. Chem. Phys.*, **52**, 6331. See GOODMAN.

SAROLÉA-MARHOT, L., *Bull. Ac. Roy. Belg. (Cl. Sc)*, **40**, 1120 (1954). [225]

SARWINSKI, R. E., TOUGH, J. T., GUO, H. M., and EDWARDS, D. O., *Phys. Rev. Lett.*, **27**, 1259 (1971). See GUO.

SAUPE, A., and MAIER, W., *Z. Naturf.*, **13a**, 564 (1958); *Z. Naturf.*, **14a**, 882 (1059); *Z. Naturf.*, **15a**, 287 (1960). See MAIER.

SAVILLE, G., ROWLINSON, J. S., and CHAPELA, G. A., *Chem. Soc. Faraday Disc.*, No. 59, 22 (1975). See CHAPELA.

SAVILLE, G., THOMPSON, S. M., and ROWLINSON, J. S., (private communication, 1978). See THOMPSON.

SAVILLE, G., THOMPSON, S. M., ROWLINSON, J. S., and CHAPELA, G. A., *J. Chem. Soc., Faraday Trans. II*, **73**, 1133 (1977). See CHAPELA.

SCATCHARD, G., *J. Phys. Chem.*, **66**, 618 (1962). [3]

SCHERAGA, H. A., and LENZ, B. R., *J. Chem. Phys.*, **58**, 5296 (1973). See LENZ.

SCHIFF, D., and VERLET, L., *Phys. Rev.*, **160**, 208 (1967). [185]

SCHMIT, J., and LUCAS, A. A., *Solid State Commun*, **11**, 415 (1972). [145, 146, 158]

SCHREIBER, D. E., BARKER, J. A., and ABRAHAM, F. F., *J. Chem. Phys.*, **62**, 1958 (1975). See ABRAHAM.

336

SCHRIEFFER, J. R., BARDEEN, J., and COOPER, L. N., *Phys. Rev.*, **108,** 1175 (1957). *See* BARDEEN.

SCHUCHOWITSKY, A., *Acta Physiochem. URSS*, **19,** 176 (1944). [123]

SCRIVEN, L. E., and DAVIS, H. T., *J. Chem. Phys.*, **69,** 5215 (1978). *See* DAVIS.

SEMENCHENKO, V. K., *Surface Phenomena in Metals and Alloys.* Pergamon, Oxford. [141]

SESSLER, A. M., and EMERY, V. J., *Phys. Rev.*, **119,** 43 (1960). *See* EMERY.

SHASHIDHAR, R., and KRISHNASWAMY, S., *Mol. Cryst. Liq. Cryst.*, **35,** 253 (1976); *Mol. Cryst. Liq. Cryst.*, **38,** 711, (1977). *See* KRISHNASWAMY.

SHEN, S. Y., GASPARINI, F. M., EKARDT, J. R., and EDWARDS, D. O., *J. Low. Temp. Phys.*, **13,** 437 (1973). *See* GASPARINI.

SHEN, S. Y., GASPARINI, F. M., TAM, C. P., EDWARDS, D. O., FATOUROS, P., IHAS, G. G., and MROZINSKI, P., *Phys. Rev. Lett.*, **34,** 1153, (1975). *See* EDWARDS.

SHIH, Y. M., and WOO. C.-W., *Phys. Rev. Lett.*, **30,** 478 (1973). [187, 193]

SHOEMAKER, P. D., PAUL, G. W., and MARC DE CHAZAL, L. E., *J. Chem. Phys.*, **52,** 491 (1970). [28, 29]

SHVETS, A. D., ESEL'SON, B. N., and IVANTSOV, V. G., *Zh. Eksp. Teor. Fiz.*, **44,** 483 (1963) (*Sov. Phys. JETP*, **17,** 330 (1963)). *See* ESEL'SON.

SILBERBERG, A., *J. Phys. Chem.* **66,** 1872 (1962); *J. Phys. Chem.*, **66,** 1884 (1962); *J. Phys. Chem.*, **46,** 1105 (1967); *J. Phys. Chem.*, **48,** 2835 (1968). [219, 224, 225, 230]

SILBERBERG, A., and PRIEL, Z., *Polym. Prepr. Amer. Chem. Soc. Div. Polym. Chem.*, **11,** 1405 (1970). *See* PRIEL.

SIMHA, R., FRISCH, H. L., and EIRICH, F. R., *J. Phys. Chem.*, **57,** 584 (1953). [219, 225]

SINGH, A. D., *Phys. Rev.*, **125,** 802 (1962). [177]

SINGH, Y., and ABRAHAM, F. F., *J. Chem. Phys.*, **67,** 537 (1977). [41]

SINGH, Y., and ABRAHAM, F. F., *J. Chem. Phys.*, **67,** 5960 (1977). [44]

SINGWI, K. S., and VASHISTA, P., *Phys. Rev.*, **B6,** 875 (1972). *See* VASHISTA.

SIOW, K. S., and PATTERSON, D., *J. Phys. Chem.*, **77,** 356 (1973). [229, 230]

SLATER, J. C., *The Self-consisten, Field for Molecules and Solids: Quantum Theory of Molecules and Solids*, Vol. 4. McGraw-Hill, New York, 1974. [148]

SMITH, E. B., *Rep. Prog. Chem.*, **63,** 13 (1966). [90]

SMITH, E. R., and PERRAM, J. W., *Chem. Phys. Lett.*, **35,** 138 (1975); *Chem. Phys. Lett.*, **39,** 328 (1976); *Phys. Lett. A.*, **59,** 11 (1976). *See* PERRAM.

SMITH, J. R., *Phys. Rev.*, **181,** 522 (1969). [142, 150, 154]

SMITH, W. R., *Canad. J. Phys.*, **52,** 2022 (1974). [93]

SMITH, W. R., *Chem. Phys. Lett.*, **40,** 313 (1976). [93]

SMITH, W. R., HENDERSON, D., and BARKER, J. A., *J. Chem. Phys.*, **55,** 4027 (1971). [36]

SMITH, W. R., MADDEN, W. G., and FITTS, D. C., *Chem. Phys. Lett.*, **36,** 195, (1975). [93]

SMITH, W. R., MADDEN, W. G., and FITTS, D. D., *Mol. Phys.*, **35,** 1017 (1978). *See* MADDEN.

SMITH, W. R., NEZEBA, I., MELNYK, T. W., and FITTS, D. D., *Faraday Disc. Chem. Soc.*, **66/8** (1978). [93]

SODA, T., ANDERSON, P. W., MOREL, P., and BRUECKNER, K. A., *Phys. Rev.*, **118,** 1442 (1960). *See* BRUECKNER.

SOUTHIN, R. T., and CHADWICK, G. A., *Scripta. Metall.*, **3,** 541 (1969). [164]

SOVEN, P., and KALKSTEIN, D., *Surf. Sci.*, **26,** 85 (1971). *See* KALKSTEIN.

SPROW, F. B., and PRAUSNITZ, J. M., *Trans. Faraday Soc.*, **62,** 1097, (1966). [28, 280]

SPROW, F. B., and PRAUNITZ, J. M., *Trans. Faraday Soc.*, **62,** 1105 (1966). [28, 133]

SPURLING, T. H., and MASON, E. A., *J. Chem. Phys.*, **46,** 322 (1967). [104]

SRINIVASAN, G., and JOHNSON, M., *Phys. Lett.*, **43A,** 427 (1973). *See* JOHNSON, M.

SRINIVASAN, G., and JOHNSON, M., *Physica Scripta*, **10,** 262 (1974). *See* JOHNSON, M.

STANSFIELD, D., *Proc. Phys. Soc.*, **72,** 854 (1967). [277]

STEELE, W. A., and SWEET, J. R., *J. Chem. Phys.*, **47,** 3029 (1967). *See* SWEET.

STELL, G., and BLUM, L., *J. Stat. Phys.*, **15,** 439 (1976). *See* BLUM.

STELL, G., RASAIAH, J. C., and NARANG, H., *Mol. Phys.*, **27,** 1393 (1974). [99]

STELL, G., and SULLIVAN, D. E., *J. Chem. Phys.*, **67,** 2567 (1977). *See* Sullivan. [70]

STILLINGER, F. H., and BEN-NAIM, A., *J. Chem. Phys.*, **47,** 4431 (1967). [198, 199, 203, 206, 213, 214, 217, 309]

STILLINGER, F. H., and BEN-NAIM, A., *Structure and Transport Processes in Water and Aqueous Solutions.* (Ed. Horne, R. A.), Wiley–Interscience, New York, 1972. *See* BEN-NAIM.

STILLINGER, F. H., and BUFF, F. P., *J. Chem. Phys.*, **39,** 1911 (1963). *See* BUFF.

STILLINGER, JR., F. H., BUFF, F. P., and LOVETT, R. A., *Phys. Rev. Lett.*, **15,** 621 (1965). *See* BUFF.

STILLINGER, JR., F. H., BUFF, F. P., and LOVETT, R. A., *Phys. Rev. Lett.*, **14,** 491 (1965). *See* BUFF.

STILLINGER, JR., F. H., HANKINS, D., and MOSKOWITZ, J. W., *J. Chem. Phys.*, **53,** 4544 (1970). *See also* Erratum, *J. Chem. Phys.*, **59,** 995 (1973). *See* HANKINS.

STILLINGER, F. H., and BUFF, F. P., *J. Chem. Phys.*, **25,** 312 (1956). *See* BUFF.

STILLINGER, F. H., and RAHMAN, A., *J. Chem. Phys.*, **60,** 1545, (1974). [200, 215]

STOGRYN, A. P., and STOGRYN, D. E., *Mol. Phys.*, **11,** 371 (1966). *see* STOGRYN, D. E.

STOGRYN, D. E., and STOGRYN, A. P., *Mol. Phys.*, **11,** 371 (1966). [104]

STRATTON, R., *Phil. Mag.*, **44,** 1236 (1953). [142]

STREHLOW, M., *Z. Elektochem.*, **56,** 119 (1952). [217]

SULLIVAN, D. E., and STELL, G., *J. Chem. Phys.*, **67,** 2567 (1977). [45, 68, 70]

SURIATECKI, W. J., *Proc. Phys. Soc.*, **A64,** 226 (1961). [142]

SUZUKI, K., and WASEDA, Y., *Phys. Stat. Solidi*, **49,** 643 (1972). *See* WASEDA.

SUZUKI, K., and WASEDA, Y., *Phys. Stat. Solidi*, **57,** 351, (1973). *See* WASEDA.

SWAIN, R. C., and MCBAIN, J. W., *Proc. Roy. Soc.*, **A154,** 608 (1936). *See* MCBAIN.

SWEET, J. R., and STEELE, W. A., *J. Chem. Phys.*, **47,** 3029 (1967). [287]

SYKES, M. F., *J. Chem. Phys.*, **39,** 410 (1963); DOMB, C., and SYKES, M. F., *J. Math. Phys.*, **2,** 63 (1961). [299]

TAM, C. P., EDWARDS, D. O., FATOUROS, P., IHAS, G. G., MROZINSKI, P., SHEN, S. Y., and GASPARINI, F. M., *Phys. Rev. Lett.*, **34,** 1153, (1975). *See* EWARDS.

TELO da GAMA, M. M., and EVANS, R., *Mol. Phys.*, **38,** 687 (1979). [317, 320]

TER HAAR, D., *Elements of Statistical Mechanics.* Rinehard, New York, 1954, pp. 147–155. [180]

THOMPSON, S. M., *Faraday Disc. Chem. Soc.*, **66,** (1978). [286]

THOMPSON, S. M., LENG, C. A., and ROWLINSON, J. S., *Proc. Roy. Soc.*, **A352,** 1 (1976). *See* LENG.

THOMPSON, S. M., LENG, C. A., and ROWLINSON, J. S., *Proc. Roy. Soc.*, **A358,** 267 (1977). *See* LENG.

THOMPSON, S. M., ROWLINSON, J. S., CHAPELA, G. A., and SAVILLE, G., *J. Chem. Soc., Faraday Trans. II*, **73,** 1133 (1977). *See* CHAPELA.

THOMPSON, S. M., ROWLINSON, J. S., and SAVILLE, G., (private communication, 1978). [28]

THOULESS, D. J., *Ann. Phys. (N.Y.)*, **10,** 553 (1960). [196]

THROOP, G. J., and BEARMAN, R. J., *J. Chem. Phys.*, **44,** 1423 (1966). [120, 128, 131]

TONKS, L., *Phys. Rev.*, **50,** 955 (1936). [32]

TOSI, M. P., and MARCH, N. H., *Ann. of Physics*, **81,** 414 (1973). *See* MARCH.

TOSI, M. P., and MARCH, N. H., *Nuovo Cimento*, **15B,** 308 (1973). [140]

338

Tosi, M. P., and March, N. H., *Atomic Dynamics in Liquids*, p. 260. London, Macmillan, 1976. See March.

Tosi, M. P., Parrinello, M., and March, N. H., *Nuovo Cimento*, **23B,** 135 (1974) [140]

Tough, J. T., Guo, H. M., Edwards, D. O., and Sarwinski, R. E., *Phys. Rev. Lett.*, **27,** 1259 (1971). See Guo.

Toxvaerd, S., *J. Chem. Phys.*, **55,** 3116 (1971). [29, 36, 37, 38]

Toxvaerd, S., *J. Chem. Phys.*, **57,** 4092 (1972). [29, 34, 37, 48, 49, 50]

Toxvaerd, S., *Prog. Surf. Sci.*, **3,** 189 (1972). [48]

Toxvaerd, S., *Mol. Phys.*, **26,** 91 (1973). [50, 59]

Toxvaerd, S., *Statistical Mechanics*. (Ed. Singer, K.), Specialist Periodical Reports, Chem. Soc., (1975), Vol. 2, Ch. 4, p. 256. [273]

Toxvaerd, S., *J. Chem. Phys.*, **62,** 1589 (1975). [264, 268, 269, 270]

Toxvaerd, S., *J. Chem. Phys.*, **64,** 2863 (1976). [52, 53]

Trapnell, G. M. W., and Hayward, D. O., *Chemisorption*. Butterworths, 2nd edition, 1964. See Hayward.

Triezenberg, D. G., and Zwanzig, R., *Phys. Rev. Lett.*, **28,** 1183 (1972). [72, 302]

Turn, C. H., *Ph.D. Thesis*, Univ. Florida, Gainesville (1976). [103]

Upstill, C. E., and Evans, R., *J. Phys. C.*, **10,** 2791 (1977). [38, 39]

van der Waals, Z. *Phys. Chem.*, **13,** 657 (1894). (English Translation, *J. Stat. Phys.*, **20,** 197 (1979)). [320]

Vargaftik, N. B., *Tables on the Thermophysical Properties of Liquids and Gases*. Wiley, New York, 2nd Ed., 1975. [306]

Vashista, P., and Singwi, K. S., *Phys. Rev.*, **B6,** 875 (1972). [157]

Verlet, L., *Phys. Rev.*, **159,** 98 (1967). [267]

Verlet, L., *Phys. Rev.*, **165,** 201 (1968). [52, 110]

Verlet, L., and Hansen, J.-P., *Phys. Rev.*, **184,** 151 (1969). See Hansen.

Verlet, L., and Schiff, D., *Phys. Rev.*, **160,** 208 (1967). See Schiff.

Verlet, L., and Weis, J. J., *Phys. Rev.*, **A5,** 939 (1972). [36]

Verschaffelt, J. E., *Bull. Sci. Acad. Roy. Belge.*, **22,** 373 (1936). [3]

Verwey, E. J. W., *Rec. trav. Chim. Pay Bas.*, **61,** 564 (1942). [217]

Viceeli, J. J., Buff, F. P., Lovett, R., and De Haven, P. W., *J. Chem. Phys.*, **58,** 1880 (1973). See Lovett.

Wainwright, T. E., and Alder, B. J., *J. Chem. Phys.*, **31,** 459 (1959). See Alder.

Waisman, E., Henderson, D., and Lebowitz, J., *Mol. Phys.*, **32,** 1373 (1976). [67, 68]

Wallis, R. F., *Progress in Surface Science*. (Ed. Davison, S. G.,), Oxford, Pergamon, Vol. 4, p. 253, 1975. [145]

Waseda, Y., and Suzuki, K., *Phys. Stat. Solidi*, **49,** 643 (1972). [151, 160]

Waseda, Y., and Suzuki, K., *Phys. Stat. Solidi*, **57,** 351 [151, 160]

Watabe, M., and Maregawa, *2nd Intl. Conf. on Liquid Metals*. Taylor & Francis, London, 1973, p. 133. [140]

Watts-Tobin, R. J., *Phil. Mag.*, **8,** 333 (1963). [199]

Webb, W. W., and Huang, J. S., *J. Chem. Phys.*, **50,** 3677 (1966). See Huang.

Webb, W. W., and Wu, E. S., *Phys. Rev.*, **A8,** 2065 (1973). See Wu.

Weeks, J. D., *J. Chem. Phys.*, **67,** 3106 (1977). [79, 80, 83, 260, 261, 262, 263, 275, 278]

Weeks, J. D., Anderson, H. C., and Chandler, D., *J. Chem. Phys.*, **56,** 3812 (1972). See Anderson.

Weeks, J. D., Chandler, D., and Anderson, H. C., *J. Chem. Phys.*, **54,** 5237 (1971). [25]

Weis, J. J., and Verlet, L., *Phys. Rev.*, **A5,** 939 (1972). See Verlet.

Welsh, W. J., and Fitts, D. D., *Chem. Phys.*, **26,** 379 (1977). See Fitts.

Wertheim, M. S., *J. Chem. Phys.*, **65,** 2377 (1976). [56, 77, 260, 263, 275, 291]

WHEATLEY, J., *Rev. Mod. Phys.*, **47**, 415 (1975). [196]

WHITE, D. W. G., *Trans. Metall. Soc.*, *A.I.M.E.*, **236**, 796 (1966). [161, 162, 165, 167]

WHITE, D. W. G., *Metals, Materials and Metallurgical Reviews*, July, (1968). [3, 165]

WHITE, D. W. G., *J. Inst. Metals*, **99**, 287 (1971). [161, 162, 164, 167]

WHITE, L. R., and PERRAM, J. W., *Faraday General Discussion, Physical Adsorption in Condensed Phases;* April, 1975. *See* PERRAM.

WIDOM, A., *Ph.D. Thesis*, Cornell University, 1968. [183]

WIDOM, A., *Phys. Rev.*, **A1**, 216 (1970). [176]

WIDOM, B., 'Surface Tension of Fluids', In *Phase Transitions and Critical Phenomena*. (Eds. Dorub, C., and Green, M. S.), Academic Press, New York, 1972, Vol. 2, Ch. 3. [275]

WIDOM, B., and EGELSTAFF, P. A., *J. Chem. Phys.*, **53**, 2667 (1970). *See* EGEL-STAFF.

WIDOM, B., and FISK, S., *J. Chem. Phys.*, **50**, 3219 (1969). *See* FISK.

WIDOM, B., and LONGUET-HIGGINS, H. C., *Mol. Phys.*, **8**, 549 (1964). *See* LONGUET-HIGGINS.

WIKBORG, E., and INGLESFIELD, J. E., *Sol. Stat. Commun.*, **16**, 335 (1975). [146]

WILETS, L., and BERG, R. D., *Proc. Phys. Soc.*, **A68**, 229, (1955). *See* BERG.

WILSON, J. R., *Metall. Rev.*, **10**, 381 (1965). [161]

WOO, C.-W., *European Physical Society Topical Conference: Liquid and Solid Helium*. Haifa, Israel, 1974. [196]

WOO, C.-W., *The Physics of Liquid and Solid Helium, Part I*. (Eds. Bennemann, K. H., and Ketterson, J. B.), Wiley, New York, 1976, Ch. 5., p. 482. [178, 196]

WOO, C.-W., and SHIH, Y. M., *Phys. Rev. Lett.*, **30**, 478, (1973). *See* SHIH.

WOOD, W. W., *Physics of Simple Liquids*. (Eds. Temperley, H. N. V., Rowlinson, J. S., and Rushbrooke, G. S.), North-Holland Publ. Co., Amsterdam, 1968, Ch. 5. [259]

WU, E. S., and WEBB, W. W., *Phys. Rev.*, **A8**, 2065 (1973). [76]

WYLLIE, G., and HUANG, K., *Proc. Phys. Soc.*, **A62**, 180 (1949). *See* HUANG.

YANG, A. J. M., FLEMING, P. D., and GIBBS, J. H., *J. Chem. Phys.*, **64**, 3722 (1976). [75, 300, 316, 319]

YVON, J., *La Théorie Statistique des fluids et l'equation d'état*. Herman & Cie, Paris, 1935. [42, 45]

YVON, J., *Proc. IUAP Symposium on Thermodynamics*, p. 9; Brussels, 1948. [72]

ZADUMKIN, S. N., *Russian J. Phys. Chem.* **33**, 539 (1959). [167]

ZADUMKIN, S. N., and PUGACHEVICH, P. P., *Proc. Acad. Sci. USSR Physics & Chemistry Section*, **146**, 743 (1962). [67]

ZADUMKIN, S. N., *Fiz. Metal. i Metalloved.*, **11**, 3 (1961); **11**, 331 (1961) (*Phys. Metals Metallog.*, **11**, 3 (1961); **11**, 11 (1961). [144, 45]

ZADUMKIN, S. N., and ZVYAGINA, V. YA., *Izvest. Akad. Nauk. S.S.S.R.* (*Metally*), **4**, 58 (1966). [165]

ZADUMKIN, S. N., and ZVYAGINA, YA V., *Russian Met. and Fuels*, **4**, 20 (1966). [165]

von ZAWIDGKI, Z., *Physikal. Chem.*, **35**, 129 (1900). [122]

ZIMAN, J. M., *Advan. Phys.*, **13**, 89 (1964). [151]

ZINOV'EVA, N. K., and BOLDAREV, S. T., *Zh. Eksp. i Teor. Fiz.*, **56**, 1089 (1969); *Sov. Phys. JETP*, **29**, 585 (1969). [195]

ZVYAGINA, V. YA., and ZADUMKIN, S. N., *Russian Met. & Fuels*, **4**, 20 (1966). *See* ZADUMKIN.

ZVYAGINA, V. YA, and ZADUMKIN, S. N., *Invest. Akad. Nauk. S.S.S.R.* (*Metally*), **4**, 58 (1966). *See* ZADUMKIN.

ZWANZIG, R. W., *J. Chem. Phys.*, **22**, 1420 (1954). [34, 93]

ZWANZIG, R. W., and TRIEZENBERG, D. G., *Phys. Rev. Lett.*, **28**, 1183 (1972). *See* TRIEZENBERG.

Subject Index

Page references in *italics* refer to tables or figures in which experimental data or the results of theoretical calculations are presented. Chemical elements are listed under their conventional symbols only if they are the subject of special comment in the text.